Information Systems and Modern Society:

Social Change and Global Development

John Wang
Montclair State University, USA

Information Science
REFERENCE

Managing Director:	Lindsay Johnston
Editorial Director:	Joel Gamon
Book Production Manager:	Jennifer Yoder
Publishing Systems Analyst:	Adrienne Freeland
Assistant Acquisitions Editor:	Kayla Wolfe
Typesetter:	Christy Fic
Cover Design:	Jason Mull

Published in the United States of America by
Information Science Reference (an imprint of IGI Global)
701 E. Chocolate Avenue
Hershey PA 17033
Tel: 717-533-8845
Fax: 717-533-8661
E-mail: cust@igi-global.com
Web site: http://www.igi-global.com

Library of Congress Cataloging-in-Publication Data

Information systems and modern society: social change and global development / John Wang, editor.
    p. cm.
 Includes bibliographical references and index.
 Summary: "This book is a comprehensive collection of research on the emergence of information technology and its effect on society, focusing on the advancements made throughout social changes and the application of information systems"-- Provided by publisher.
 ISBN 978-1-4666-2922-6 (hbk.) -- ISBN 978-1-4666-2923-3 (ebook) -- ISBN 978-1-4666-2924-0 (print & perpetual access) 1. Information technology. 2. Information technology--Social aspects. 3. Management information systems. I. Wang, John, 1955-
 T58.5.I5275 2013
 303.48'33--dc23
                        2012039282

British Cataloguing in Publication Data
A Cataloguing in Publication record for this book is available from the British Library.

The views expressed in this book are those of the authors, but not necessarily of the publisher.

Table of Contents

Section 2
Information Systems and Information Technologies Issues in Nonprofits

Section 3
Informatics and Semiotics in Organisations

Detailed Table of Contents

Section 1
Social and Personalized Learning in Web-Based Environments

Chapter 1
Malinka Ivanova, Technical University of Sofia, Bulgaria
Anguelina Popova, University of Utrecht, The Netherlands

This paper presents the results of an exploratory study examining bachelor degree students' experiences of learning with a new generation learning management system Edu 2.0 combined with Web 2.0 applications. The authors discuss students' perceptions of formal and informal activities within this environment as captured through a collection of surveys, activities' tracking, and assessment. The main functional characteristics and available social tools of Edu 2.0 are examined in the context of students learning support. A model of Learning area is developed to analyze the formal and informal learning flows from the point of view of learning enhancement.

Chapter 2
Yuri Nishihori, Sapporo Otani University, Japan
Chizuko Kushima, Tsuda College, Japan
Yuichi Yamamoto, Hokkaido University, Japan
Haruhiko Sato, Hokkaido University, Japan
Satoko Sugie, Hokkaido University, Japan

The main objective of this project is to design and implement Web-based collaborative environments for a global training based on a multiple perspective assessment for future and novice ALTs (Assistant Language Teachers) who will come to Japan from various parts of the world. The system was created in order to give better chances to acquire professional knowledge in advance with support from experienced senior teachers, both Japanese teachers and former ALTs. Computer Support for Collaborative Learning (CSCL) was adopted as a platform for their online discussion with much focus on multiple perspective assessment to support social and personalized aspects such as individual accountability and contribution to the collaboration. Initial results are reported using the analysis of system design and the Web-based questionnaire answered by the participants involved in this knowledge-building forum.

# Chapter 3

This paper illustrates the use of WELSA adaptive educational system for the implementation of an Artificial Intelligence (AI) course which is individualized to the learning style of each student. Several of the issues addressed throughout this paper are describing similar approaches existing in literature, how the AI course is created, and what kind of personalization is provided in the course including the underlying adaptation mechanism. The authors also focus on whether the course is used effectively by the stakeholders (teachers and students respectively). Results obtained in the paper confirm the practical applicability of WELSA and its potential for meeting the personalization needs and expectations of the digital native students.

# Chapter 4

In today's knowledge-based economy, having proper expertise is crucial in resolving many tasks. Expertise Finding (EF) is the area of research concerned with matching available experts to given tasks. A standard approach is to input a task description/proposal/paper into an EF system and receive recommended experts as output. Mostly, EF systems operate either via a content-based approach, which uses the text of the input as well as the text of the available experts' profiles to determine a match, and structure-based approaches, which use the inherent relationship between experts, affiliations, papers, etc. The underlying data representation is fundamentally different, which makes the methods mutually incompatible. However, previous work (Watanabe et al., 2005a) achieved good results by converting content-based data to a structure-representation and using a structure-based approach. The authors posit that the reverse may also hold merit, namely, a content-based approach leveraging structure-based data converted to a content-based representation. This paper compares the authors' idea to a content only-based approach, demonstrating that their method yields substantially better performance, and thereby substantiating their claim.

# Chapter 5

SOCIALX is a web application supporting social-collaborative and cooperative aspects of e-learning, such as sharing and reuse of (solutions to) single exercises, and development of projects by group-work and social exchange. Such aspects are supported in the framework of a reputation system, in which learners participate. We describe design and motivational issues of the system, show implementation details and describe a small-scale experimentation that helped evaluating the effectiveness of the system as well as the best lines for further development. In the system, learner's reputation is computed, presented and maintained during her/his interactions with the system. The algorithm to compute reputation can be configured by the teacher, by tuning weights associated to various aspects of the interactions. To enhance collaboration on exercises, we support contextual (to the exercise) micro-forum and FAQs, together with a currency-based concretization of the perceived usefulness of questions/answers. Group responsibilities, peer-assessment and self-evaluation are supported by group-based projects with self/

peer-evaluated phases: Different stages of a project are assigned to different groups; a stage-deliverable is both self-evaluated (at submission) and peer-evaluated (by the group receiving it for the next stage). This paper is an extension of the original version, published on the International Journal of Information Systems and Social Change: mainly two sections were added, describing the latest improvements and some experimentation.

Section 2
Information Systems and Information Technologies Issues in Nonprofits

Chapter 6

Louis-Marie Ngamassi Tchouakeu, Prairie View A&M University, USA

Edgar Maldonado, The Pennsylvania State University, USA

Kang Zhao, Iowa State University, USA

Harold Robinson, The Pennsylvania State University, USA

Carleen Maitland, The Pennsylvania State University, USA

Andrea H. Tapia, The Pennsylvania State University, USA

Humanitarian nongovernmental organizations (NGOs) are increasingly collaborating through inter-organizational structures such as coalitions, alliances, partnerships, and coordination bodies. NGO's information technology coordination bodies are groups of NGOs aimed at improving the efficiency of ICT use in humanitarian assistance through greater coordination. Despite their popularity, little is known about these coordination bodies, specifically the extent to which they address inter-organizational coordination problems. This paper examines coordination problems within two humanitarian NGO's information technology coordination bodies. Based on data collected through interviews, observation, and document analysis, despite positive attitudes toward coordination by members, seven of eight widely accepted barriers to coordination still exist among members of these coordination bodies. Further, in a comparison of mandate-oriented, structural and behavioral coordination barriers, research finds mandate issues were most significant and structural factors were found in the greatest numbers. Findings suggest that effective humanitarian NGO's information technology coordination bodies must pay attention to both organizational design and management issues, although the former are likely to have a greater impact on coordination.

Chapter 7

Geoffrey Greenfield, University of Queensland, Australia

Fiona H. Rohde, University of Queensland, Australia

During the past decade there has been an increasing interest in research within Not-for-Profit (NFP) organisations. Research has indicated that there are a number of characteristics that make NFPs different from other organisations. This paper considers whether workers within the NFP sector have different attitudes to technology and whether such differences affect the measures used within technology acceptance models. An exploratory study of workers within two NFPs indicated that workers within the service delivery functions of NFPs have different attitudes to technology than workers within the standard business functions of a NFP organisation e.g., marketing. These attitudes affected their perceptions of the use of and ease of technology.

# Chapter 8

Semantic Web services (SWS) have attracted increasing attention due to their potential to automate discovery and composition of current syntactic Web services. An issue that prevents a wider adoption of SWS relates to the manual nature of the semantic annotation task. Manual annotation is a difficult, error-prone, and time-consuming process and automating the process is highly desirable. Though some approaches have been proposed to semi-automate the annotation task, they are difficult to use and cannot perform accurate annotation for the following reasons: (1) They require building application ontologies to represent candidate services and (2) they cannot perform accurate name-based matching when labels of candidate service elements and ontological entities contain Compound Nouns (CN). To overcome these two deficiencies, this paper proposes a query-based approach that can facilitate semi-automatic annotation of Web services. The proposed approach is easy to use because it does not require building application ontologies to represent services. Candidate service elements that need to be annotated are extracted from a WSDL file and used to generate query instances by filling a Standard Query Template. The resulting query instances are executed against a repository of ontologies using a novel query execution engine to find appropriate correspondences for candidate service elements. This query execution engine employs name-based and structural matching mechanisms that can perform effective and accurate similarity measurements between labels containing CNs. The proposed semi-automatic annotation approach is evaluated by employing it to annotate existing Web services using published domain ontologies. Precision and recall are used as evaluation metrics. The resulting precision and recall values demonstrate the effectiveness and applicability of the proposed approach.

# Chapter 9

There was a time in the history of GM when it was the largest corporation in the US. The history of GM also shows that it was the single largest employer in the world. The announcement of GM's bankruptcy on June 1, 2009 shocked the world and had a tremendous impact on the United States economy. Looking back at the history of GM, there were many indicators which suggested the fate of the company. There were several internal factors that answer the question, what went wrong with GM. These internal factors are management arrogance, not meeting customer demands, the costs and demands of unions, poor forecasting, and internal controls on accounting standards. Similarly, there were several external factors that answer the same question, which include increased competition and loss of market share, rising gas prices and environmental friendliness, and the costs and burdens of meeting government regulations and restrictions. This paper will explore and answer the following questions: What are the fundamental causes of GM's problems? What can be learned from GM's mistakes and experiences? How and why an industrial icon came to ruin?

Section 3
Informatics and Semiotics in Organisations

Information systems actability theory builds on a communicative action perspective on IS. Information systems are seen as instruments for technology mediated work communication. Human actors are communicating (i.e. sending and/or receiving messages) through an information system. Information systems actability emphasises pragmatic dimensions of information systems. The paper presents 19 actability criteria divided into three groups: 1) criteria concerning user-system interaction, 2) criteria concerning user-through-system-to-user communication, and 3) criteria concerning information system's contribution to workpractice processes. These actability criteria should be possible to use in design and evaluation of information systems.

# Chapter 11

Information and Communication Technology has the potential of benefiting citizens, allowing access to knowledge, communication and collaboration, and thus promoting the process of constitution of a fairer society. The design of systems that make sense to the users' community and that respect their diversity demands socio-technical views and an in-depth analysis of the involved parties. The authors have adopted Organizational Semiotics and Participatory Design as theoretical and methodological frames of reference to face this challenge in the design of an Inclusive Social Network System for the Brazilian context. This paper presents the use of some artifacts adapted from Problem Articulation Method to clarify concepts and prospect solutions. Results of this clarification fed the Semantic Analysis Method from which this paper presents and discusses an Ontology Chart for the domain and the first signs of the inclusive social network system.

Improvements in electronics and computing have increased the potential of monitoring and surveillance technologies. Although now widely used, these technologies have been known to cause unintended effects, such as increases in stress in those being observed. Further advancements in technology lead people towards the 'pervasive era' of computing, where a new means of monitoring ubiquitously becomes possible. This monitoring differs from existing methods in its distinct lack of physical boundaries. To address the effects of this kind of monitoring, this paper proposes a model consisting of a series of factors identified in the monitoring and pervasive literature believed to influence behaviour. The model aims to understand and predict behaviour, thereby preventing any potential undesirable effects, but also to provide a means to analyse the problem. Various socio-technical frameworks have been proposed to guide research within ubiquitous computing; this paper uses the semiotic framework to analyse the model in order to better understand and explain the behavioural impact of ubiquitous monitoring.

Chapter 13

Soheil Ghili, Sharif University of Technology, Iran
Hengameh Shams, Sharif University of Technology, Iran
Madjid Tavana, La Salle University, USA

This paper develops a mathematical model of innovation in technology with two main characteristics. First, it discusses the endogenously made decision on not only how much to innovate, but also, how much to imitate. Second, it demonstrates that the decision to innovate or imitate are not mutually exclusive and a firm can innovate and imitate simultaneously. A mathematical model is presented, and the authors explain the barriers to innovation development and diffusion. The model is further used to investigate the effectiveness of two technology innovation and imitation policies. It is shown that an intellectual property right (IPR) policy will better function if the price of innovation is set to a level lower than the cost of innovation. The concept "superfluous innovation" (innovations whose costs are higher than their benefits) is also proposed and developed through investigating the policy of levying subsidies on innovation.

Section 4
Technology Trends and Critical Social Challenges

Chapter 14

Melih Kirlidog, Marmara University, Turkey & North-West University, South Africa
Aygul Kaynak, Marmara University, Turkey

Technology Acceptance Model (TAM) is an important tool to understand the dynamics of acceptance of Information Systems in an organization. The model posits that perceived ease of use and perceived usefulness are key factors in the adoption. This study extends TAM for investigating the user rejection of technology by reversing the two key factors into perceived difficulty of use and perceived uselessness. The study was conducted by surveying the customers of an e-banking application in Turkey who disuse the system. The results reveal important hints for the organization that wants to get an insight into the causes of the system disuse.

Chapter 15

Biresh K. Sahoo, Xavier Institute of Management, Bhubaneswar, India
Dieter Gstach, Vienna University of Economics & Business Administration, Austria

Two alternative estimation models, i.e., a translog cost function and data envelopment analysis (DEA) based on a cost model are compared and contrasted in revealing scale economies in the Indian commercial banking sector. The empirical results indicate that while the translog cost model exhibits increasing returns to scale for all the ownership groups, the DEA model reveals economies of scale only for foreign banks, diseconomies of scale for nationalized banks, and both economies and diseconomies of scale for private banks. The divergence of the results obtained from these two estimation models should concern model builders. From an empirical perspective the definition of scale economies through a constant input mix is very restrictive. The DEA cost model is much more flexible in this respect: It neither requires the restrictive assumptions that the unit factor prices are always available with certainty, nor that these prices are exogenous to the firms. However, the very volatile nature of the banking industry might question the validity of the empirical estimates in this deterministic setting. Therefore, further research is required to examine the bank performance behavior using both SFA and chance constrained DEA for the comparison in a stochastic setting.

# Chapter 16

Inspired by patterns of behavior generated in social networks, a prototype of a new object was designed and developed for the World Wide Web – the stigmergic hyperlink or "stigh". In a system of stighs, like a Web page, the objects that users do use grow "healthier", while the unused "weaken", eventually to the extreme of their "death", being autopoieticaly replaced by new destinations. At the single Web page scale, these systems perform like recommendation systems and embody an "ecological" treatment to unappreciated links. On the much wider scale of generalized usage, because each stigh has a method to retrieve information about its destination, Web agents in general and search engines in particular, would have the option to delegate the crawling and/or the parsing of the destination. This would be an interesting social change: after becoming not only consumers, but also content producers, Web users would, just by hosting (automatic) stighs, become information service providers too.

This paper studies the differential practices of change management in organizations of western origin and compares it with the best practices prevalent in Indian organizations, with special emphasis on social and cultural challenges faced in these countries. Since Enterprise Resource Planning (ERP), as part of an information and communication technology (ICT) initiative, is frequently associated with organization change and transformation in relation to its adaptation, it has been used as the context in this study. The impact of social factors and cultural challenges on change management processes and elements are compared and contrasted using multiple case studies from USA, Canada, European (Western/Eastern) and Indian organizations who have adopted ERP technologies. The conceptual framework highlights cultural and social factors that affect ERP implementation, and offers suggestions to researchers to empirically test these influences using sophisticated analytical methods and develop change strategies and practices in response to these challenges. Further, it also draws attention to the need for a contemporary, result-oriented, quantitatively measurable framework of change management at the individual and enterprise levels. It is expected that such an approach would result in better buy-in from all stakeholders in terms of increased accountability.

Preface

"Information Systems and Modern Society: Social Change and Global Development" belongs to the *Advances in Information Systems and Social Change Series* book project. There are four sections and 17 chapters in this book.

SECTION 1: SOCIAL AND PERSONALIZED LEARNING IN WEB-BASED ENVIRONMENTS

Section one consists of 5 chapters. In Chapter 1, Malinka Ivanova and Anguelina Popova explore "Formal and Informal Learning Flows Cohesion in Web 2.0 Environment." The current state of the Web is recognized as "social" because of the multiple possibilities for participation, stimulation, and facilitation of interactions in groups, communities, networks, and among people with similar interests. Knowledge exchange is easier than ever allowing access to tools, services, and content. Social software and social media already have demonstrated their implications on the implementation of new methods for learning in both formal and informal educational settings. Web 2.0 can therefore be presented as a platform for building, running and combining services and applications. These social and technological changes have influenced Learning Management Systems (LMSs) development, proposing improved mechanisms for authoring, communication and collaboration. The features of contemporary LMSs 2.0 are conducive for the realization of multiple methods, directions, and channels for learning, encouraging informal conversation, generating collaborative content, and sharing information and knowledge.

This exploratory study examines the affordances of a free hosted LMS 2.0 for learning enhancement in order to reveal the most effective approaches to its use. The research contributes to a better understanding of formal and informal learning flows in Web 2.0 environment, including LMS 2.0 and applications beyond it, that can help lead to improvements in students' learning experiences. The activities of students are examined and their analysis is used as a basis for a model showing the preferred places and applications for formal and informal learning. This research explores the advantages of new generation LMSs combined with Web 2.0 technologies to enhance the openness of the learning process (media, content, human capabilities), bring about dynamics and interactivity in formal and informal ways, and elicit opportunities for self-regulated learning. The findings suggest that informal learning activities are performed in all parts of the learning areas organized under LMS 2.0, including personal spaces, course learning places, public libraries, and learning communities. The results suggest the positive effects of social and technological changes using the Web on student engagement and achievements. The facilitating role of informal media to bring about better learning is noted. The employment of formal learning flow strongly dominates the course learning section, and in the public repository which collects learning resources in the course domain. It is observed that several formal activities occur beyond the LMS 2.0 frame, in other Web 2.0 applications and social media. For some students, this step transfers the abstract concepts into real world scenarios, helping them see patterns and movements that explain ideas and their importance. These informal learning flows can complement and enrich the formal learning process, and formal learning flows can be premises for informal learning to occur. Further, the parallel performance of formal and informal learning activities could be used for learning optimization.

In Chapter 2, Yuri Nishihori, Chizuko Kushima, Yuichi Yamamoto, Haruhiko Sato, and Satoko Sugie introduce "Global Teacher Training Based on a Multiple Perspective Assessment: A Knowledge Building Community for Future Assistant Language Teachers." The main objective of this research is to design and implement Web-based collaborative environments to support global training based on multiple perspective assessment for both future and novice ALTs (Assistant Language Teachers) who will come to Japan from various parts of the world. The system was created in order to improve the ability to acquire professional knowledge in advance from experienced senior teachers, both Japanese teachers (JTEs) and former ALTs. The use of Computer Support for Collaborative Learning (CSCL) was adopted as the platform for their online discussion. Named "Forest Forum," it mainly focuses on a multiple perspective assessment to help support social and personalized aspects such as individual accountability, and contributions to collaboration. It also enables future and novice ALTs all over the globe to undergo pre-training and join a community of teachers before starting to work.

This research makes a major contribution to both the creation of knowledge, and the visualization of each member's contribution. Inside the Forest Forum, the actual knowledge building can be seen in the comment box, under which various topics have been visualized as trees. Contributions from each member can be visualized as flowers on the trees. When the reply to a topic is posted, a flower is displayed in the topic tree on the reply screen as well as on the home page. Whenever a reply to the same topic is posted, the comment boxes extend downward in order of posting, and the color of the comment box becomes the same as that of the flower. Therefore, one can see the degree of members' contributions at a glance. Eventually, the more trees and flowers there are on the screen, the more active is the discussion currently within the community.

The initial results are reported using system design analysis design and a Web-based questionnaire answered by the participants involved in this knowledge-building forum. 16 ALTs and 14 JTEs enrolled in the forum, and data was collected by an anonymous online questionnaire. The results indicate that not only future and novice ALTs, but also experienced teachers demonstrated greater gains in their professional development. Both future and novice ALTs came to realize the development of their professional identity with the help of experienced teachers. Experienced teachers themselves gained a deeper awareness of their professional knowledge and identity. Moreover, most of members exhibited a willingness and engagement with the community. All in all, it can be concluded that a global online discussion forum with a multiple perspective assessment can work as an effective pre-training system offered to prospective ALTs.

According to Elvira Popescu and Costin Badica, the advent and omnipresence of information systems have been revolutionizing and changing not only the way we communicate, access information, and conduct businesses, but also the way we learn. Chapter 3, entitled "Creating a Personalized Artificial Intelligence Course: WELSA Case Study" illustrates the use of WELSA, an adaptive educational system for the implementation of an Artificial Intelligence (AI) course, which is individualized to the learning style of each student. Several of the issues addressed throughout this chapter describe similar approaches existing in the literature, such as how the AI course is created, and what kind of personalization is provided in the course including the underlying adaptation mechanism. The authors also focus on whether the course is used effectively by the stakeholders (teachers and students respectively). The results presented in the chapter confirm the practical applicability of WELSA and its potential for meeting the personalization needs and expectations of digital native students.

Neil Rubens, Toshio Okamoto, and Dain Kaplan demonstrate "Expertise Learning & Identification with Information Retrieval Frameworks" in Chapter 4. In today's knowledge-based economy, having the proper expertise is crucial to resolving many tasks. Expertise Finding (EF) is an area of research concerned with matching available experts to given tasks. The traditional approach to expertise finding is typically a slow and burdensome process, involving directly contacting individuals who are familiar with the areas for which expertise is required, and then relying on their ability to provide appropriate referrals. This research looks at the use of computers to help mitigate this burden to a considerable degree. As a result of this technological support, expert finding systems (EFS) have started to gain acceptance and are being deployed in a variety of areas, for example the Taiwanese National Science Council has utilized EFS to find reviewers for grant proposals, and Australia's Department of Defense has deployed a prototype EFS to better utilize and manage its human resources.

Mostly, EF systems operate either via a content-based approach, which utilize the text of the input along with the text of the available experts' profiles to determine a match, or a structure-based approach, which analyze the inherent relationships between experts, affiliations, papers, etc. (such as are available in citation networks) to find likely experts. Both approaches have strengths and shortcomings. In an attempt to leverage the advantages of both, there has been work on combining the two; but the coupling has been loose – independently implementing both approaches and then combining their outputs. Since this increases the effort needed to implement and maintain both systems, they do not improve the representations of either, but only result in a combining of outputs at its final stage.

Recent work has demonstrated the merits of converting a content-based representation to a structure based one (e.g. semantic networks), and applying structure-based analysis methods to it. The authors show that a reverse approach can be equally if not more effective – meaning the technique of converting a structure-based representation to a content-based one. In addition to using content-based analysis methods (on initially structured data), the proposed conversion allows for easily combining both representations, by simply appending the structural contents to existing contents; unlike structure-based representations where the merging of structures is far from trivial. Finally, the proposed method allows for an integrated implementation with content-based (information retrieval) frameworks that provide easy deployment and scalability; this is something that structure-based approaches often lack.

In Chapter 5, Andrea Sterbini and Marco Temperini present "SocialX: An Improved Reputation Based Support to Social Collaborative Learning through Exercise Sharing and Project Teamwork." Collaborative learning is an all-important learning methodology, supporting learners in developing their knowledge, skills, and critical thinking, for both individual and group work. More recently the interest of researchers has been extended to comprise the social dimension of e-learning, where social exchange and interaction in a community of students support learning effectively. Since it helps to enable monitoring (as well as evaluating and supporting) a learner's participation through instructional activities in a social-collaborative learning system, a reputation system (a tool quite often found in social networks) can be quite effective in the learning environment.

This chapter profiles the SocialX system, in which the authors aim to integrate the group-collaborative and the social-collaborative dimensions of e-learning. The authors propose a comprehensive bouquet of modules that allow instructors to employ learning activities based on collaborative development of exercise-tasks (home-work) and of group projects, together with the management of micro-forums dedicated to single tasks, all connected in a reputation system equipped to represent the learners' behavior and outcomes. Peer-assessment and self-assessment are supported, as components of the reputation system. The reader will have the opportunity to evaluate the design issues of the system, and to look, in particular, at those features that make the system suitable for the following: 1) training applicative skills, dealing with the applied counterpart to theoretical, and may be more individual-focused; 2) training group-work capabilities, where, in particular, such skills are used in a project-development environment where social-collaborative features are available; 3) deploying learning experiences in a social-collaborative framework, including peer and self-assessment, towards the development of high cognitive levels of skill.

Prior to this work, the authors designed and developed a system for personalized course construction and delivery, including the management of a repository of standard learning objects. The authors are focused currently on the fusion between Lecomps and SocialX, in a framework supporting the Vygotskij pedagogical theory and implementing its characteristic Zone of Proximal Development. The result of these efforts is a personalized course construction system in which the course can comprise both individual and social collaborative learning activities.

SECTION 2: INFORMATION SYSTEMS AND INFORMATION TECHNOLOGIES ISSUES IN NONPROFITS

Section two has 4 chapters. In Chapter 6, Louis-Marie Ngamassi Tchouakeu et al. work on "Exploring Barriers to Coordination between Humanitarian NGOs: A Comparative Case Study of two NGO's Information Technology Coordination Bodies." Taking the context of the massive international response to humanitarian crises such as the South Asian Tsunami in 2004, the Hurricane Katrina in 2005 and the Haiti earthquake in 2010, this chapter examines and explores the importance of humanitarian inter-organizational collaboration and coordination. Since humanitarian nongovernmental organizations (NGOs) are increasingly collaborating through inter-organizational structures such as coalitions, alliances, partnerships, and coordination bodies, these NGOs have information technology coordination bodies as groups of NGOs aimed at improving the efficiency of ICT use in humanitarian assistance endeavors through greater coordination. Despite their popularity, little is known about these coordination bodies, especially the extent to which they address inter-organizational coordination problems.

This chapter examines the effectiveness of humanitarian NGOs' information technology coordination bodies in addressing inter-organizational coordination problems. To guide the study, the authors employ an analytic framework that enables organizing a myriad of well-known inter-organizational coordination barriers into three categories, which have been recognized as factors for successful coordination among organizations in coordination bodies. The analytic framework is applied to data collected through interviews, observation, and document analysis, from two coordination body case studies that revealed fifteen different barriers to coordination among humanitarian NGOs. Findings suggest that despite positive attitudes toward coordination by members, seven of eight widely accepted barriers to coordination still exist among members of these coordination bodies. Furthermore, in a comparison of mandate-oriented, structural and behavioral coordination barriers, the research found that mandate issues were the most significant, and structural factors were found in the greatest numbers. Findings also suggest that effective humanitarian NGO information technology coordination bodies must pay attention to both organizational design and management issues, although the former are likely to have a greater impact on coordination.

One of the greatest contributions of this research is to point out that the value of creating IT issue focused coordinating bodies in the humanitarian relief sector is that it reduces or eliminates the barrier to coordinate around resources. The authors believe that the coordinating bodies create a structure and mechanism for the home organizations and outside donors to channel funding, staff, and supplies to create collaborative IT projects that may have been impossible within any single NGO. Given the other (seven) significant barriers that still exist in this sector with IT coordinating bodies, it is significant when one can see the diminished effects of one barrier. This study has several limitations that may not allow the claim to made that all coordinating bodies, or even all IT-focused coordinating bodies, help to resolve resource barriers to coordination. However, the implications are that through the use of a well-structured coordinating body with the appropriate mandates and culture, the result may be that it helps to facilitate coordination around IT issues across organizations, at least in the area of resources.

In Chapter 7, Geoffrey Greenfield and Fiona H. Rohde focus on "Technology Acceptance: Are NFPs or their Workers Different?" Workers' attitudes towards technology, particularly those within not-for-profit (NFP) organizations, have received less attention and focus despite the growing interest in the NFP sector and the potential differences between for-profit and NFP organizations and their employees. Irrespective of firm type, the change unleashed within a firm through the implementation of new technology ultimately requires that users accept the technology. Therefore, an understanding of workers' attitude to technology allows for the improved likelihood of a successful implementation of technology in organizations.

The main contribution of this to technology acceptance research is the contrast in attitudes toward technology between NFP employees within a traditional business area, i.e., the marketing division of an NFP, or workers in the more socially sensitive area, such as social work. The present research examines whether the two groups of people who have entered different careers, have different attitudes to technology. These different attitudes to technology may in turn affect the variables contained in technology acceptance models. Data was collected using a survey instrument in two NFP organizations. In particular, the authors explored whether models developed through research focusing on the for-profit sector are equally applicable to NFP organizations and their workers. In doing this, the authors explore the relationship between attitude, perceived usefulness, and perceived ease of use in relation to technology for workers within NFP organizations.

The Technology Acceptance Model (TAM) is well accepted as a model to predict acceptance of new technology by individuals, though the universality of the TAM model, to fit across all situations, has recently been called into question. One of the main research findings was, for successful deployment of technology, organizations need to understand the attitude of their workers towards technology. Additionally, workers' underlying view of technology may drive some other decisions they make such as career choice, and ultimately their acceptance of technology. From a practical perspective, by gaining a better understanding of workers attitudes to technology, organizations can better tailor technology deployment to suit their worker's needs thus improving the likelihood of a successful implementation. That is, when deploying the same technology within an organization, different groups of employees within the same organization may have different attitudes, thereby requiring firms to consider different users when implementing new technologies.

In Chapter 8, Mohammad Mourhaf Al Asswad, Sergio de Cesare, and Mark Lycett conduct "A Query-based Approach for Semi-Automatic Annotation of Web Services." Semantic Web Services (SWS) are very important components of the future intelligent Web because they can automate the discovery and composition of current syntactic Web services. A major issue that prevents a wider adoption of SWS is related to the manual nature of the semantic annotation task. Manual annotation is a hard, time-consuming and error-prone process and thus its automation is highly desirable. Few approaches have been proposed to semi-automate the annotation task, as they are difficult to use and cannot perform accurate annotation for the following reasons: (1) They require building application ontologies to represent candidate services; and (2) they cannot perform accurate similarity measurements between service elements and ontological entities, especially those contain compound nouns (CNs). In response to calls for a more effective SWS annotation system, the authors propose a novel annotation approach that can facilitate semi-automatic annotation of Web services. This approach is simpler and more effective than existing annotation approaches. It is simpler because it does not require building application ontologies to represent WSDL files of candidate services. Instead, candidate service elements are extracted from a WSDL file and then used to generate query instances by filling a Standard Query Template. The resulting query instances are then executed against a repository of ontologies using a novel query execution engine to find appropriate correspondences for candidate service elements. This query execution engine employs name and structural matching mechanisms that can effectively measure similarities between service elements and ontological entities containing CNs.

The proposed semi-automatic annotation approach is evaluated by employing it to annotate existing Web services using published domain ontologies. Precision and recall are used as evaluation metrics. The evaluation results demonstrate the effectiveness and applicability of the proposed approach since almost complete and clean annotation results, in relation to manual results, can be obtained. Some service elements cannot be annotated using the selected ontologies, however, due to issues related to candidate WSDL files and coverage of domain ontologies. This problem is called the low percentage problem. To alleviate the low percentage problem, future research should focus on developing effective and automatic ontology extension mechanisms that can extend existing ontologies with appropriate correspondences for service elements that do not have matches in ontologies. Moreover, text analysis techniques should be used to automate the extraction of service elements from candidate WSDL files. Automating the extraction process may lead to a fully automatic service annotation process.

In Chapter 9, Yanli Zhang et al. dissect "A Broken Supply and Social Chain: Anatomy of the Downfall of an Industrial Icon." There was a time in the history of GM when it was the largest corporation in the US and in the world. The announcement of GM's bankruptcy on June 1, 2009 shocked the world and had a tremendous impact on the United States economy. This chapter explores the following questions: What were the fundamental causes of GM's problems? What can be learned from GM's mistakes and experiences? How and why did an industrial icon come to ruin? The authors analyze the reasons that led GM into dire straits during the 2008-09 financial crises. The authors argue that the stage for the difficult situation GM fell into was actually set decades ago. GM, for far too long, had been ignoring factors that were directly responsible for the success or failure of the company. A major factor was that GM had been extremely slow in reacting to the competition from Japanese automakers and the threat of more fuel-efficient cars, and it was more focused on thinking on what they could produce, rather than what consumers wanted to buy.

The chapter goes on to analyze in detail several of the internal and external factors that led to GM's financial woes. They point out the following internal factors: 1) arrogant leadership – Executives at GM were accustomed to thinking that whatever is good for GM is good for the United States; 2) unhappy customers – GM had been losing the loyalty of younger generations because it had not been listening to customer demands, 3) "Generous Motors" – GM's pensions and benefits were too generous to be affordable; 4) poor forecasting – GM never undertook scenario planning to anticipate market shifts; and 5) casual controls – GM failed to have a handle on its internal financial control and ran out cash to carry out the operations. External factors include 1) severe competition – foreign companies flooded the US market and changed the rule of the game with their quality and cost; 2) relentless environment – gas price continuously increasing put GM in a disadvantaged position; and 3) intense regulations – government regulation added to GM's cost. In conclusion, the authors argue that if GM had responded to its problems earlier, its financial troubles could have been avoided. The authors went on further to warn that this is what could happen to any company if it rests on its laurels and fails to adapt to change.

SECTION 3: INFORMATICS AND SEMIOTICS IN ORGANISATIONS

Section three consists of four chapters. Göran Goldkuhl proposes "Actability Criteria for Design and Evaluation: Pragmatic Qualities of Information Systems" in Chapter 10. Information systems actability theory (ISAT) is a conceptualisation of information systems emphasizing their pragmatic dimensions. It can be seen as a practical theory aiming to support the design and evaluation of IS. As a practical theory, ISAT comprises a conceptualisation of IS and several models. ISAT comprises also normative criteria of pragmatic character (quality ideals).

Information systems actability theory builds on a communicative action perspective on IS. Information systems are seen as instruments for technology mediated work communication. Human actors are communicating (i.e. sending and/or receiving messages) through an information system. Sending a message through an IS means performing a communicative action. In effect, the IS affords a communicative action repertoire to its users. This repertoire enables and constrains the users in their communicating. Such a communicative action perspective does not, however, dismiss the perspective of a human utilising and interacting with an IT artifact. ISAT comprises a number of aspects of human-computer interaction. However, usage of an information system is considered to be a user-via-system-to-user communication. ISAT gets its current theoretical backing from theories and knowledge traditions like pragmatic philosophy, speech act theory, classical semiotics, social action theories, affordance theory, semiotic HCI engineering, conversation analysis, discourse theory, and activity theory.

The chapter presents 19 actability criteria divided into three groups: 1) criteria concerning user-system interaction, 2) criteria concerning user-through-system-to-user communication, and 3) criteria concerning an information system's contribution to work practice processes. The first group is concerned with fundamental interaction criteria and navigation criteria and it consists of the following criteria: Clear action repertoire, intelligible vocabulary, action transparency, clear feedback, easy navigation, action stage overview, conceptual consistency, and action accessibility. The second group is concerned with reading and formulation criteria and it consists of the following criteria: clear and accessible work practice memory, information accuracy, actor clarity, intention clarity, satisficing communication needs, relevant communication demands, work practice memory addition, addressee relevant communication and addressee adapted communication. The third group consists of only one criterion: subsequent action support. These criteria should be possible to use in the design of information systems. The functions of the criteria are here to express possible quality ideals to strive for. The criteria can be used as inspiration for hen designing the system based on domain-specific goals. They can also be used in the formative evaluation of design proposals during the IS development process. These criteria should also be possible to use in post-evaluations of information systems.

In Chapter 11, Vânia Paula de Almeida Neris et al. concentrate on "Collective Construction of Meaning and System for an Inclusive Social Network." This chapter addresses the problem of designing an inclusive social network information system which makes sense to the user community and also respects their diversity. To reach this audacious challenge, the authors have adopted Organizational Semiotics and Participatory Design as theoretical and methodological frames of reference. The research took place in Brazil, and the authors provide an overview of the scenario of vast diversity found in Brazil regarding societal issues such as poverty, illiteracy, and lack of access to technologies, just to mention a few examples. Considering the challenging scenario – and one that is not exclusive to of Brazil, – the authors introduce the domain of online inclusive social networks as a platform that may contribute for the constitution of a fairer society.

The research examined a group of thirty people participating in workshops in which organizational semiotics artifacts and methods were collaboratively used, such as stakeholder analysis, semantic analysis, and use of an evaluation frame. The authors argue in favor of a socio-technical view for designing inclusive systems with the involved parties, and show an approach to accomplish it. The results from the workshops were analyzed revealing people involved or affected by the system, requirements based on problems and solutions discussed with the participants, and also perceptions of the meaning of an inclusive social network in that context. Finally, the authors present an insightful discussion about how each of the aspects discussed in the workshop influenced direct or indirectly the resulting inclusive social network system. Besides the design solution and lessons learned that could inspire other user interface designers, the authors have articulated theory and practice. Some important organizational semiotics methods and artifacts were revisited to allow collective participation of users from the local community. Moreover, another interesting contribution is the system itself that proposes ways to cope with low literacy, communication among people with different functional conditions, and system maintenance by the community; which is significantly different from mainstream social network systems.

In Chapter 12, Keiichi Nakata and Stuart Moran discuss "A Semiotic Analysis of a Model for Understanding User Behaviors in Ubiquitously Monitored Environments." Ubiquitous computing as a field is developing rapidly, driven by advances in mobile, embedded, wireless, and sensor technologies. A key enabler of the interactive aspects of this technology is an unprecedented degree of data collection specifically about aspects of its users. This data is used to provide a multitude of novel services and applications to users. However, one of the problems for data collection in ubiquitous computing is the act of collecting data itself. Existing research has shown that the observation/monitoring of people can cause changes in their behavior. Hence, when users are observed they may be affected, making any data collected about them potentially inaccurate. This could render any services or applications provided by the system ineffective. The act of monitoring and the resulting impact on system services may also lead to other undesirable effects such as stress and distrust. This is particularly disconcerting when considering the scale of ubiquitous computing. This motivated the development of a predictive model, grounded in the principles outlined by the Theory of Planned Behavior and the relationship between the system characteristics of monitoring systems, and user perceptions of those characteristics. By predicting the potential undesirable effects of ubiquitous monitoring, prior to the development and deployment of systems, it gives developers the opportunity to improve their designs to minimize these effects. The model is designed around three conceptual layers: physical, technical, and social, which have several parallels with those used in semiotic analysis. As such, semiotics was placed as an analytical perspective to gain useful insights into the problem.

The key contributions of this chapter center on the proposed model and its examination through a semiotic lens. The model provides a systematic means for analyzing and explaining user behavior in terms of both attitudes and perceptions of users in relation to systems characteristics. This is in direct contrast to other similar predictive models that do not explicitly detail the specific characteristics of a system that influences behaviors. A secondary examination of the model through semiotic framework explains how different people would interpret a monitoring device in different ways based on the unique perspective of treating such a device as a sign. This is central to the argument in the model that user attitudes and behaviors toward a system are grounded in the way a user perceives the system; and that differences in social norms that govern its interpretation and ways to interact with it lead to different, sometimes unexpected, behaviors. Overall, this chapter contributes to framing and deepening the understanding of the ways in which users can be influenced by ubiquitous monitoring systems.

In Chapter 13, Soheil Ghili, Hengameh Shams, and Madjid Tavana study "Innovation or Imitation: Some Economic Performance and Social Welfare Policy Perspectives." The tradeoff between innovation development and innovation diffusion has long been a controversial subject widely studied in the technology change and industry performance literature. This tradeoff arises from the fact that preventing free availability of existing innovative discoveries to all producers, although beneficial from an ex post efficiency standpoint, often fail to provide the ex-ante incentives for further innovation. The problem of how firms decide on innovation-related issues is of much interest and importance. In its simplest form, the appropriability problem is concerned about a firm deciding on whether or not to innovate - or how much to innovate - based on the extent to which the innovation is appropriable.

This chapter studies the interdependency between innovation and imitation and shows that a firm can be both an innovator and an imitator at the same time. While this two-fold role of a firm has been addressed sporadically in the field literature, this research considers the interdependency between innovation and imitation. In this chapter a mathematical model is constructed, addressing the question of how much to imitate in the context of the two-fold role of a firm (innovator and imitator) in a strictly competitive game setting. Other investigated issues are how firms decide on how much to innovate and imitate, and how their decisions affect social welfare under different conditions of imitability. It has also studied the effects of two widely noted policies: first, the intellectual property right (IPR) and second, the policy of treating innovation as a public good.

One of the main research findings was that imitable innovation (having low imitation cost) can be considered as a public good. By simulating the model, it was observed that in this case, tax-subsidy policy might result in more innovation development and more social welfare. Meanwhile, this policy does not function properly in the case of non-imitable innovations. It was also illustrated that IPR policy leads to more innovation development in two ways. First, it guarantees that the firm's innovations will not be easily imitated (it appropriates innovation). Secondly, it guarantees that if the firm's innovations are imitated, it will earn a profit from selling that innovation and also an equal cost will be incurred to the imitating competitor. It was also shown that the latter function is not fulfilled when the intellectual property is priced high. Another significant insight gained from this study was that an increase in imitation cost causes the effect of the tax-subsidy policy gradually become less effective and even harmful to welfare since it causes a large amount of superfluous innovation to be developed.

SECTION 4: TECHNOLOGY TRENDS AND CRITICAL SOCIAL CHALLENGES

Section four consists of four chapters. In Chapter 14, Melih Kirlidog and Aygul Kaynak initiate a "Project Technology Acceptance Model and Determinants of Technology Rejection." Technology Acceptance Model (TAM) posits that usage of an information system is determined by behavioral intention and that behavioral intention is also determined by the person's attitude towards the use of the system and also by his perception of its utility. If the individual thinks that usage of the system was affecting his/her performance, the system is evaluated as useless by them. Therefore, TAM hypothesizes a direct link between perceived usefulness and perceived ease of use.

The main contribution of this research is to show whether there is such direct relationship between usefulness and ease of use. While doing this the authors employed the negative or reverse of these ideas and observed the relationship between difficulty of use and uselessness on user rejection of technology. They conducted a survey on the customers of an Internet banking application in Turkey who disused the system. One of the research findings was that where a survey question was related to usefulness in a negative way, question about easiness of use has also been answered negatively. Survey results were collected and evaluated using SPSS. The reliability and variance analyses supported the validity of this study's TAM constructs which are perceived uselessness and perceived difficulty of use in evaluating the surveyed Internet banking application. According to their results, one of the main study findings was that TAM is useful for understanding the user behavior for not adopting the system as much as TAM is useful for adopting an information system. Collected data also showed that personal privacy and involvement both had a significant effect on a customer's decision. Collection of data taken almost 2 months, and users were reluctant to participate survey. However the authors convinced a majority of users to do the survey by committing that these will be used for improvement of the system. The results of the research were shared with this bank. The bank considered these and took necessary steps in order to improve and enhance current its Internet banking application.

In Chapter 15, Biresh K. Sahoo and Dieter Gstach pinpoint "Scale Economies in the Indian Commercial Banking Sector: Evidence from DEA and Translog Estimates." To promote efficiency and competition in the Indian financial sector, the Reserve Bank of India initiated in 1992 a number of reforms. In this scenario the authors believe that banks are in the pursuit of enlarging their sizes using available scale economies in order to enhance their asset base and profit so as to meet global standards. This chapter applies and compares two alternative approaches – the translog cost model and the DEA cost model, to analyze the impact of competition on scale economies' performance across the entire spectrum of ownership groups of the Indian commercial banking sector. This will enable one to investigate the economic linkage between ownership and performance in the light of the property right hypothesis and public choice theory. The common premise underlying both the approaches is the deterministic nature of the observations.

The empirical results indicate that while the translog model exhibits increasing returns to scale for all the ownership groups, the DEA model reveals economies of scale only for foreign banks, diseconomies of scale for nationalized banks, and both economies and diseconomies of scale for private banks. The divergence of the results obtained from these two estimation models may arise from two sources: one is the model set up, and the other is the assumed deterministic nature of production, employed in both the approaches. From a theoretical perspective, the use of DEA is advantageous as the econometric method can confound the effects of misspecification of functional form with scale economies, flexible functional forms are susceptible to multicollinearity, and theoretical restrictions imposed to have a well-behaved production technology may be violated. From an empirical perspective, the definition of scale economies through a constant input mix employed in the translog model is restrictive. The DEA model is, however, flexible in this respect: it neither requires the restrictive assumptions that the unit factor prices are always available with certainty, nor that these prices are exogenous to the firms. This is so because the cost setup in the DEA model assumes that firms not only have control over the mix and quantities of inputs used but also exercise control over input prices. Using several different aspects of production planning process, the DEA cost model imputes a multi-factor perspective in its scale estimates to track overall performance. However, the very volatile nature of the banking industry might question the validity of the empirical estimates in this deterministic setting. Therefore, further research is required to examine the bank performance using both SFA and chance constrained DEA for the comparison in a stochastic setting.

In Chapter 16, Artur Sancho Marques and José Figueiredo seek "Stigmergic Hyperlinks: A New Social Web Object." Relatively unnoticed, at the core of every hypertext system is the hyperlink, so ubiquitous and so simple to use that hardly anyone ever glances at it long enough to question: could it be different, serve more than a navigation purpose, serve a better informed understanding of the Web and help rethink some other information systems? The authors glanced at the common hyperlink and gave it a vitality by calling this new Web object species as the "stigmergic hyperlink," or stigh, inspired in Biology's "stigmergic" behaviors, i.e. indirect communication via some mark (stigma) of work (ergon) over a shared medium, e.g. termites' pheromones while building a nest. Stighs look exactly like regular hyperlinks, but thrive when used and fade when neglected, relatively to the usage of their siblings, eventually to a terminal point when their vitality nulls. This attribute and associated behaviors create some interesting possibilities and applications. Web users perform like insects, leaving a mark when clicking hyperlinks.

The most immediate contribute of stighs is as an automatic and decentralized solution to the broken links problem. By using stighs instead of regular hyperlinks, content authors put in place pointers that, if their linked destinations disappear, on time the corresponding stighs will self-terminate after being ignored relatively to the hyperlinks that remained valid, assuming that users see less value in broken resources and hence stops following them. On the same principle, one other application is as the basic building blocks of recommender systems, where the vitality reading is at the root of the classifications. Other strictly usage based applications include real-time pricing systems for hard to price items, such as digital stock photos.

It is important to notice that there is neither central intelligence nor direct communication involved, contrary to other approaches. Each stigh is autonomous, reading its local environment and responding to it, in a bio-inspired "stigmergic" fashion. Direct applications are demonstrable since the authors did materialize the stigh into an existing Web object, first using the C# language and requiring .NET compatible Web servers, and currently on open source using PHP. The authors also coded some artificial users for stress testing. Furthermore, they have been researching other less direct applications, some dependent on factors out of their control, namely an alternative PageRank measure, computed with effective usage instead of just with structural data, and new business models that could support new relationships between content providers and infomediaries, by using stighs as objects that could be extended to serve specialized "Deep Web" search methods for their linked destinations.

In Chapter 17, Sapna Poti, Sanghamitra Bhattacharyya, and T.J. Kamalanabhan face "Social and Cultural Challenges in ERP Implementation: A Comparative Study across Countries & Cultures." Change management in the context of Enterprise Resource Planning (ERP) is not given importance to the extent that would be expected. This research is pivotal in nature because of the fact that it highlights how different countries approach the subject of change during an ERP implementation. In certain cases while the variables studied are the same, the way each country approaches the variable varies, and one needs to understand the subtle differences of approach. In this research, data was collected by way of case recordings based on open ended questions. Further, personal interviews of respondents were undertaken to elicit qualitative information that could support the questionnaire responses. In addition to direct interviewing, company documents, e-mails, and group discussions were used to shape and triangulate findings.

The impact of social factors and cultural challenges on change management processes and elements are compared and contrasted using multiple case studies from USA, Canada, European (Western/Eastern) and Indian organizations who have adopted ERP technologies. The conceptual framework highlights cultural and social factors that affect ERP implementation. These factors should be studied prior to implementation of an ERP. Further organizations should customize their change initiatives as per the influence of the factors in their country and culture. Finally, the research framework proposed suggests a focus on certain key elements of change processes and strategies that could help address these social and cultural challenges faced by organizations in specific countries. If organizations ignore the pulse of culture and do not tweak the change processes to combat the situation they may not be satisfied with the success rate of the ERP implementation. Such organizations will also find themselves handling huge resistance to the change. Hence, this study brings in the required awareness to manage an ERP change.

John Wang
Montclair State University, USA

Jeffrey Hsu
Fairleigh Dickinson University, USA

Section 1
Social and Personalized Learning in Web–Based Environments

Chapter 1
Formal and Informal Learning Flows Cohesion in Web 2.0 Environment

Malinka Ivanova
Technical University of Sofia, Bulgaria

Anguelina Popova
University of Utrecht, The Netherlands

ABSTRACT

This paper presents the results of an exploratory study examining bachelor degree students' experiences of learning with a new generation learning management system Edu 2.0 combined with Web 2.0 applications. The authors discuss students' perceptions of formal and informal activities within this environment as captured through a collection of surveys, activities' tracking, and assessment. The main functional characteristics and available social tools of Edu 2.0 are examined in the context of students learning support. A model of Learning area is developed to analyze the formal and informal learning flows from the point of view of learning enhancement.

INTRODUCTION

In spite of recent debates at conferences, web sites and blogs (Stiles, 2007; ALT-C 2009 conference, 2009; Goldsmiths University London Blog, 2009) about the end of the Learning Management Systems (LMSs), universities and training organizations still successfully (Rankine et al., 2009;

Sterbini & Temperini, 2009) empower LMSs to support a high-quality education delivery in a blended-learning model (Ellis & Calvo, 2007) or in distance learning. The core components of the LMSs enable students to access the course content, as well as to participate in the course activities in formal learning flows. Web 2.0 technologies and eLearning 2.0 strategies influence LMSs development and implementation. Today's LMS rapidly adapt to meet the needs for social and informal

DOI: 10.4018/978-1-4666-2922-6.ch001

learning. Research shows that some corporate and open source solutions are moving fast in this direction. For example, Blackboard 9 is extended in the highlights of new Web 2.0 and social learning capabilities including blogs, journals and enhanced group tools; notification dashboards highlighting time sensitive information and alerts; and a completely redesigned, customizable Web 2.0 user interface. The open source LMS ATutor adds a social networking module "ATutor Social" that allows ATutor users to connect with each other. They can gather contacts, create a public profile, track network activity, create and join groups, and customize the environment with any of the OpenSocial gadgets available all over the Web.

In the context of the expansive adoption of social software, "traditional LMS" evolve to a new generation of LMSs 2.0. These are extended into building networks within a course, or between training institutions registered in the system and among all the registered students, educators and professionals. The LMSs 2.0 represent a new way of thinking teaching and learning that has profound implications not only in terms of traditional concepts of authority and value, but also on the opportunities presented for developing and sustaining communities of practice, content generation by educators and learners and the aggregation of resources (TimeCruiser Computing Corporation White Paper, 2008). The content provided can come from mixed resources.

In the perspective of eLearning 2.0 ecosystems in general, LMS 2.0 environments are still an underexplored segment of eLearning. Elearning 2.0 ecosystems focuses on collaborative and open learning techniques, where learners are not at the end of the learning chain but actively participate in the learning process as authors, co-authors and contributors of knowledge and their products are based on collective intelligence and personal progress. eLearning 2.0 ecosystems capture the learning space as a mashable space for personal activities and for collaboration and communication with other learning communities. These character-

istic grows the interest for informal learning. There are different resources as evidence that the learners are engaged in a wide range of technology-based informal learning at home and the community of practice (Cranmer, 2006; Gray, 2004).

It is a challenge to explore and organize the tools and approaches for learning based upon new social computing capabilities within the LMSs 2.0. Arrays of learning activities are facilitated not only within the specified course environment, but also beyond its borders (McLoughlin & Lee, 2008). Students have the opportunity to interact with the educator and classmates in a given course as well as with peers and experts from specific groups and communities. As the options grow for different learning flows to coexist within the online learning environment, so does the necessity for identifying the types of these flows that are the most valued by students (Canzi et al., 2003). In the same way, a new model for assessment of students' learning and their achievements in Web 2.0 environment is needed. Recently, educators try to weave the informal learning forms and informal media in a formal learning process. For example, using wiki for constructivist learning environments forming, as it facilitates collaboration (Notari, 2006); utilizing blog for group work, for sharing course related resources, and for submission of students' assignments and home work (Luján-Mora, 2006); microblog to train communicative and culture competence anytime anywhere without face-to-face interaction (Borau et al., 2009), start pages for self-organizing learning and personal development (Ivanova, 2009).

The focus of this work is on formal and informal usage of the next generation Learning Management Systems (LMSs) based on Web 2.0 technologies (LMS 2.0) and Web 2.0 applications for enhancing learning. The aim of the exploratory study is to examine the learning experiences and flows occurring in Edu 2.0 Learning Management System and beyond it: in Web 2.0 environment, gained by 110 bachelor degree students (in a two years project) concerning the various learning activities that they undertake during a course in Computer Graphics.

The following four primary research questions guided this research: (1) How could the affordances of LMS 2.0 for learning enhancement be used in the most effective way? (2) How could LMS 2.0 social capabilities be integrated into a learning process? (3) How could students' learning experience be improved combining formal and informal learning flows? (4) How could student-generated artifacts be incorporated in a formal learning program?

The paper presents a model describing the formal and informal learning flows in Edu 2.0 which took place during the course. The meaning of the learning flows in terms of learning enhancement is analyzed in the light of performed students' activities and survey results.

BACKGROUND

Edu 2.0 Learning Management System Affordances for Learning

The course Computer Graphics is implemented within the free web hosted learning management system Edu 2.0. The course incorporates a variety of content and learning management tools typical for the standards of LMS. It also integrates Web 2.0 tools and services which enable students to be authors, co-authors and developers, and to interact online to communicate, collaborate and share information in a wide range of formats in formed groups, community and networks. The learning area spreads out upon four learning spaces, briefly described below.

Personal Learning Space

The personal learning space provides support for students to set and manage their own learning via: (a) a calendar allowing them to plan and manage learning activities and meetings, record class events and assignments deadlines; (b) a personal blog stimulating analytical thinking and problem reflecting; (c) a portfolio is used for storage of the contributed artifacts and also as a digital record of the learning process; (d) image and resource sharing as a reflective activity on the subject matter; (e) contact with classmates, friends and community members to exchange ideas or get advice, (f) game section affording learning by contributing with questions, learning from mistakes. The personal space in Edu 2.0 is extended with the use of a start page as a mashable and flexible environment for aggregating resources, widgets and various applications.

Course Learning Place

There is a standard *Lessons* section, where activities related to lectures and laboratory practices are formed in sequential or hierarchical structures. The content in this section is aggregated from different social media using the web as a platform: PPT presentations from Slideshare, videos from YouTube, podcast programmes, widgets and other applications. The learning activities then can be completed in other Edu 2.0 sections as well as outside the system. The *Resources*, *Assessment* and *Feeds* sections have standard functions, but with possibilities for embedding content and linking to various social sites. The Course calendar is an important part for arranging meetings, consultations, and events. Collaborative work, project-based work, group work is organized in the Collaboration and communication section via course blog, wiki, debates, chat and forum. The course blog allows each course member to write and share a secure web journal that is used as a digital record of the learning process. An ebook with project information, project examples and evaluation criteria is created within the wiki. Every student can extent the ebook with his/her contribution and get acquainted in working with the wiki. The debate function is not yet used in the Computer Graphics course, but it is an important tool for encouraging critical thinking. The communication process among students and educators is facilitated by chat, forum and internal email.

Public Resource Repository

Within the lessons, every link to public web resources added by educators is automatically submitted to Edu 2.0 moderators for addition to the library. Also, every registered user - student or educator can manually submit a resource. The library grows together with the contributions of the community. In this case the learning resources can easily be found through a powerful search engine with advanced options for searching by subject, grades, contributor's name, language, object type and format. The resource repository is user-centered: students dynamically use existing and add new content; it provides a multimedia experience, accumulating various media formats: blogs, ebooks, presentations, surveys, audio/video, etc.; it is socially engaging: students can add to favorites and vote about resources; it is community innovative - it could be modified by community participants.

Participation in the Community

The community section provides direct access to registered schools, groups, and members from over the world in Edu 2.0. Social networking features allow participants to create and view social profiles, create friends lists, and directly message each other, and also create groups with special interests. By connecting to a different part of this community, students meet particular needs and goals by expressing personal opinions, asking for support or specific information and by sharing ideas and experience.

Informal Learning and LMS 2.0

There are typically three subsystems of educational activities: formal, non-formal and informal. For-

mal learning is curriculum-based and takes place within an educational institution. Non-formal learning refers to all organized learning programs that take place outside the formal school system, that are usually short-term and voluntary, and might provide with a certain certification, or not. As for informal learning, since its nature is not captured in any particular form neither follows a pre-described path, definitions differ. Livingstone (1999) defines informal learning as "any activity involving the pursuit of understanding, knowledge or skill which occurs outside the curricula of educational institutions, or the courses or workshops offered by educational or social agencies." Informal learning includes self-directed learning, incidental learning, and socialization. Informal learning is usually seen as learning occurring in relation to practice-based activities (e.g., work). Marcia Conner in her website (Conner, 2009) defines informal learning as follows: "Informal learning describes a lifelong process whereby individuals acquire attitudes, values, skills and knowledge from daily experience and the educational influences and resources in his or her environment, from family and neighbors, from work and play, from the market place, the library and the mass media". With the emerging of Web 2.0 technologies, understanding of informal learning expands beyond the practice context, to take place in virtual environments and among online communities of interests and networks. There are different articles that report about increased learners' engagement in informal learning at home and the community (Jokisalo & Riu, 2009; Chang et al., 2009).

The informal learning based on Web 2.0 exists and functions like a "university in shadow" with possibilities of a borderless digital learning environment. In the context of LMSs of second generation (e.g., augmented with social software applications), repositioning informal learning seems necessary. Social software affords both individual and social aspects of informal learning. Formal assessment in LMSs 2.0 on the other side extends beyond the formal learning settings. This study sees informal learning in LMS 2.0 as learning occurring in informal settings, made meaningful by the process of its integration into formal learning assessment.

METHODOLOGY

The methodology includes steps related to information gathering about students' activities and their role in the learning process through surveys, monitoring and tracking students' behavior and assessment results.

Students' learning activities and their arrangement in flows were tracked via the platform and assessed throughout two academic years. Typical learning flows scenario is examined as evidence of the formal and informal learning activities. Additionally, data was gathered through a survey questionnaire on students' experiences and perceptions of formal and informal activities facilitating their learning within the Edu 2.0 environment and beyond its borders. The survey questionnaire asked students about their previous experience of Web 2.0 applications, their familiarity with Web 2.0 tools before the beginning of the course, the frequency of Web 2.0 usage, the Web 2.0 tools used in Edu 2.0, the time spent in different sections of the Learning area.

RESULTS

Learning Flows Scenario in Computer Graphics Course

Computer Graphics have become common in many aspects of scientific research as well as in our daily life. The high visual impact has encouraged its usage in computational biology, physics, medicine, CAD/CAM/CAE, digital art, information visualization, scientific visualization, video games, movies, virtual reality, web design, education.

Applied Computer Graphics is a unique part of Computer Science education which bridges mathematics, physics, art, and engineering techniques.

At lectures time, students learn mathematical representation of object's generation, transformation and modification via matrix apparatus, and also fundamental algorithms for clipping, shading, mapping, coloring, some more advanced techniques including raytracing and radiosity. In laboratory practices they use professional graphics package 3DStudioMax for models and scenes constructing. The coursework is focused on applying the acquired knowledge and skills during lectures and practices into realistic visual solution.

Understanding and transferring the theory into realistic three dimensional scenes or object models requires precise modeling, arranging and rendering of objects, light, effects and cameras. This is possible after detailed examination, visual mapping and understanding of the real spatial approaches. The process is characterized with analytical thinking, analyzing in the context of subject matter and creating unique artifacts. Deep and meaningful formal learning with its main forms of interaction student-educator, student-student, and student-content (Anderson, 2003) is facilitated by using informal learning flows. One scenario often exploited by students in their individual projects, which demonstrates the cohesion of formal and informal learning flows, is the following:

- The students are introduced to the state-of-the-art of problems in the area of Computer Graphics and they are aware of the potential projects' topics. They have to identify a challenging problem and start the project. During the course, the students build their Personal Learning Environment (PLE) utilizing search engines, RSS feeds, video.

- During work on the project, the students use printed books and tutorials in electronic format and in form of wiki pages.
- When they choose a more complex project they need additional knowledge and skills to complete it. Then they read forums, participate in groups with special interests and discuss their problems.
- Also, the students often use the internal Edu20 email messaging to contact educators and peers, to explain the problems and to look for solution and advice.
- The educators provide them with immediate feedback and recommendations and according to these students do the following:
  ○ Update their Personal Learning Environments according to their emergent goals and interests, including:
    - Add new education resources – open courseware lectures and tutorials, videos and presentations, podcast programmes;
    - Expand their social networking and add new contacts of learners/educators/experts working in the similar areas;
    - Add other tools, applications and services to support their learning and the specific project.
- As they move forward to complete the project, they frequently access the course learning materials and manuals to complete several tasks: to remember theory, practical techniques or path for model implementation.
- Also, they take self-tests available in Edu 2.0 to identify their knowledge state.
- After they complete and defend their projects, the projects are posted on the course web site and are available for online and offline course discussions.

In this scenario, there is a combination of *formal learning* solutions: reading course materials and tutorials and participation in virtual coaching sessions and *informal learning* flows, integrating searching for books, videos, reading RSS feeds, communicating with peers and experts via social networks, participating in discussions and groups.

As this scenario shows, there is a great amount of learning going on outside the formal structures designed for the course of Computer Graphics. In the case of students' projects, informal learning can both support and extend the formal learning of learners when they choose to learn independently.

Formal and Informal Learning Flows

In this paper the flows are described in terms of connectivity (one-way, two-way, static connections, dynamic connections, etc.). Learning flows address connectivity among nodes (student, educator, content, class, group, community member) that are part of the learning process. Formal learning flow occurs in planned learning that derives from activities within a structured learning setting. Informal learning flow occurs in a variety of places: groups, community of practice, networks, social sites and working with Web 2.0 tools.

Edu 2.0 proposes an environment for getting the advantages of formal and informal learning and achieving effective learning as a final aim. The observed learning flows are presented in Figure 1.

The study shows informal learning flows appearing at all four parts of the Learning area. The personal learning space is used by students for management of their own goals, resources, artifacts and relationships. According to their present interests they consume information or contribute to the public repository or they have the possibility to look for ideas and advice from community members. In the course learning place students in an informal way communicate with peers and educators or participate in self-formed and regulated groups (e.g., to solve a problem or work on a project). Formal learning flow strongly domi-

Figure 1. Formal and informal learning flows

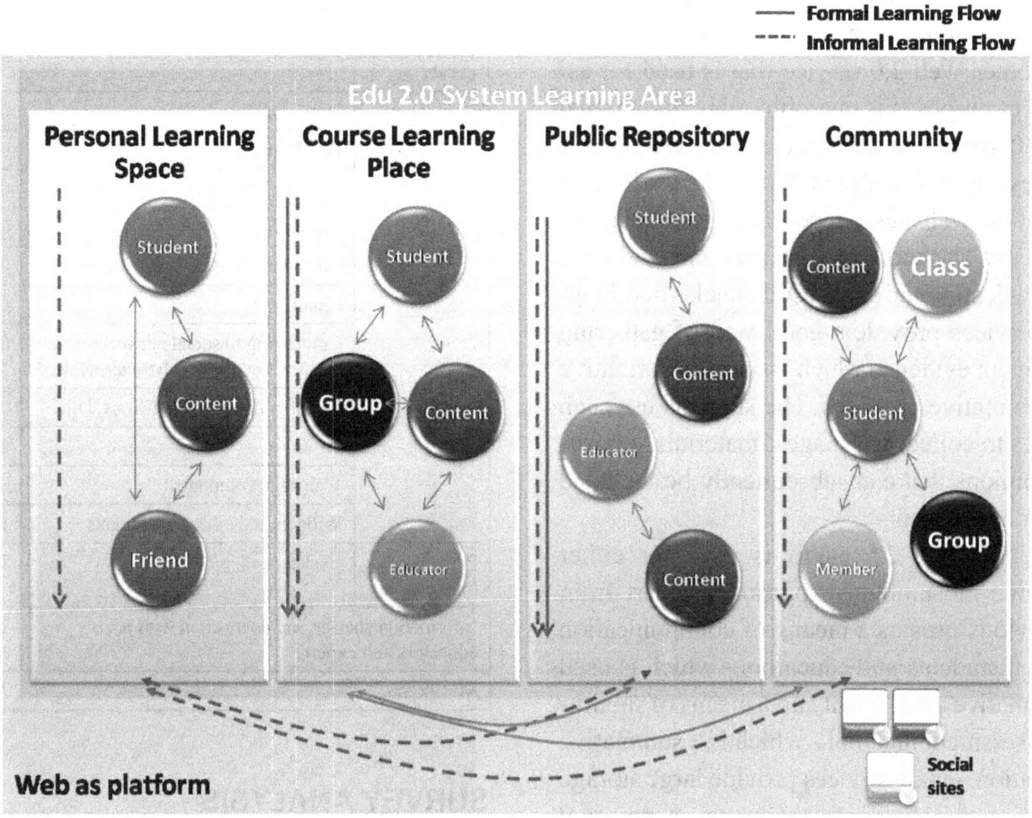

nates in the course learning section with units of learning (UoLs) organized in simple sequencing or in hierarchical structures and where instructions for learning activities performing take place.

Formal work occurs mainly in the public repository interacting with learning resources in the course domain. But several formal activities occur beyond the Edu 2.0 frame, in other Web 2.0 applications (in informal media), for example video watching in YouTube or working with Glogster. While participating in the course students create and share their artifacts, add resources of interest, communicate via forum, blog, wiki, so the course is dynamically extended and reshaped. The educator has to foresee such a design of UoLs that allow their extending and supplementing according to the learning activities performed by students without breaking the learning strategy.

Assessment 2.0

During the course, students built a PLE via a start page Netvibes/iGoogle/Pageflakes preparing their own design about numbers of panels, kind of components, customized look and information sharing. The main components that they use are for planning learning activities: creating a list of activities, using a simple text editor, access to blogs and wikis; RSS syndicating information of rich media sources; exploring and researching via search engines and using additional widgets, for example polls and analytical tools; collaboration and networking: sharing of information and knowledge, connecting to social networks; personalization of feel and look as well as using widgets, for example for quizzes and surveys. The shared panels and components with suitable information are under formative assessment: the given marks are as Table 1 shows. The given points form 5% from the final course mark.

Another Web 2.0 service that is used for assessment purposes is microblogging via internal Edu 2.0 statement field and via Twitter. These services allow sharing and discussing the links to interesting websites, videos and other learning resources. The tools are used for fast and easy communication and as a FAQ channel. Microblogging services provide a good way of gathering assessment evidence which is used for formative and summative purposes. The students use such services to collect web pages, materials, advises and opinions that can subsequently be included in their assessments.

Internal Edu 2.0 email is used for either formative or summative purposes. Apart from the obvious uses as a means of communication between students and educators - which is used for formative assessment and a means of submitting assessment material - which is a summative application, email services provide large storage capacities that facilitate their use as personal portfolios.

The PLE components, microblog and participation in the community are utilized for peer assessment, in which students comment on and judge their colleagues work. In our case the peer assessment has a role in formative assessment, but it can also be used as a component in a summative assessment package. It is a technique that increases the ability of the learner to make independent judgments of their own and others' work. In terms of summative assessment, studies have found student ratings of their colleagues to be both reliable and valid.

Table 2 summarizes the evidence for students' participation in the course Computer Graphics and the way they influence assessment process.

Table 1. Assessment of PLE building

Personal Learning Environment building		Points
Number of panels	3-7	1
Components used	Hyperlink	0,5
	HTML module	0,5
	RSS feed module	0,5
	Search engine	0,5
	Calendar	0,5
	Email	0,5
	Connecting to social networks Twitter, Facebook, MySpace, Ning	0,5
	Social bookmarks: Delicious, Diigo	0,5
	To-do list	0,5
	Web page component	0,5
	Widgets and other components	1
Personalized look and feel	Background, picture	1
Information sharing and connection with peers, educators and experts		2
Max points		*10*

SURVEY ANALYSIS

The study sample includes all students enrolled in the course Computer Graphics - 110 students in 5 groups (two years project), average age 22 years, female/male: 30/70%.

The course starts with a survey about students' previous experience with LMS and Web 2.0 applications. The students can choose more than one option from the survey's answers. The results point that YouTube and Facebook are well known applications and they are used by respectively 30% and 25% students (see Figure 2). After them are Netvibes, Flickr, Vimeo and Delicious with 11%, 9%, 9%. Most of the students have several favorite Web 2.0 applications and services. According to their LMS usage, nobody at this moment has no Web 2.0 learning experience.

Table 2. Assessment based on participation

Web service	Uses in a learning process	Formative	Summative	Peer
PLE Netvibes/iGoogle/ Pageflakes	Evidence PLE bulding Evidence collection	yes	yes	yes
Microblog	Link sharing, discussion, FAQ	yes		yes
Email	Communication between students and educators	yes	yes	
Calendar	Assessment scheduling		yes	
Course Blog	Reflection, analytical thinking		yes	
Photo storage	Evidence storage		yes	
Portfolio	Evidence storage	yes	yes	
Wiki	Projects work		yes	
Create contacts, participation in community	Evidence discovery Peer support Reflection			yes

The survey asked students to rate their familiarity with five different types of Web 2.0 tools. The distribution of the self rating scores (by percentage) is shown in Table 3. To summarize the findings from Figure 3, the self-rating scores on familiarity with blog, wiki, games, RSS feed and social networking tools indicated that: (1) The majority of male and female students say that they do not know much about blogs. (2) The students are more familiar with wikis than blogs. There is no sizeable difference among male and female students according their knowledge and experience with wikis. (3) In the area of internet gaming, 21% t of the male respondents and 7% of the female that they are very well familiar with online games. (4) In the area of RSS feeds knowledge males rate themselves much higher than females. 31% of the females surveyed are reported they do not know about RSS technology. (5) The female students rated themselves much higher using social networking software such as Facebook, MySpace and others than males.

The students were also asked to indicate the frequency (one time/day, 2-5 times/day, 1-3 times/ week, several times/month) of Web 2.0 tools usage such as blogs, wikis, RSS feeds, games and social networking tools. The results are summarized in Table 4 and the conclusions are the following: (1) The high number indicating non-usage or low-usage and participation of blogs correlates with respondents' self-rating of low knowledge about blogs. This is consistent among both male and female students. (2) The data suggests that more of the students are involved in some form

Table 3. Student familiarity with Web 2.0 tools before course starting

	Very good		Good		Average		Poor		Don't know	
	Male	Female	Male	Female	Male	Female	Male	Female	Male	Female
Blog	7%	8%	10%	7%	11%	15%	34%	21%	38%	49%
Wiki	14%	16%	19%	17%	17%	27%	28%	34%	22%	6%
Social network	27%	31%	21%	28%	10%	17%	15%	11%	27%	13%
RSS feed	18%	14%	21%	11%	13%	9%	31%	35%	17%	31%
Games	21%	7%	19%	10%	20%	23%	25%	38%	15%	22%

Figure 2. Experience of web 2.0 applications usage

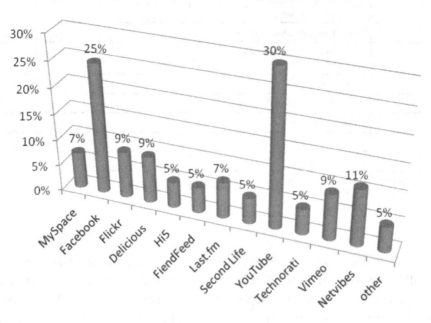

of wiki activity – reading/writing with different frequency. Among those who use wikis, the majority often: several times/week consult themselves with wiki information and learning resources. Frequency of wiki usage is similar for both male and female students. (3) The male students dominate the online internet gaming. 21% play several times per week, 28% play games several times in a month. 13% of the female students play online games 1-3 times/week and 21% of them play games several times in a month. (4) Social networking tools usage is slightly dominated by female students. 37% of the females and 33% of the males use social networks several times in a month.

At the end of the course data was collected in the form of surveys with questions about students' participation and preferences about formal and informal learning activities in Edu 2.0. The results summarized below (see Figure 3) show that students spent most of their time in Edu 2.0 in the

personal learning space and slightly less in the course learning section.

At this moment the possibility for networking with other peers and professionals over the world and group participating beyond the course boundaries is underused. Only a small number of students are involved in such networking. In contrast to this networking among students from different groups and among students and educators is well developed, and this facilitates informal learning.

The analysis of the Web 2.0 tools used in Edu 2.0 system points that students are engaged in the learning process through file adding, sharing and managing, with intercourse and intergroup messaging, and through wiki content creation (see Figure 4). Also, students are active in activities/ events planning, posting reflections in the forum, RSS subscription and news reading.

Survey results on students' familiarity to wiki, blog, RSS feed and social networks at the end of the course are shown in Table 5. They indicated

Figure 3. Time spent in different sections of the learning area

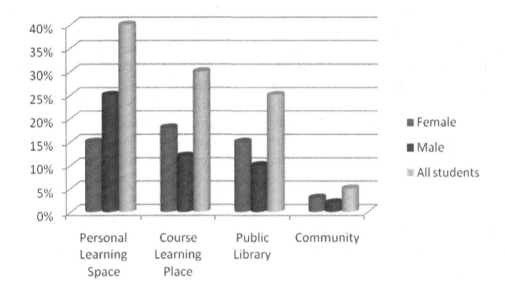

that: (1) A high percent of the students who were actively involved in the course, know these Web 2.0 technologies better at the end of the course. (2) At the end of the course nearly the half of the students reported they were familiar with all tools in Edu 2.0 and other Web 2.0 tools that they have used this semester, such as SladeShare, Scribid, and so on. (3) Social networking relationships between students and students-educators from this course are well developed. More students participate in social networks other than Edu 2.0 such as Facebook, Ning, Flickr, Delicious, Twitter.

The 50 students (participated in the course this year) were asked: "If the Web 2.0 tools are valuable for their learning experience?" According to

Figure 4. Used web 2.0 tools in edu 2.0

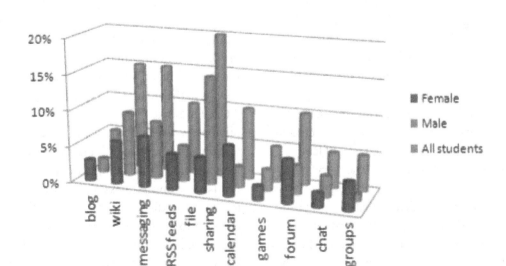

the findings, the students perceive the Web 2.0 as enhancements to their learning experience: (73% male and 70% female agree/strongly agree), represented in Table 6. Also, they were asked: "If the Web 2.0 tools help them for better understanding of course concepts?" The students responded that the Web 2.0 tools increased their understanding of course concepts (69% male and 70% female agree/strongly agree). Most students agreed/strongly agreed that they were pleased with their performance on the Web 2.0 tools usage.

THE IMPACT OF THE PROPOSED TECHNOLOGICAL SOLUTION ON STUDENTS' ACHIEVEMENTS

The results of this study show the positive effects of technology on student engagement and achievements and also the facilitating role of informal media for better learning. According to the curriculum of the course Computer Graphics formative quizzes (at the start of course, during it and at its end and summative quizzes) and also summative quizzes for self-testing are planned. Many of formative quizzes and all self-quizzes are performed in the Edu 2.0 environment. The pre-test results of students using Edu 2.0 are the same (with a little bit of difference) as students that did not work in eLearning system (the students from 2 years ago): excellent-6%, very good-13%, good-30%, average-46%, poor-5%. This means that one small part of the students is introduced with the vision and main topics at the beginning of the course Computer graphics.

The formative quizzes during the semester show how students learn new concepts, facts and topics. The results of the new/old way of learning are as follows: excellent-19%/9%, very good-25%/16%, good-31%/29%, average-16%/31%, poor-9%/15%. This indicates that the proposed possibilities of such eLearning environment offering new forms of content availability, communication channels and a wider range of online activities, led to the increased motivation for learning and engaged the students' attention on the course topics.

At the end of semester, the results show increasing marks of the scale excellent-very good: excellent-29%/15%, very good-36%/26%, good-25%/35%, average-7%/15%, poor-3%/9%.

Also, the functions of Edu 2.0 for self-testing are exploited by 67% of students that have regularly passed the created tests on different topics receiving the objective and fast evaluation of their current knowledge.

CONCLUSION AND FUTURE WORK

The positive characteristics of Web 2.0 technologies do not stand alone from other considerations such as the social and cultural settings in which learning is situated. Traditionally, students are immersed in formal learning forms. Also, most of them have used social media without understanding or intentionally using its power for learning. Participation in the course focused their attention on: (1) How to use social software in formal learning and (2) How informal interactions could facilitate formal learning. Students demonstrated engagement in Web 2.0 technologies for authoring, social networking and community participation.

The exploration study indicates that openness of the learning process (media, content, human capabilities), dynamics and interactivity in formal and informal ways and opportunities for self-regulated learning are advantages of the new generation of LMSs combined with Web 2.0 technologies The study points at the complimentary adhesion between different types of learning flows. It highlights the ways informal learning flows can complement and enrich the formal learning process and points out that formal learning flows can be premises for informal learning to occur. Students' learning experience was enriched during the course not only because formal and informal learning activities are performed in parallel, but also because students could interact with and within Edu 2.0 which is a flexible and mashable environment where a wide range of Web 2.0 tools and services are available. The created virtual learning space continues to be used as a knowledge source and communication tool after the end of the course itself. Several of the artifacts added by students are incorporated in the course' formal learning - videos, images, links, sites, widgets, interactive posters.

Future tasks will treat the following problems:

- Experimenting with engaging socially-oriented scenarios which incorporate the most important tools for students and suitable for engineering education;
- Trying to facilitate some educators' activities by forming self-regulated students' groups;

Table 4. Frequency of Web 2.0 tools usage

	one time/day		2-5 times/day		1-3 times/week		several times/month		Non-usage	
	Male	Female	Male	Female	Male	Female	Male	Female	Male	Female
Blog	-	-	-	-	4%	6%	12%	13%	84%	81%
Wiki	7%	6%	2%	3%	36%	37%	23%	24%	32%	30%
Social network	3%	5%	7%	9%	23%	29%	33%	37%	34%	20%
RSS feed	1%	-	-	-	4%	5%	45%	37%	50%	58%
Games	8%	5%	2%	1%	21%	13%	28%	21%	41%	60%

Table 5. Web 2.0 Tools Knowledge at the end of the course

	Very good		Good		Average		Poor		Don't know	
	Male	Female	Male	Female	Male	Female	Male	Female	Male	Female
Wiki	48%	52%	29%	27%	14%	13%	6%	7%	3%	1%
Blog	43%	50%	29%	25%	17%	17%	8%	6%	3%	2%
Social network	41%	49%	31%	30%	18%	19%	9%	1%	1%	0%
RSS feed	38%	35%	26%	26%	28%	27%	6%	9%	2%	3%
Games	21%	7%	19%	10%	20%	23%	25%	38%	15%	22%

Table 6. Web 2.0 tools and learning experience

	Strongly disagree		Disagree		Neutral		Agree		Strongly agree	
	male	female	male	female	male	female	male	female	male	female
Web 2.0 tools are valuable for my learning experience	3%	2%	1%	3%	23%	25%	44%	43%	29%	27%
Web 2.0 tools help me for better understanding of course concepts	5%	7%	9%	10%	17%	13%	39%	40%	30%	30%

- Exploring in more details the influence of informal media on knowledge creation and organization;
- Investigating how social networking and participation in communities will contribute to the enhancement of the learning activities' performance;
- Trying to formalize the learning flows and knowledge receiving with a suitable notation system.

REFERENCES

ALT-C 2009 conference, Manchester, UK. (2009, September 8-10). *The VLE is Dead*. Retrieved December 9, 2009, from http://www.alt.ac.uk/altc2009/index.html

Anderson, T. (2003). Getting the mix right again: An updated and theoretical rationale for interaction. *The International Review of Research in Open and Distance Learning, 4*(2). Retrieved December 9, 2009, from http://www.irrodl.org/index.php/irrodl/article/view/149/230

Borau, K., Ullrich, C., Geng, J., & Shen, R. (2009). Microblogging for Language Learning: Using Twitter to Train Communicative and Cultural Competence. In *Proceedings of the ICWL 2009 conference* (pp. 78-87). Retrieved December 9, 2009, from http://www.carstenullrich.net/pubs/Borau09Microblogging.pdf

Burden, K., & Atkinson, S. (2008). Evaluating pedagogical affordances of media sharing Web 2.0 technologies: A case study. In *Proceedings of the ascilite*, Melbourne, Australia. Retrieved December 14, 2009, from http://www.ascilite.org.au/conferences/melbourne08/procs/burden-2.pdf

Canzi, A., Folcio, A., Milani, M., Radice, S., Santangelo, E., & Zanoni, E. (2003). *The management of flows of distance communication between tutors and students in the context of an English course in blended learning held at the Università degli Studi of Milan*. Retrieved December 9, 2009, from http://www.ctu.unimi.it/pdf/Edmedialucerna_2003.pdf

Chang, R. L., Stern, L., Sondergaard, H., & Hadgraft, R. (2009). Places for learning engineering: A preliminary report on informal learning spaces. In *Proceedings of the Research in Engineering Education Symposium*, Palm Cove, QLD. Retrieved December 14, 2009, from http://rees2009.pbworks.com/f/rees2009_submission_86.pdf

Chang, R., Stern, L., Sondergaard, H., & Hadgraft, R. (2009). Places for learning engineering: A preliminary report on informal learning spaces. In *Proceedings of the Research in Engineering Education Symposium*, Palm Cove, QLD. Retrieved December 12, 2009, from http://rees2009.pbworks.com/f/rees2009_submission_86.pdf

Conner, M. (2009). *Marcia Conner Web site*. Retrieved December 12, 2009, from http://marciaconner.com/

Cranmer, S. (2006). Children and young people's uses of the internet for homework' learning. *Media and Technology, 31*(3), 301–315. doi:10.1080/17439880600893358doi:10.1080/17439880600893358

Downes, S. (2005). E-Learning 2.0. *Stephen's Web*. Retrieved December 9, 2009, from http://www.downes.ca/post/31741

Duval, E. (2007). *Snowflakes Effect: Open learning without boundaries*. Retrieved December 9, 2009, from http://ariadne.cs.kuleuven.be/mediawiki2/index.php/SnowflakeEffect

Ellis, A., & Calvo, A. (2007). Minimum Indicators to Assure Quality of LMS-supported Blended Learning. *Educational Technology and Society, 10*(2), 60-70. Retrieved December 14, 2009, from http://www.ifets.info/journals/10_2/6.pdf

Goldsmiths University London Blog. Learning Technologist jottings at Goldsmiths Blog. (2009). *The VLE is Dead. Debate at ALT-C 2009*. Retrieved December 9, 2009, from http://celtrecord.wordpress.com/2009/09/08/the-vle-is-dead-debate-at-alt-c-2009/

Gray, B. (2004). Informal Learning in an Online Community of Practice. *Journal of Distance Education, 19*(1), 20–35.

Ivanova, M. (2008). Knowledge Building and Competence Development in eLearning 2.0 Systems. In *Proceedings of the I-KNOW '08 conference*, Graz, Austria (pp. 84-91).

Ivanova, M. (2009). Use of Start Pages for Building a Mashup Personal Learning Environment to Suport Self-Organized Learners. *Serdica Journal of Computing*, 227-238.

Johnson, L., Levine, A., & Smith, R. (2009). *The Horizon Report*. Austin, TX: The New Media Consortium. Retrieved December 9, 2009, from http://www.nmc.org/pdf/2009-Horizon-Report.pdf

Jokisalo, E., & Riu, A. (2009). *Informal learning in the era of Web 2.0*. Retrieved December 14, 2009, from http://www.elearningeuropa.info/files/media/media19656.pdf

Livingstone, D. (1999). Exploring the icebergs of adult learning: Findings of the first Canadian survey of informal learning practices. *CJSAE, 13*(2), 49–72.

London Knowledge Lab. (2008). *Education 2.0? Designing the web for teaching and learning*. Retrieved December 9, 2009, from http://www.tlrp.org/pub/documents/TELcomm.pdf

Luján-Mora, S. (2006). *A Survey of Use of Weblogs in Education. Formatex*. Retrieved December 14, 2009, from http://www.formatex.org/micte2006/pdf/255-259.pdf

McLoughlin, C., & Lee, M. J. W. (2008). Future learning landscapes: Transforming pedagogy through social software. *Innovate Journal of Online Education, 4*(5). Retrieved December 9, 2009, from http://www.jeffrudisill.com/Content/Student%20Centered%20Learning/Future%20Learning%20Landscapes%20Exhibit%202.pdf

Notari, M. (2006, August 21-23). How to use a wiki in education:"Wiki based effective constructive learning". In *Proceedings of the WikiSym '06*, Odense, Denmark. Retrieved December 14, 2009, from http://www.wikisym.org/ws2006/proceedings/p131.pdf

O'Reilly, T. (2005). *What is Web 2.0. Design Patterns and Business Models for the Next Generation of Software*. Retrieved December 9, 2009, from http://oreilly.com/pub/a/web2/archive/what-is-web-20.html?page=1

Rankine, S. Malfroy, & Ashford-Rowe. (2009). Benchmarking across universities: A framework for LMS analysis. In *Proceedings of the ascilite Auckland 2009*. Retrieved December 9, 2009, from http://www.ascilite.org.au/conferences/auckland09/procs/rankine.pdf

Redecker, C. (2009). *Review of Learning 2.0 Practices: Study on the Impact of Web 2.0 Innovations on Education and Training in Europe* (Tech. Rep.). JRC Scientific. Retrieved December 9, 2009, from http://ftp.jrc.es/EURdoc/JRC49108.pdf

Sterbini, A., & Temperini, M. (2009). Collaborative projects and self evaluation within a social reputation-based exercise-sharing system. In *Proceedings of the IEEE/WIC/ACM International Conferences on WI and IAT, 2nd International Workshop on SPeL* (Vol. 3, pp. 243-246).

Stiles, M. (2007). Death of the VLE?: A challenge to a new orthodoxy. *The Journal for the Serials Community, 20*(1), 31–36. doi:10.1629/20031d oi:10.1629/20031

TimeCruiser Computing Corporation. (2008). *LMS 2.0: How to Select an Advanced Learning System*. Retrieved December 12, 2009, from http://www.timecruiser.com/timecruiser/admin/ UserFiles/White_Paper_LMS_43008.pdf

This work was previously published in the International Journal of Information Systems and Social Change, Volume 2, Issue 1, edited by John Wang, pp. 1-15, copyright 2011 by IGI Publishing (an imprint of IGI Global).

Chapter 2
Global Teacher Training Based on a Multiple Perspective Assessment:
A Knowledge Building Community for Future Assistant Language Teachers

Yuri Nishihori
Sapporo Otani University, Japan

Yuichi Yamamoto
Hokkaido University, Japan

Chizuko Kushima
Tsuda College, Japan

Haruhiko Sato
Hokkaido University, Japan

Satoko Sugie
Hokkaido University, Japan

ABSTRACT

The main objective of this project is to design and implement Web-based collaborative environments for a global training based on a multiple perspective assessment for future and novice ALTs (Assistant Language Teachers) who will come to Japan from various parts of the world. The system was created in order to give better chances to acquire professional knowledge in advance with support from experienced senior teachers, both Japanese teachers and former ALTs. Computer Support for Collaborative Learning (CSCL) was adopted as a platform for their online discussion with much focus on multiple perspective assessment to support social and personalized aspects such as individual accountability and contribution to the collaboration. Initial results are reported using the analysis of system design and the Web-based questionnaire answered by the participants involved in this knowledge-building forum.

DOI: 10.4018/978-1-4666-2922-6.ch002

INTRODUCTION

The ALTs are young overseas graduates who assist in international exchange and foreign language instruction in local governments, boards of education, as well as junior and senior high schools in Japan. Their main job is to work on team-teaching lessons in cooperation with the JTE (Japanese Teachers of English) in English classes. There has been recently, however, a pressing need to propose more effective preparation for ALTs on a practical level in Japan, since the Ministry of Education, Culture, Sports, Science and Technology, Japan, announced in 2002 that the number of ALTs should be increased to the unprecedented number of over 8400. The New Course of Study, revised in 2008, has raised this need by introducing elementary school English. Even at this level, ALTs are expected to teach basic communicative skills. To cope with this rapid increase, however, surprisingly few projects have been carried out for professional training in the recent years within the educational arena of ELT (English Language Teaching) in Japan.

This paper discusses effective ways to meet the above-mentioned pressing need, with regard to appropriate preparation based on the questionnaire survey given to 119 ALTs and 119 JTEs in Japan (Kushima & Nishihori, 2006). ALTs have little job training on a practical level at the orientation sessions offered at present, and virtually no individual preparation is provided for them before coming to Japan (see Figure 1).

ALTs answered the question of what they thought was "Necessary information for ALTs before coming to Japan" (see Table 1).

ALTs viewed the most necessary preparations for them as being "awareness of the actuality of team-teaching lessons in Japan," "awareness of the purpose of team-teaching as part of English lessons in Japan", and "what a Japanese teacher's job actually entails" (see Table 1). With regard to what ALTs actually prepared, the information above shows us that the response rate for the answer "Collecting information about team-teaching lessons" was low, 4.8% (see Figure 1). This leads us to the most likely explanation that ALTs only recognized the importance of this after arriving at their assigned school. Namely, they could not accurately envisage how their actual job was going to be.

To help ameliorate this situation, our system creates a knowledge building community, where individual accountability of knowledge development and contributions can be illustrated from multiple perspectives. This forum offers a good opportunity for future ALTs to undergo pre-training and join the teachers' community in advance.

Figure 1. What ALTs actually prepared before coming to Japan

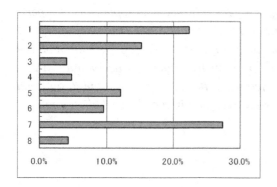

1. Collecting information about experiences of those who were ALTs
2. Learning Japanese
3. Training as a language teacher
4. Collecting information about team-teaching lessons
5. Collecting information about language education
6. Collecting information about Japanese education
7. Collecting information about Japanese life and culture
8. Other

Table 1. Necessary information for ALTs before coming to Japan

	ALT	JTE
1.Japanese education administration system and Japanese schooling system	8.6%	11.5%
2. Japanese school management system	6.2%	9.7%
3. Standard Japanese classroom management	10.8%	11.3%
4. Subjects which Japanese students take and their respective syllabi	7.9%	3.8%
5. Language policy in Japanese public education	7.9%	4.9%
6. The purpose of team-teaching lessons as part of English lessons in Japan	14.1%	14.3%
7. The actuality of team-teaching lessons in Japan	15.9%	13.0%
8. What a Japanese teacher's job actually entails	13.0%	13.8%
9. Japanese students' daily routine	10.4%	15.1%
10. Other	5.5%	2.2%

LITERATURE REVIEW

Since the advent of the Japan Exchange and Teaching (JET) Programme in 1987, there has been much research data accumulated with regard to the teaching methods of team-teaching lessons which have been conducted in many junior and senior high schools in Japan. A number of articles have reported on the effectiveness of introducing ALTs into Japanese schools, as well as effective team-teaching lesson plans (Tajino, 2000). The main interest was placed on finding an effective method for Japanese teachers to collaborate with ALTs. Through this process, however, some problems were revealed in relation with the way in which the JET Programme is run; the entrance examination system, the inadequacy of ALTs' and JTEs' language skills, and cultural friction. In particular, it has been pointed out that there are also differences in educational philosophy which impedes the actual recognition of the ALT's role by JTEs and ALTs (Fenton-Smith, 2000).

In order to solve these problems and make team-teaching lessons more successful, some researchers were of the opinion that in-service training should be enhanced. Scholefield's survey (1996) showed that ALTs had little preparation for their jobs on a practical level. He proposed that ALTs become more familiar with designated vocabulary lists for each year level and with the Japanese education system in general. Gillis-Furutaka (1994) and Crooks (2001) pointed out

that the success of team-teaching in the JET Programme can be enhanced by professional joint training for ALTs and JTEs.

According to another survey (Kushima & Nishihori, 2006), pre and in-service training and guidance is at present offered for ALTs by MEXT (the Ministry of Education, Culture, Sports, Science and Technology), CLAIR (Council of Local Authorities for International Relations), and the Ministry of Foreign Affairs, but this training only focuses on general guidance of how to live in Japan and to start work as an ALT. Thus, it should be noted that there exists no systematic professional training for participants in the program. It is also reported that ALTs are not provided with background information on the Japanese educational system, nor on the actuality of Japanese students' lives or even their academic level. Some ALTs positively evaluated the present training, while others complained of the timing or the contents of the training.

Some researchers also raised issues with the present service training. Gillis-Furutaka (1994) insisted that ALTs should take TESL/FL courses before entering the classrooms. Fanselow (1994) discussed that friction might arise from different views on the actual role of the assistant between both JTEs and ALTs. Moreover, he pointed out that having people with no preparation assisting others could be disadvantageous.

Around the turn of the century, the situation was much improved by the use of ICTs over the information network. In particular, professional training utilizing the Internet started being viewed as a method which could be highly evaluated thanks to its improvements on teacher training. Under these circumstances in Japan, our project organized a global pre-service training for ALTs who were from different countries all across the globe. This kind of Web-based training is useful since it is not bound by normal difficulties taking to do with time and distance.

In contrast to this, however, it was pointed out by Farrell (2000) that asynchronous Web-based training does have its disadvantages in that students tend to have less interaction in this type of activity. Much focus is placed on interaction (Ferry et al., 2000; Vygotsky, 1978) in the discussion forum, since this offers learners the opportunity to analyze information, explore ideas, and share feelings among learners and instructors in an e-learning environment (Khan, 2005). In the case of these types of learning communities, collaboration and knowledge building are considered to be essential elements. In its essential part, knowledge building is viewed as the creation of knowledge as a social product (Scardamalia & Bereiter, 1994), and a number of studies have been carried out concerning knowledge building communities (Costa, 2010; Halonen et al., 2010; Li, 2004).

In the field of professional teacher development, recent studies have focused on the teacher community of practice supported by networked environments (Lave & Wenger, 1999). There have been interesting case studies conducted within an Asian context (Chai & Tan, 2005; Tan & Kwok, 2005). Our study was conducted to further extend this type of training to a global context. Furthermore, assessments of collaborative activities have been analyzed with much focus on recent studies. In particular, our concern is how to include an evaluation of individual knowledge development as well as the individual input (Palloff & Prat, 2005) to the collective knowledge development.

SYSTEM DESIGN PRINCIPLES FOR THE ONLINE DISCUSSION PLATFORM

In our research project, collaborative space ontology (Takeuchi et al., 2006) was applied to the design and construction of a Web-based training platform, which was implemented in a network environment. According to collaborative space ontology, collaborative space is classified into two areas: one for practice to create community knowledge, and the other for education to pass it on to the next generation. Our system is designed to amalgamate both functions within the same space. In this sense, our project is a challenge to realize learning environments in which a learner or novice teacher can draw on the expertise of other experienced learners and senior teachers effectively in an online forum to create and disperse community knowledge.

Several viewpoints have been taken into consideration with the design of the platform in our project, and they have been realized within the Forest Forum (see Figure 2) as follows (Kushima, Obari, & Nishihori, 2008):

1. The Forest Forum is an online discussion forum which develops a 'knowledge building community' (Bereiter & Scardamalia, 1993) as a group of people who work on team-teaching lessons in English classes in Japanese schools or are going to.
2. The discussions center on vocational education for ALTs.
3. Future ALTs will gain comprehension from discussions among JTEs and experienced ALTs as mentors.
4. The level of ALTs' understanding of their profession is evaluated.
5. The discussion level and the degree to which mentors contribute are visualized.

In the Forest Forum, future or novice ALTs ask questions about their jobs, while JTEs or ex-

Figure 2. Paradigm of the Forest Forum

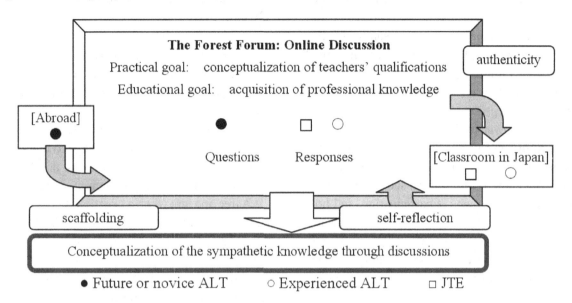

perienced ALTs answer the questions. Future or novice ALTs can develop an understanding of the values, norms, and habits of the teaching profession in Japan through interaction with experienced teachers. In addition, all the participants come to share the ideas of their profession and conceptualize teachers' qualifications. Through discussions, they can build professional knowledge and conceptualize sympathetic knowledge. Moreover, it is said that a good online discussion forum offers three functions: authenticity, scaffolding, and self-reflection (Miyake, 1997). Authenticity is related to applying learners' knowledge to dealing with things. Scaffolding refers to interactive spaces such as this forum built by advanced learners to guide novices. Learners observe the perspectives of others through the forum, thereafter reflecting on them and revising their knowledge to a higher level. These three qualities are clearly included in the Forest Forum.

MULTIPLE PERSPECTIVE ASSESSMENT

In this project, we applied multiple perspective assessment (Oosterhof, Conrad, & Ely, 2008) to design and construct a Web-based training platform, which was implemented in a network environment. In a multiple assessment, the instructor, learners, and peers provide collective input to determine the level of knowledge developed by a learner. In essence, our system is designed to visualize individual knowledge development and contributions to a collaborative product. It has been suggested that future research should look at ways to improve the combination of these types of assessment in the online learning environment (Conrad, 2009). Our project is a challenge to realize learning environments in which a learner or novice teacher can draw on the expertise of other experienced learners and senior teachers effectively in an online forum to create and disperse community knowledge.

With the design of this platform, several viewpoints have been taken into consideration. The actual knowledge building can be seen in the comment box under a certain topic which has been visualized as a tree in this Forest Forum. Contributions from each member can be visualized in this forest which takes the form of a visual display of the actual community.

Each member is indicated by the color of the group they represent in the comment box, and also by a flower in the tree (see Figure 3). The learning process is visualized by the assessment of a discussion which is made by a future or novice ALT. A wealth of information such as the nature of the discussion, who is participating, and how the discussion is being evaluated can be made at a glance by following this process. We can witness this creation of novel learning experiences once we are in this forum.

TRAINING IN A LEARNING COMMUNITY

Forest Forum is an online community whose members are composed of future and novice ALTs, JTEs, and former ALTs. Its aim is to support young people who are interested in working as an ALT to prepare for their professional career. Members get together to grow a topic tree by having future and novice ALTs posting questions, and former ALTs and JTEs answering them.

A tree appears when a question is posted. The newest tree always appears on the upper-left. Clicking on the title will take a member to the detailed question page. A flower will appear when a comment on a question is posted. The number of flowers is equivalent to the number of comments. The color shows the status of the respondent. The flower will blink when the future ALT evaluates the comment positively by clicking the "good" button. When the discussion reaches the stage of eight comments, a butterfly will appear. When

Figure 3. Visualization of a community

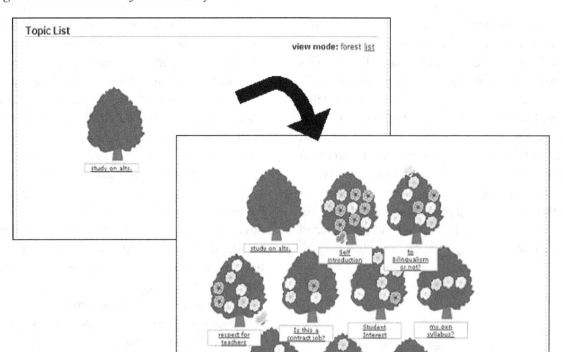

the discussion reaches the next stage of sixteen comments, a second butterfly will appear. When the future ALT clicks the "satisfied" button for the final evaluation of the topic, a rainbow-colored butterfly will appear.

KNOWLEDGE BUILDING PROCESS

The knowledge building process can be visualized through the discussion flow in the Forum. Contributions by each participant can be easily traced by the color of a flower, which is decided according to their member status. Furthermore, the color is the same as that of the background in the text posted by the various participants (see Figure 4).

Interaction is visualized in the discussion flow following each contribution from start to finish (Nishihori, 2007).

- **Starting Stage:** A future ALT asks a question about the ALT's job.

- **Responding Stage:** A JTE or an experienced ALT answers the question, selecting three keywords relating to the answer.
- **Interacting Stage:** Participants further interact by asking and answering additional questions. During the course of the discussion, the future ALT evaluates answers to the original question.
- **Closing Stage:** The future ALT evaluates the whole discussion when she/he finds the answer to the initial question satisfactory, and then closes the topic.

In the Forest Forum, each tree has a discussion flow in sequence. Once a particular tree, i.e., a topic, is selected, the details of the future ALT's question appear in the text as shown in the paradigm (see Figure 4).

EVALUATION AND REFLECTION IN THE TRAIN OF THOUGHT

It is of great importance to provide learners with meta-awareness centered on the discussion. In

Figure 4. Visualization of the discussion and the degree of contribution

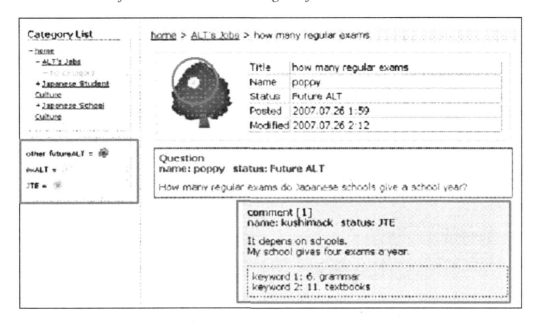

the knowledge building community, this can be achieved by designing a collaborative space to offer possibilities for interaction and reflection over the course discussion. Our project offers this opportunity through its design, which also enables reflection on the actual content during the course of interactive discussion.

The "good" button plays a role for evaluation during the course of discussion in that a future ALT can evaluate any answer made by senior experienced teachers (see Figure 5). In this way, novice teachers can acquire meta-awareness for their profession. By clicking the "good" button, a "blinking" flower will appear on both the tree and the comment box so that evaluation can be visually identified by other members in the forum community.

The above evaluation process is incorporated in the discussion flow to make the train of thought in the forum. The contributions of each member, which are displayed in their assigned colors, can be visualized at a glance in the course of the discussion (see Figure 6). The main stream of the flow consists of the comments by a future or novice ALT. These are shown against a white colored background and are placed toward the left edge of the discussion forum, so that they can

be physically identified as playing a major role in the discussion.

Other contributions can easily be seen by various colored texts posted by senior teachers. In this way, an amalgamation of knowledge can be viewed and traced as a whole and at a glance.

The ultimate stage is marked by closing the topic (see Figure 7). When future ALTs need no more answers and want to close the topic, they make a final evaluation on the whole discussion, based on a 3-point scale; "satisfied," "so-so," and "not satisfied." If they select "satisfied," a rainbow-colored butterfly will fly around the topic tree. They can also comment on their evaluation in the box.

DATA COLLECTIONS & RESEARCH FINDINGS

The Forest Forum was established in August, 2007, and a variety of discussions were observed over 68 days. 16 ALTs and 14 JTEs enrolled in the Forest Forum, most of them accessing the forum from their own computers. Data was collected by an anonymous online questionnaire after the above-mentioned discussion period.

Figure 5. The "good" button

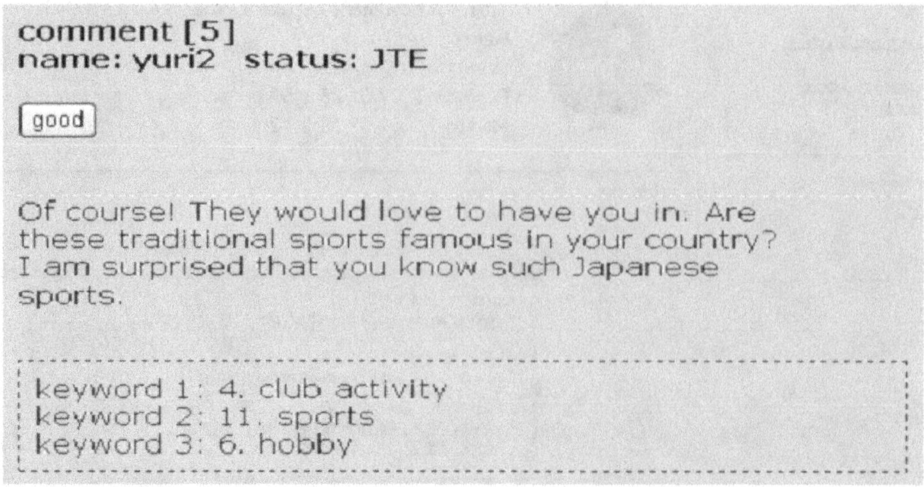

```
comment [5]
name: yuri2   status: JTE

[good]


Of course! They would love to have you in. Are
these traditional sports famous in your country?
I am surprised that you know such Japanese
sports.

keyword 1: 4. club activity
keyword 2: 11. sports
keyword 3: 6. hobby
```

Figure 6. The train of thought

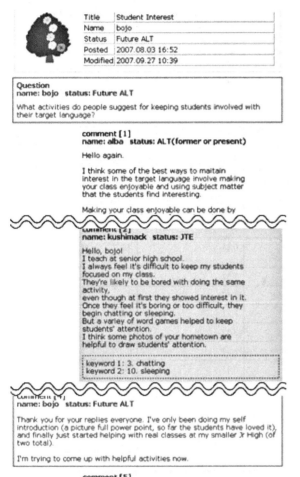

We measured whether or not future or novice ALTs were able to develop an understanding of the nature of teachers' work by comparing three kinds of numerical index in the questionnaire items: "comprehension of keywords," "satisfaction with response," and "sense of belonging to the Forest Forum." "Keywords" means important words regarding ALTs' work as shown in Table 2. The mean score of the degree of comprehension and selection frequency of each keyword were obtained from seven futures or novice ALTs.

The correlations of these two items were measured by using Spearman rank correlations. There was no statistically significant correlation between them, since the correlation was rs=0.075. Although the keywords of "club activity" and "school uniform," for example, were unmentioned in the Forest Forum, their correlations (both are 3.71%) were comparatively high. The correlation is the average computed from nine respondents, and thus it is necessary to look more closely at the numeric values of each respondent one by one. In fact, one ALT respondent fully understood "flexibility" after several Japanese teachers chose it as the keyword or gave explanations of the word in the forum.

On the other hand, there was a significant correlation to be found among "comprehension of keywords," "satisfaction with response," and "sense of belonging to the Forest Forum" (see Figure 8) (Kushima, 2008)

Figure 7. Closing the topic

Table 2. Degree of comprehension and selection frequency of keywords

Category	Keyword	Compre-hension	Frequency
ALT's Jobs	ALTs and JTEs role	3.71	6
	assistant	3.57	3
	communicative ability	3.43	8
	flexibility	3.43	7
	grammar	3.43	0
	informant	3.29	0
	model reading	3.43	0
	praise students	3.57	5
	team-teaching	3.71	6
	textbooks	3.29	0
	Mean	3.49	3.50
	Standard Deviation	0.15	3.27
Japanese Student Culture	cell phones	3.57	0
	chatting	3.57	4
	club activity	3.71	0
	friendship	3.57	0
	hobby	3.57	0
	music	3.43	0
	personality	3.29	5
	school uniform	3.71	0
	sleeping	3.71	1
	sports	3.43	0
	Mean	3.56	1.00
	Standard Deviation	0.14	1.89
Japanese School Culture	bureaucracy	3.43	1
	classroom environment	3.43	6
	collectivism	3.57	1
	communication	3.43	0
	discipline	3.43	2
	individualism	3.43	1
	Japanese school management system	2.86	5
	Japanese schooling system	3.14	4
	school event	3.33	0
	staff room environment	3.29	1
	Mean	3.33	2.10
	Standard Deviation	0.20	2.13
(Mean) (Standard Deviation)		3.27	2.23
		0.81	2.49

Note: degree of comprehension 4 (fully understood),3 (almost understood), 2 (not so understood), and 1 (not understood at all)

Figure 8. Mean and standard deviation of "comprehension of keywords," "satisfaction with response," and "sense of belonging to the Forest Forum"

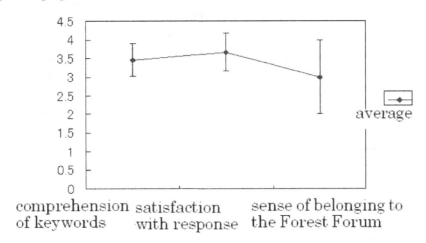

Table 3 shows correlation analysis of future or novice ALTs' "comprehension of keywords," "satisfaction with response," and "sense of belonging to the Forest Forum." We can see that the three factors correlate with one another. Namely, future or novice ALTs became satisfied with the responses, and at the same time they developed an understanding of the nature of teachers' work. Consequently, they were able to develop a sense of belonging to the community. However, the correlation coefficient in the middle line is a little lower than the others. This implies that ALTs are likely to be satisfied with the whole reply even though they have not sufficiently understood their profession yet.

In order to validate the perceived sense of belonging, "access frequency" to the Forum was examined along with two other items: "satisfaction with response" and "sense of belonging to the Forest Forum." Out of seven future or novice ALTs in Table 3, there were six who responded to the question: "How often do you see the Forest Forum?" The answer choices were (1) Every day, (2) Every time the answer to your own question is posted, (3) Once a week, (4) A few times a month, (5) Once a month, and (6) Other. Table 4 shows correlation analysis of future or novice ALTs' "frequency," "satisfaction with response," and "sense of belonging to the Forest Forum." A strong correlation can be seen in each combination, but the number of respondents is limited; hence, we need to collect more data in the future.

Table 3. Correlation analysis of "comprehension of keywords," "satisfaction with response," and "sense of belonging to the Forest Forum"

	N	r	t	P	t(0.975)
C X S	7	.90*	4.10	.01	2.78
C X B	7	.78*	2.75	.04	2.78
S X B	7	.92*	4.67	.01	2.78

Note: C = degree of comprehension of keywords, S = degree of satisfaction with response, B = degree of sense of belonging to the Forest Forum.

*p < .05.

Table 4. Correlation analysis of "frequency," "satisfaction with response," and "sense of belonging to the Forest Forum"

	N	r	t	P	t(0.975)
F X B	6	.90*	4.11	.01	2.78
F X S	6	.88*	4.67	.02	2.78

Note: F = frequency, B = degree of sense of belonging to the Forest Forum, S = degree of satisfaction with response

*p < .05.

CONCLUSION AND FURTHER CONSIDERATIONS

This paper presented a description of the system design which was constructed for a Web-supported learning community to help future and novice ALTs with support from experienced Japanese teachers and former ALTs. The system was designed with much consideration on the interactive nature of discussion in a Web-based forum. It is an online community where novice and experienced teachers can share professional knowledge and expertise in advance of ALTs coming to Japan. The platform for their discussion was designed with much focus on multiple perspective assessment. Even novices can not only visualize their contribution to the community, but evaluate the discussion by closing the discussion. Experienced teachers can offer their answers and advice to them while watching the process of discussion, paying particular attention to how the novice members are forming their knowledge and evaluating the discussion. The online community can be visualized thanks to this process of multiple perspective assessment. In this way, this method can support social and personalized aspects as individual accountability and contribution to the collaboration.

The data was collected in order to analyze its effectiveness in providing future and novice ALTs with better chances to acquire professional knowledge at the initial stage of their development as ALTs. An anonymous online questionnaire was conducted after the discussion period. We measured whether or not future or novice ALTs were able to develop an understanding of the nature of teachers' work by comparing three kinds of numerical index in the questionnaire items: "comprehension of keywords," "satisfaction with response," and "sense of belonging to the Forest Forum." There was a significant correlation among "comprehension of keywords," "satisfaction with response," and "sense of belonging to the Forest Forum." In order to validate the perceived sense of belonging, "access frequency" to the Forum was

examined along with two other items: "satisfaction with response" and "sense of belonging to the Forest Forum." A strong correlation was noticed in each combination. The results indicate that both groups of teachers demonstrated greater gains for their professional development. The findings will encourage the researchers in refining knowledge building communities such as this one for future professional development.

We plan to further develop this Forest Forum to support online training in Web-based collaborative environments. In order to make it successful, evaluation is of critical importance, as shown by the amount of the data we gained in order to make a substantial conclusion. Our future work lies with analyzing this system with a much more detailed survey which looks at learners' awareness toward collaboration.

ACKNOWLEDGEMENT

This study is supported by Grant-in-Aid for Scientific Research (B) No.18320086 from Japan Society for the Promotion of Science (JSPS).

REFERENCES

Bereiter, C., & Scardamalia, M. (1993). *Surpassing ourselves: An inquiry into the nature and implications of expertise*. Chicago: Open Court.

Chai, C. S., & Tan, S. C. (2005). Fostering learning communities among teachers and students: potentials and issues. In Looi, C. K., (Eds.), *Towards sustainable and scalable educational innovations informed by the learning sciences*. Amsterdam, The Netherlands: IOS Press.

Conrad, R. M. (2009). Assessing collaborative learning. In Rogers, P., (Eds.), *Encyclopedia of distance learning* (pp. 89–93). Hershey, PA: IGI Global.

Costa, C. (2010). Lifelong learning in Web 2.0 environments. *International Journal of Technology Enhanced Learning, 2*(3), 275–284. doi:10.1504/IJTEL.2010.033582

Crooks, A. (2001). Professional development and the JET program: Insights and solutions based on the Sendai City Program. *JALT journal, 23*, 31-46.

Fanselow, J. F. (1994). JET as an exercise in program analysis. In Wada, M., & Cominos, A. (Eds.), *Studies in team teaching* (pp. 201–216). Tokyo, Japan: Kenkyusha.

Farrell, J. N. (2000). Long live c-learning: The advantages of the classroom. *Training & Development, 54*(9), 43–46.

Fenton-Smith, B. (2000). Foreign teachers in Japanese secondary schools: Why aren't they happier? *Kanda university of international studies, 12*, 409-426.

Ferry, B., Kiggins, J., Hoban, G., & Lockyer, L. (2000). Using computer-mediated communication to form a knowledge-building community with beginner teachers. *Journal of Educational Technology & Society, 3*(3).

Gillis-Furutaka, A. (1994). Pedagogical preparation for JET programme teachers. In Wada, M., & Cominos, A. (Eds.), *Studies in team teaching* (pp. 29–41). Tokyo, Japan: Kenkyusha.

Halonen, R., Thomander, H., & Laukkanen, E. (2010). DeLone & McLean IS success model in evaluating knowledge transfer in a virtual learning environment. [IJISSC]. *International Journal of Information Systems and Social Change, 1*(2), 36–48.

Khan, B. H. (2005). *Managing e-learning strategies*. Hershey, PA: Information Science Publishing.

Kushima, C. (2008). *Research on the system design for supporting ALT's job preparation: a global learning community ALTs and JTEs are building*. Unpublished doctoral dissertation, Hokkaido University, Sapporo, Japan.

Kushima, C., & Nishihori, Y. (2006). Reconsidering the role of the ALT: Effective preparation for ALTs based on the questionnaire survey. *Annual review of English language education in Japan, 17*, 221-230.

Kushima, C., Obari, H., & Nishihori, Y. (2008). Fostering global teacher training: the design and practice of a web-based discussion forum as a knowledge building community. In *Proceedings of WorldCALL, 2008*, 236–239. Retrieved from http://www.j-let.org/~wcf/proceedings/proceedings.pdf.

Lave, J., & Wenger, E. (1999). Legitimate peripheral participation in the communities of practice. In McCormic, R., & Paechter, C. (Eds.), *Leraning and knowledge* (pp. 21–35). Thousand Oaks, CA: Sage.

Li, Q. (2004). Knowledge building community: keys for using online forums. *Research & practice to improve learning, 48*(4), 24-28.

Miyake, N. (1997). *The Internet children*. Tokyo, Japan: Iwanami Shoten.

Nishihori, Y. (2007). The design of web-based collaborative environments for global teacher training: A knowledge building community for future assistant language teachers in Japan. In *Proceedings of the 12th Conference of Pan-Pacific Association of Applied Linguistics* (pp. 168-171).

Oosterhof, A., Conrad, R. M., & Ely, D. P. (2008). *Assessing learners online*. Upper Saddle River, NJ: Pearson Education.

Palloff, R., & Pratt, K. (2005). *Collaborating online: learning together in community*. San Francisco, CA: Jossey-Bass.

Scardamalia, M., & Bereiter, C. (1994). Computer support for knowledge-building communities. *Journal of the Learning Sciences, 3*(3), 265–283. doi:10.1207/s15327809jls0303_3

Scholefield, W. F. (1996). What do JTEs really want? *JALT journal, 18*, 7-25.

Tajino, A., & Tajino, Y. (2000). Native and non-native: What can they offer? *ELT Journal, 54*(1), 3–11. doi:10.1093/elt/54.1.3

Takeuchi, M., Hayashi, Y., Ikeda, M., & Mizoguchi, R. (2006). A collaborative learning design environment to integrate practice and learning based on collaborative space ontology and patterns. In *Proceedings of the 8th international conference on intelligent tutoring systems* (pp. 187-196).

Tan, C., & Kwok, P. (2005). *Knowledge building in inter-school learning communities: Reflections from a case on project learning in Hong Kong. Towards sustainable and scalable educational innovations informed by the learning sciences.* Amsterdam, The Netherlands: IOS Press.

Vygotsky, L. S. (1978). *Mind in society*. Cambridge, UK: Harvard University Press.

This work was previously published in the International Journal of Information Systems and Social Change, Volume 2, Issue 1, edited by John Wang, pp. 16-30, copyright 2011 by IGI Publishing (an imprint of IGI Global).

Chapter 3
Creating a Personalized Artificial Intelligence Course:
WELSA Case Study

Elvira Popescu
University of Craiova, Romania

Costin Bădică
University of Craiova, Romania

ABSTRACT

This paper illustrates the use of WELSA adaptive educational system for the implementation of an Artificial Intelligence (AI) course which is individualized to the learning style of each student. Several of the issues addressed throughout this paper are describing similar approaches existing in literature, how the AI course is created, and what kind of personalization is provided in the course including the underlying adaptation mechanism. The authors also focus on whether the course is used effectively by the stakeholders (teachers and students respectively). Results obtained in the paper confirm the practical applicability of WELSA and its potential for meeting the personalization needs and expectations of the digital native students.

INTRODUCTION

The advent and omnipresence of information systems have been revolutionizing and changing not only the way we communicate access information and conduct businesses, but also the way we learn. In the world of pervasive Internet, learners

are also evolving: the so-called "digital natives" want to be in constant communication with their peers, they expect an individualized instruction and a personalized learning environment. In this context, we present such an adaptive educational system, called WELSA, illustrating it with a course module on "Artificial Intelligence". According to Brusilovsky and Millan (2007), adaptation can be done with respect to various factors, such as

DOI: 10.4018/978-1-4666-2922-6.ch003

knowledge, interests, goals, background, individual traits and context of work. In this paper we base our adaptation on one of the students' individual traits, namely their learning style (i.e., a specific manner of approaching a learning task, the preferred learning strategies activated in order to fulfill that task).

Motivation

Our endeavor was motivated by several aspects.

First, many educational psychologists support the use of learning styles, claiming that they have an important effect on the learning process (Popescu, 2010a); however this is not to say that the domain is free from controversies (Coffield et al., 2004).

Secondly, during the past several years, quite a few researchers dedicated their time to the development of learning style based adaptive educational systems (LSAES), as we will see in the next section. Most of them reported positive experimental results with their systems, finding improvements in student learning gain and/or satisfaction (Bajraktarevic et al., 2003; Carver et al., 1999; Graf et al., 2009; Lee et al., 2005; Limongelli et al., 2009; Papanikolaou et al., 2003; Sangineto et al., 2008; Triantafillou et al., 2004; Wang et al., 2008). Once again, contrary results have also been reported, with (Brown et al., 2009) being a representative study in this respect.

Thirdly, due to the huge expansion of the Web, the amount of information made available in current e-learning systems is very large, definitely larger than what could be presented by traditional teaching means. While being a positive aspect, this availability can also have a downside - it could easily become overwhelming for the students. It is therefore of a particular importance to filter the content in order to avoid cognitive overload of the learners. Furthermore, it is important to decide how to best present this content and in what sequence (the navigation type).

WELSA Overview

The e-learning platform used in our study is called WELSA (Web-based Educational system with Learning Style Adaptation). More details about the system and the principles behind it can be found in (Popescu et al., 2009). Basically, WELSA's main pedagogical goal is to provide an educational experience that best suits the learning preferences of each student, in terms of perception modality, way of processing and organizing information, as well as motivational and social aspects. All these preferences are condensed in a so-called Unified Learning Style Model (ULSM). A detailed description of the ULSM components, together with its rationale and its advantages in Web-based learning settings over traditional learning style models are provided in (Popescu, 2010a).

WELSA is composed of three main modules:

- An authoring tool for the teachers, allowing them to create courses conforming to the internal WELSA format (XML-based representation).
- A data analysis tool, which is responsible for interpreting the behavior of the students and consequently building and updating the learner model, as well as providing various aggregated information about the learners.
- A course player (basic learning management system) for the students, enhanced with two special capabilities: i) learner tracking functionality (monitoring the student interaction with the system); ii) adaptation functionality (incorporating adaptation logic and offering individualized course pages).

The rest of the paper is structured as follows: the next section includes a review of other courses deployed using related LSAES (i.e., courses adapted to students' learning styles). The following two sections present an AI course (inspired from

(Poole et al., 1998) classical textbook) deployed in WELSA; first the system is seen through the eyes of the teacher, who also plays the role of course author; next the system is seen through the eyes of the student, who has to learn the adapted course. Subsequently, both the authoring and the adaptation approaches are validated by means of experimental studies. The last section contains some conclusions and future research directions.

RELATED WORKS

In what follows we will give an overview of similar works reporting on the implementation of personalized courses with respect to learning styles; the adaptation techniques used are presented, together with evaluation data where available. A summary of the reviewed papers is included in Table 1.

- Carver et al. (1999) devised a hypermedia course on "Computer Systems", individualized for 3 dimensions of the Felder-Silverman model (FSLSM) (Felder &

Silverman, 1988): sensing/intuitive, visual/verbal, sequential/global.

The course includes a large variety of multimedia educational resources: hypertext, audio files, graphic files, digital movies, instructor slideshows, lesson objectives, note-taking guides, a virtual computer simulation tool, quizzes in a 3D gaming engine etc. For each category of resources, the teacher has to mention its suitability (support) for each learning style (by rating it on a scale from 0 to 100). When a student logs into the course, a CGI executable loads the student profile (i.e., his/her learning style as resulted from answering a dedicated questionnaire); it then computes a unique ranking of each category of resources, by combining the information in the student's profile with the resource ratings. Next the CGI dynamically creates an HTML page containing an ordered list of the educational resources, from the most to the least effective from the student's learning style point of view. So Carver et al. (1999) don't actually propose a fully integrated adaptive educational system, but just a CGI script

Table 1. Overview of papers on personalized courses with respect to learning styles

Paper	Course subject	Learning style model	Adaptation techniques	Experimental validation
(Carver et al., 1999)	Computer Systems	FSLSM (sensing/intuitive, visual/verbal, sequential/ global)	Fragment sorting	Yes
(Bajraktarevic et al., 2003)	Geography	FSLSM (sequential/global)	Customize system's interface	Yes
(Papanikolaou et al., 2003)	Computer Architecture	Honey and Mumford model	Fragment sorting	Yes
(Triantafillou et al., 2004)	Multimedia Technology Systems	Witkin's field dependence /field independence	Conditional text and page variants	Yes
(Cha et al., 2006)	Heritage Alive of an Old Temple	FSLSM	Customize system's interface	No
(Graf et al., 2009)	Object Oriented Modeling	FSLSM (active/reflective, sensing/intuitive, sequential/global)	Fragment sorting	Yes
Current	*Artificial Intelligence*	*ULSM*	*Fragment sorting and adaptive annotation*	*Yes*

that applies the fragment sorting technique on an already existing hypermedia course.

The informal experimental evaluations showed the benefits of using this hypermedia course: an increase in the students' learning gain as well as a reduction in the students' requests for additional instruction outside the classroom. However, the learning gain was not evenly distributed among students, with best learners benefiting more from the courseware and weakest learners benefiting less.

- Bajraktarevic et al. (2003) devised a Geography course for 14-year old students, individualized according to the FSLSM sequential / global preference. Namely, the course content is presented in a specific layout: pages for global students contain diagrams, table of contents, overview of information, summary, while pages for sequential learners only include small pieces of information, and Forward and Back buttons. Just as in the previous case, Bajraktarevic et al. (2003) don't propose a fully integrated adaptive educational system, but just user interface templates tailored for sequential/global students.

The empirical study involved 22 14-year old students, who achieved significantly higher scores while browsing the session that matched their learning styles; however, no significant difference between browsing times for the matched and mismatched groups were found.

- Papanikolaou et al. (2003) devised a course module on "Computer Architecture", using the INSPIRE educational platform. The system uses adaptive presentation techniques to adapt the learning content to the 4 learning styles in Honey and Mumford model (2000): Activist, Pragmatist, Reflector and Theorist. All learners are presented with the same knowledge mod-

ules, but their order and appearance (either embedded in the page or presented as links) differs for each learning style. Thus for Activists (who are motivated by experimentation and challenging tasks), the module "Activity" appears at the top of the page, followed by links to examples, theory and exercises. In case of Pragmatists (who are motivated by trying out theories and techniques), the module "Exercise" appears at the top of the page, followed by links to examples, theory and activities. Similarly, in case of Reflectors the order of modules is: examples, theory, exercises, and activities, while in case of Theorists the order is: theory, examples, exercises and activities. The system offers also the students the possibility to choose their preferred order of studying. Furthermore, INSIPRE includes also adaptation strategies based on the students' knowledge level, in the form of curriculum sequencing and adaptive navigation support.

The empirical study involved 23 undergraduate learners of the Department of Informatics and Telecommunications of the University of Athens. According to the opinion surveys, learners were quite satisfied with being offered several types of educational material in a specific order, which apparently facilitated their study.

- Triantafillou et al. (2004) devised a course on "Multimedia Technology Systems" for fourth-year undergraduate students. The course was implemented using the AES-CS adaptive educational system. The platform makes use of both adaptive presentation technique and adaptive navigation support to individualize the information and the learning path to the field dependence (FD)/ field independence (FI) characteristic of the students (Witkin, 1962).

Specifically, AES-CS uses conditional text and page variants to present the information in a different style: from specific to general in case of FI learners (who have an analytic preference) and from general to specific in case of FD learners (who have a global preference). AES-CS offers also two control options: program control for FD learners, by means of which the system guides the learner through the learning material; learner control for FI learners, by means of which the learners can choose their own learning paths, through a menu. Since FD learners benefit more from instructions and feedback, an additional frame at the bottom of the page is used to provide them with explicit directions and guidance. This frame is missing in case of FI learners, who prefer few instructions and feedback. Similarly, in case of self-assessment tests, the feedback provided for FI learners is less extensive than in case of FD learners. Another feature offered for FD learners is an *advance* organizer (i.e., a bridging strategy offered at the beginning of a new unit, providing connections with the other units); conversely, FI learners are provided with a *post* organizer (i.e., a synopsis located at the end of a unit). Finally, FI learners are allowed to develop their own course structure, while FD learners are offered two navigational tools in order to help them structure the learning material and create the big picture: a concept map (a visual representation of the domain concepts and the relations between them) and a graphic path indicator (presenting the current, the previous and the next topic). Furthermore, AES-CS allows all students to modify the adaptation options provided by the system, making their own choices between program / learner control, minimal / maximal feedback etc. It should be mentioned that AES-CS includes also adaptation strategies based on students' knowledge level, in the form of adaptive navigation support. More specifically, it uses adaptive annotation (blue for "recommended" links and grey for "not ready to be learned" links), as well as direct guidance (the most suitable sequence of knowledge units to study).

The empirical study involving 76 undergraduate students showed a positive effect of adaptation, reflected in an increased performance (particularly in case of FD learners) and a high degree of learner satisfaction.

- Cha et al. (2006) devised a course on "Heritage alive of an old temple", individualized according to FSLSM. More specifically, the interface is adaptively customized: it contains 3 pairs of widget placeholders (text/image, audio/video, Q&A board/Bulletin Board), each pair consisting of a primary and a secondary information area. The space allocated on the screen for each widget varies according to the student's FSLSM dimension: e.g. for a Visual learner the image data widget is located in the primary information area, which is larger than the text data widget; the two widgets are swapped in case of a Verbal learner. Similarly, the Q&A Board and Bulletin Board are swapped in case of the Active versus Reflective learners.

No experimental data is available for this course.

- Graf et al. (2009) devised a course on "Object Oriented Modeling" for undergraduate students, individualized to three FSLSM dimensions (active/reflective, sensing/intuitive, sequential/global). The course was deployed in Moodle Learning Management System (Moodle, 2009), which was extended with an add-on providing the required adaptation. More specifically, it provides an individualized sequence and number of learning objects of each type (examples, exercises, self assessment tests, content objects).

The empirical study involving 147 students showed that adaptivity has the potential to support learners, having however different effects for learners with different learning styles.

As far as the course authoring process is concerned, most of the existing LSAES offer no support for the teacher, providing no dedicated authoring tool, with the notable exception of AHA! (version 3.0) (Stash, 2007). Unlike these systems, which only provide functionalities for the students, WELSA caters also for the teacher as we will show in the next section. A further difference from the above mentioned works is that our AI course is personalized according to a complex of learning preferences (distilled in ULSM), and not to one of the traditional learning style models. Finally, there are also differences in terms of the adaptation mechanism used, as shown in the next sections.

TEACHER'S PERSPECTIVE (COURSE AUTHORING)

The process of authoring adaptive hypermedia involves several steps (Stash et al., 2005):

- Creating the actual content (which should include alternatives to correspond to various learner needs, in terms of media type, instructional role, difficulty level etc).
- Creating the domain model (defining the concepts that are to be taught and the prerequisite relations between them).
- Specifying the criteria to be used for adaptation (e.g., knowledge level, goals, learning style).
- Creating the adaptation model (defining the rules for learner modeling and adaptation logic).

In case of WELSA, authors only have to create the actual content and annotate it with a predefined set of metadata (provide the static description). These metadata also include information about the hierarchical and prerequisite relations between

concepts, as we will see later on. The criteria to be used for adaptation are the learning preferences of the students, as defined in ULSM (Popescu, 2010a). Finally, the adaptation model (the dynamic description) is supplied by the application, in the form of a predefined set of adaptation rules (Popescu & Badica, 2009).

In order to support the teacher in creating courses conforming to WELSA internal format, we have designed a course editor tool, which allows authors to easily assemble and annotate learning resources, automatically generating the appropriate file structure. It should be noted that WELSA course editor does not deal with the creation of actual content (text, images, simulations etc) – a variety of existing dedicated tools can be used for this purpose (text editors, graphics editors, HTML editors etc). Instead, WELSA course editor provides a tool for adding metadata to existing learning resources and defining the course structure (specifying the order of resources, assembling learning objects in pages, sections and subsections) (Popescu et al., 2008a).

The course structure that we propose in WELSA is a hierarchical one: each course consists of several chapters, and each chapter can contain several sections and subsections. The lowest level subsection contains the actual educational resources. Each such elementary learning object (LO) corresponds to a physical file and has a metadata file associated to it (Popescu et al., 2008b). Apart from being widely used for organizing the teaching materials, this approach also insures a high reusability degree of the educational resources. Furthermore, due to the fine granularity level of the LOs, a fine granularity of adaptation actions can also be envisaged. Finally, since each LO has a comprehensive metadata file associated to it, we know all the information about the learning resource that is accessed by the learner at a particular moment, so we can perform a detailed learner tracking.

Figure 1 shows the hierarchical structure of one of the chapters in the AI course, namely the one

Figure 1. Hierarchical structure of CSP chapter (white boxes designate sections and subsections, while grey boxes designate LOs)

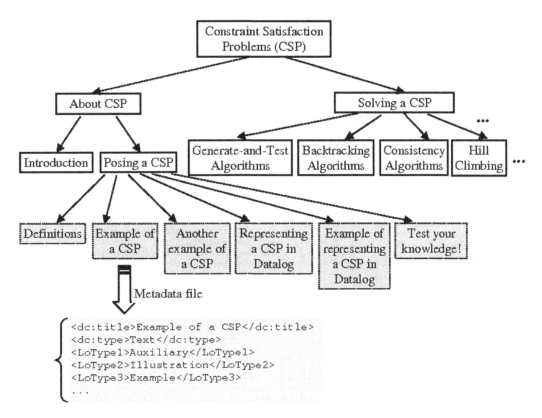

on "Constraint satisfaction problems (CSP)". The corresponding XML files can be seen in Figure 2 (XML for chapter and metadata respectively).

The teacher can define this chapter structure in a simple and intuitive way, by using the course editor, as shown in Figure 3. The corresponding XML files (i.e., those from Figure 2) are subsequently generated by the application and stored on the server.

A few explanations regarding metadata are in order. One possible approach would be to associate to each learning object the learning style that it is most suitable for. One of the disadvantages is that this approach is tied to a particular learning style. Moreover, the teacher must create different learning objects for each learning style dimension and label them as such. This implies an increase in the workload of the teacher, and also the neces-

sity that she/he possesses knowledge in the learning style theory. Instead, we propose a set of metadata that describe the learning object from the point of view of instructional role, media type, level of abstractness and formality, type of competence etc. These metadata were created by enhancing core parts of Dublin Core (DCMI, 2009) and Ullrich's instructional ontology (Ullrich, 2005), with some extensions to cover the requirements specific to learning styles.

Thus some of the descriptors of a learning object are (Popescu et al., 2008b):

- Title (the name given to the resource) → *dc:title*
- Identifier (a reference to the actual resource, such as its URL) → *dc:identifier*

Figure 2. XML files for chapter structure (left-hand side) and LO metadata (right-hand side)

```xml
<?xml version="1.0" encoding="UTF-8"?>
<Chapter xmlns:dc="http://purl.org/dc/elements/1.1/"
    xmlns:xsi="http://www.w3.org/2001/XMLSchema-instance"
    xsi:noNamespaceSchemaLocation="chapter.xsd"
    xsi:schemaLocation="http://purl.org/dc/elements/1.1/
    http://dublincore.org/schemas/xmls/qdc/2006/01/06/dc.xsd ">
    <About>
        <title>Constraint Satisfaction Problems</title>
        <creator>Elvira Popescu</creator>
        ...
    </About>
    <Content>
        <Div1>
            <Title>About CSP</Title>
            <Div2>
                <Title></Title>
                <Div3>
                    <Title>Introduction</Title>
                    <Div4>
                        <Title></Title>
                        <LO>csp_def_informal.xml</LO>
                        <LO>csp_remark_graph.xml</LO>
                    </Div4>
                </Div3>
                <Div3>
                    <Title>Posing a CSP</Title>
                    <Div4>
                        <Title></Title>
                        <LO>csp_def_formal.xml</LO>
                        <LO>csp_example1.xml</LO>
                        <LO>csp_example2.xml</LO>
                        <LO>csp_datalog.xml</LO>
                        <LO>csp_datalog_example.xml</LO>
                        <LO>csp_exercises.xml</LO>
                    </Div4>
                </Div3>
            </Div2>
        </Div1>
        <Div1>
            <Title>Solving a CSP</Title>
            <Div2>
                <Title></Title>
                <Div3>
                    <Title>Generate-and-Test Algorithms</Title>
                    <Div4>
                        <Title></Title>
                        <LO>generate_test_procedure.xml</LO>
                        <LO>generate_test_example.xml</LO>
                    </Div4>
            ...
        </Div1>
    </Content>
</Chapter>
```

```xml
<?xml version="1.0" encoding="UTF-8"?> <LO
xmlns:dc="http://purl.org/dc/elements/1.1/"
    xmlns:xsi="http://www.w3.org/2001/XMLSchema-instance"
    xsi:noNamespaceSchemaLocation="metadata.xsd"
    xsi:schemaLocation="http://purl.org/dc/elements/1.1/
    http://dublincore.org/schemas/xmls/qdc/2006/01/06/dc.xsd ">
    <title>Example of a CSP</title>
    <identifier>csp_example1.html</identifier>
    <type>Text</type>
    <format>text/html</format>
    ...
    <LoType1>Auxiliary</LoType1>
    <LoType2>Illustration</LoType2>
    <LoType3>Example</LoType3>
    <hasAbstractness>concrete</hasAbstractness>
    <hasFormalness>informal</hasFormalness>
</LO>
```

- Type (the nature of the content of the resource, such as text, image, animation, sound, video) → *dc:type*
- Format (the physical or digital manifestation of the resource, such as the media type or dimensions of the resource) → *dc:format*
- Instructional role, either i) fundamental: definition, fact, law (law of nature, theorem, policy) and process or ii) auxiliary:

evidence (demonstration, proof), explanation (introduction, conclusion, remark, synthesis, objectives, additional information), illustration (example, counter example, case study) and interactivity (exercise, exploration, invitation, real-world problem) → *LoType1, LoType2, LoType3, LoType4*
- Related learning objects: i) *isFor / inverseIsFor* (relating an auxiliary learning

Figure 3. Snapshot of WELSA authoring tool: adding chapters (main window) and editing chapter structure (dotted box)

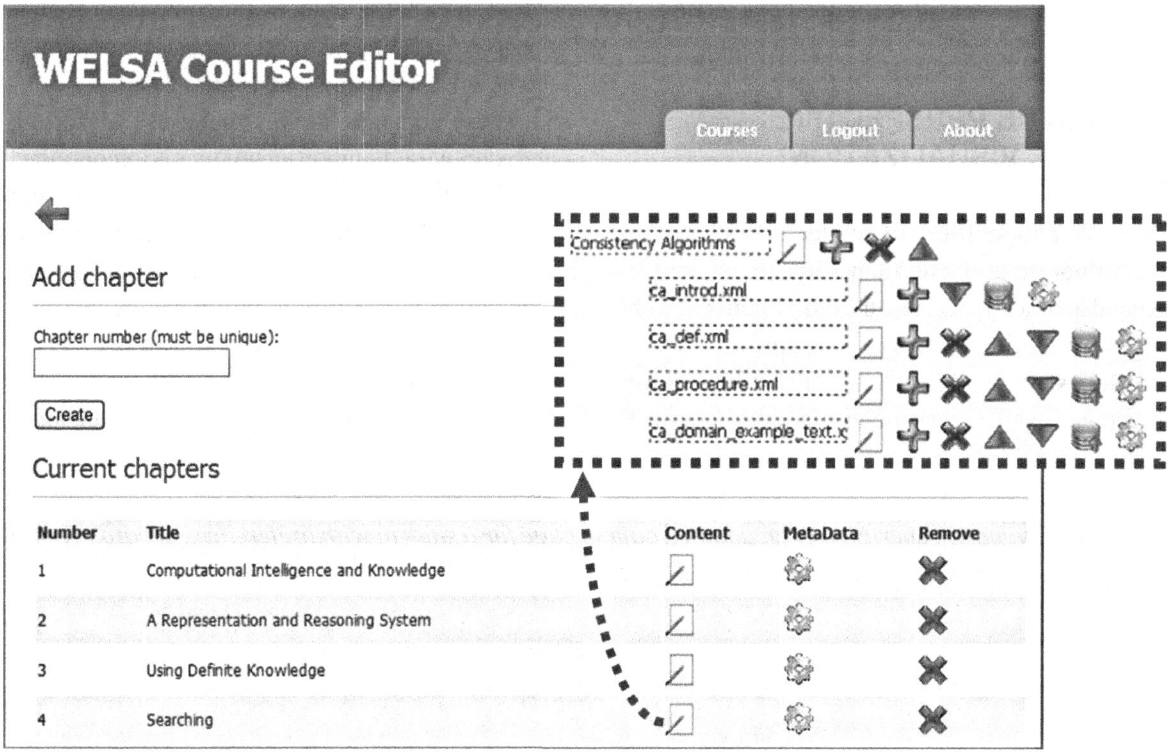

object to the fundamental learning object it completes); ii) *requires / isRequiredBy* (relating a learning object to its prerequisites); iii) *isA / inverseIsA* (relating a learning object to its parent concept); iv) *isAnalogous* (relating two learning objects with similar content, but differing in media type or level of formality).

This mechanism reduces the workload of authors, who only need to annotate their LOs with standard metadata and do not need to be pedagogical experts (neither for associating LOs with learning styles, nor for devising adaptation strategies). The only condition for LOs is to be as independent from each other as possible, without cross-references and transition phrases, to insure that the adaptation component can safely apply reordering techniques. Obviously, there are cases

in which changing the order of the learning content is not desirable; in this case the resources should be presented in the predefined order only, independently of the student's preferences (the teacher has the possibility to specify these cases by means of the prerequisites mechanism, e.g., *requires/isRequiredBy*).

Also authors should ideally provide as many equivalent LOs as possible, but represented in different media formats, different level of abstractness and formality etc. Of course, this might not be always feasible. Just as Gardner said about customizing the learning material to fit the seven intelligence types, "there is no point in assuming that every topic can be effectively approached in at least seven ways, and it is a waste of effort and time to attempt to do this" (Gardner, 1995, p. 206). However, our AI module is an example of a successful case; it was devised starting from an existing course, with little additional work from the teacher; the adaptation results were highly satisfactory, as we will see in the next section.

STUDENT'S PERSPECTIVE (COURSE VISUALIZATION)

Once the course files are created and stored by the Authoring tool, the Adaptation component is needed in order to generate the individualized web pages that will be shown to each student.

More specifically, each time an HTTP request is received by the server, the adaptation component queries the learner model database, in order to find the ULSM preferences of the current student. Based on these preferences, the component applies the corresponding adaptation rules and generates the new HTML page (see Figure 4).

These adaptation rules make use of sorting and adaptive annotation techniques, to recommend students the most suited learning objects and learning path. The popular "traffic light metaphor" is also used, to differentiate between recommended LOs (with a highlighted green title), standard LOs (with a black title) and not recommended LOs (with a dimmed light grey title). Basically, these rules are aggregated from elemen-

Figure 4. Automatic generation of an adapted course page for a student with preferences towards Verbal perception modality, Abstract concepts and Reflective observation

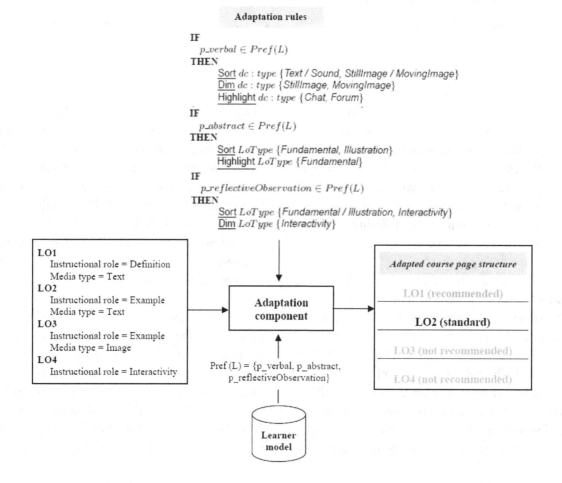

tary actions, such as annotating, inserting, eliminating, sorting or moving learning objects. They also involve the use of LO metadata, which convey enough information to allow for the adaptation decision making (i.e., media type, level of abstractness, instructional role etc).

In what follows we will show the way this adaptation mechanism is visualized by the students, in the Web browser. Let us take a student who has a preference towards *Visual* perception modality and *Concrete, practical examples* (among other ULSM characteristics). The rules in Box 1. Will be applied for this student.

Consequently, the course page on "Posing a CSP" (see Figure 5) will start with two recommended (green-titled) examples followed by a definition, since the student prefers the abstract concepts to be first illustrated to her by concrete, practical examples. Similarly, for the "Breadth-First Search Algorithm" page, the graphical animated example is placed first and marked as recommended, while the text-based one is placed second and marked as less recommended, since the student has a predominantly visual preference (see Figure 6).

COURSE VALIDATION

Student Validation

In order to validate our WELSA AI course, we tested it with 42 undergraduate students from the field of Computer Science at the University of Craiova, Romania. After following the AI course, students had to fill in a questionnaire regarding their learning experience with WELSA system. First, they were asked to assess the course content, the presentation, the platform interface, the navigation options, the communication tools and the course as a whole, on a 1 to 10 scale. The results are presented in Figure 7.

As we can see from Figure 7, the students' evaluation of the AI course and WELSA platform

Box 1.

Adaptation rule for learners with "Visual" preference
IF

$$p\_visual \in \Pr ef(L)$$

THEN
<u>Sort</u> *dc: type {StillImage/MovingImage, Text/Sound}*
<u>Dim</u> *dc: type {Text, Sound}*
Adaptation rule for learners with "Concrete" preference
IF

$$p\_concrete \in \Pr ef(L)$$

THEN
<u>Sort</u> *LoType {Illustration, Fundamental}*
<u>Highlight</u> *LoType {Illustration}*

is very positive. 71.42% of the students assessed the course content as very good (marks 9-10), 26.19% as good (marks 7-8) and only one student as average. As far as the presentation is concerned, the majority of the students (88.09%) found it very enjoyable, while the rest of 11.91% were also quite pleased with it. Students declared themselves equally satisfied with the course interface, 85.71% of them assigning it marks 9 and 10 and 14.29% marks 7 and 8. Students also appreciated positively the navigation features offered by the system, 80.95% of them giving very high marks (9-10). The lowest marks were obtained by the communication tools, with an average of only 7.54. This can be explained by the quite basic tools offered (chat and forum), while students were expecting more advanced communication tools (like audio / video conference, whiteboard, blog etc). The course as a whole received very high marks (9-10) from 85.71% of the students, the rest evaluating it as quite good (marks 7-8).

All in all, very good marks were assigned to most of the features, with only one feature (the communication tools) receiving lower (but still satisfactory) ratings. We can therefore conclude that students had a very positive learning experience with WELSA.

The main goal of our course was the provisioning of an adaptive learning experience. Therefore, evaluating the adaptivity features of the system is of a particular importance. We were first in-

Figure 5. Snapshot from WELSA AI course: a page adapted for a student with Concrete learning preference

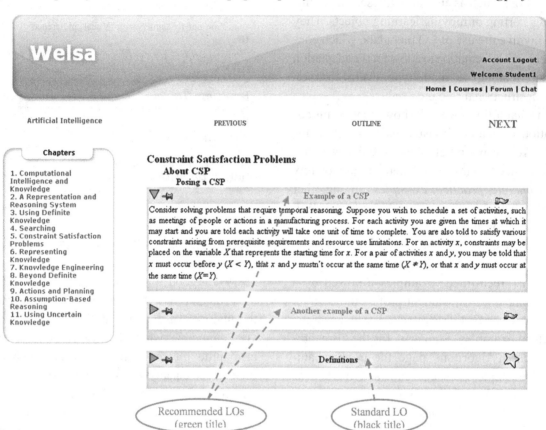

teresed in finding out the perceived degree of concordance between the course and the students' self-diagnosed learning preferences. "To which extent do you believe the course matched your real learning preferences?" was the question addressed to the students. The subjects could choose from a 5-point-scale ("Very large", "Large", "Moderate", "Small", "Very small"). The results are presented in Figure 8, showing a good correspondence between the adapted course and the students' real learning preferences.

The next survey item aimed at identifying the extent of the adaptation effect on the learning process. Students' answers to the question: "To which extent was this adaptation useful for you?" are summarized in Figure 9. As you can see, the majority of the students (80.95%) reported that the adaptation provided by the system proved useful for their learning process, at least to a moderate extent.

Finally, we were interested in students' desire to use WELSA system for other courses, on an everyday basis. The results are summarized in Figure 10. As can be seen from the figures, the large majority of the students (83.33%) are ready to adopt WELSA system for large scale use, with only 7.14% reluctant.

More evaluations regarding the adaptivity features of WELSA, coming from different experiments, were reported in (Popescu, 2010b), including comparisons between the adaptive and non-adaptive sessions as well as between matched and mismatched learners. The overall results are very encouraging, proving the positive effect that our adaptation to learning styles has on the learning process.

Figure 6. Snapshot from WELSA AI course: a page adapted for a student with Visual learning preference

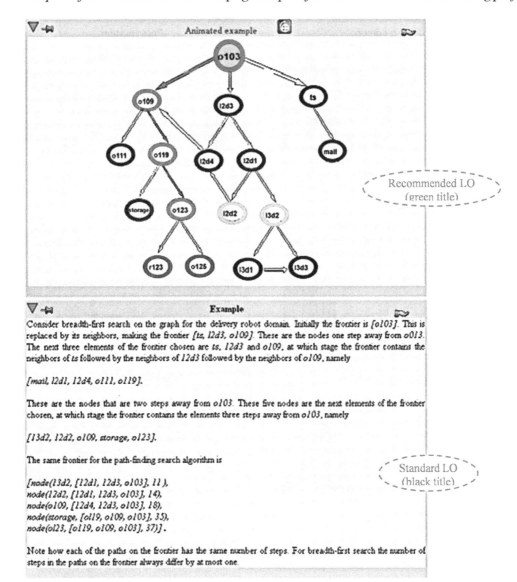

Teacher Validation

The implementation of the AI course in WELSA started from an existent learning material, inspired from the textbook of Poole et al. (1998). The authoring process was quite straightforward, requiring few additions and modifications. The authoring tool proved easy to use, both for the initial version and for subsequent editing of the course. Some additional time was required for the creation of videos, animations and interactive simulations, to support learners with visual and/or active preferences. However, as mentioned in section 3, the creation of the actual course content is outside the scope of our system: WELSA authoring tool is only concerned with the structuring and annotating of the LOs, which are presumed already available. From this point of view, the editor tool proved very handy and the authoring process was efficient and enjoyable.

Figure 7. Students' assessment of their learning experience with WELSA

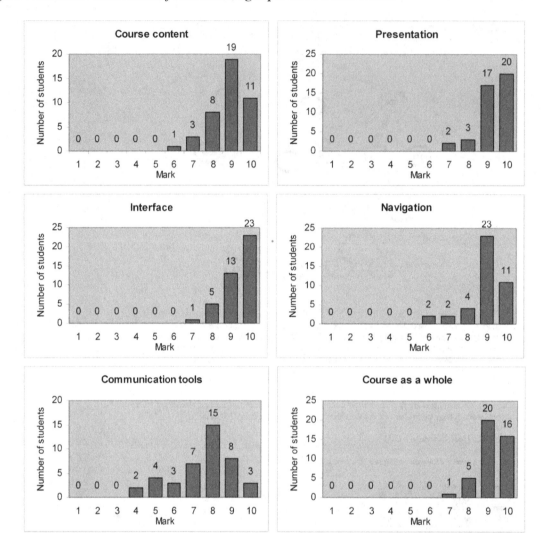

Figure 8. Perceived degree of concordance between the course and the matched students' self-diagnosed learning preferences

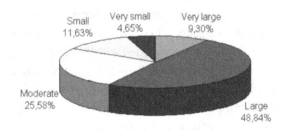

Figure 9. Perceived usefulness degree of the adaptation process

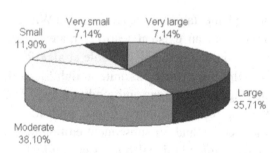

Figure 10. Students' willingness to adopt WELSA system for everyday use

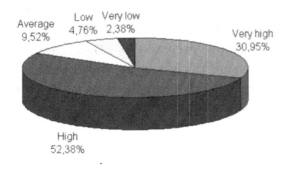

However, in order to validate the WELSA authoring tool, we needed also an independent evaluation, from teachers not engaged in the WELSA development process. We therefore performed a small experiment involving 3 professors from the Systems Engineering field. First, they went through a training session, where they were familiarized with the principles and functionalities of WELSA, including the course editor. Next, they were asked to use the authoring tool for implementing a course fragment of their choice. Finally, they had to fill in a questionnaire regarding their experience in interacting with WELSA authoring tool. The results of this questionnaire are summarized in Table 2.

We then analyzed the resulted course fragments: all three were correct and complete, with well defined hierarchical organization as well as comprehensive metadata. One thing that we noticed is that teachers tend to create their courses in a quite inflexible manner, defining strict prerequisite relations between LOs, even when this is not needed (which limits the applicability of the resource ordering technique). Perhaps more training and practice would solve this problem. Furthermore, not enough multimedia LOs were

Table 2. Teachers' assessment of their authoring experience with WELSA course editor

Question	Teacher_1	Teacher_2	Teacher_3
How much time did you spend with WELSA Authoring tool?	4h	3h	4.5h
Did you understand how to work with the tool? (Yes / No)	Yes	Yes	Yes
Was it easy to learn to work with the tool? (Very easy / Easy / Average / Difficult / Very difficult)	Very easy	Easy	Easy
Was the interface intuitive? (Very intuitive / Intuitive / Average / Not very intuitive / Not at all intuitive)	Very intuitive	Intuitive	Very intuitive
Did you experience any problems with the tool? (Yes/No) Please describe	No	No	No
Did you get the results that you expected? (Yes/No)	Yes	Yes	Yes
Are you satisfied with the resulted course? (Very satisfied / Satisfied / Neutral / Unsatisfied / Very unsatisfied)	Satisfied	Satisfied	Very satisfied
Overall impression (pleasant / unpleasant; easy / difficult)	Pleasant, easy	Pleasant, easy	Pleasant, easy
Suggestions and comments	"It would be nice to have the possibility to visualize the resulted course as you add LOs (preview)"	"My original course doesn't include any simulations / interactive resources and it would take a lot of time to make them"	"The course looks really nice and I'd like to actually use it with my students. I would probably have to add more interactive resources, though"

provided, which means that Visual learners are not well catered for; similarly, not enough opportunities for practice (simulations, interactive resources) were included, which means limited support for Active learners. However, we should point out that these limitations pertain to the course itself, not to the WELSA system.

Finally, while the 3 teachers assessed the tool as intuitive and easy to learn and encountered no problems while using it, we should not forget that they all come from a technical field. Perhaps for less technical-oriented authors a graphical tool for expressing hierarchical and prerequisite relations between LOs, as well as drag-and-drop facilities for positioning the LOs in the course, would be more welcome. Furthermore, a preview option could be added, as suggested by Teacher_1.

CONCLUSION

This paper reported a case study on the implementation of a personalized AI course using WELSA educational system. Several steps were covered: i) course authoring; ii) adaptation mechanism; iii) experimental validation. The results obtained are very encouraging, proving the practical applicability of WELSA system, both from the point of view of the students and the teachers.

However, in order to allow for generalization, more courses in various domains have to be implemented in WELSA. Furthermore, the system will have to be tested on a wider scale, with students of variable age, field of study and background knowledge, as well as with teachers having various degrees of technical experience.

Another future research direction would be to offer students a wider variety of adaptation strategies, by extending the adaptation component. Additionally, as students suggested in the opinion questionnaire, more advanced communication and collaboration tools should be incorporated, including Web 2.0 applications (blog, wiki, social bookmarking tool etc).

Further support could also be provided for the teacher (course author): adding an import / export facility to the course editor, allowing for conversion between various course formats and standards (e.g. SCORM, IMS LD etc) would be very helpful. This would allow teachers to use existing courses as they are (perhaps adding some additional metadata), which would provide for greater reuse.

ACKNOWLEDGMENT

This work was supported by the strategic grant POSDRU/89/1.5/S/61968, Project ID 61968 (2009), co-financed by the European Social Fund within the Sectorial Operational Program Human Resources Development 2007 – 2013.

REFERENCES

Bajraktarevic, N., Hall, W., & Fullick, P. (2003). Incorporating learning styles in hypermedia environment: Empirical evaluation. In *Proceedings of the Workshop on Adaptive Hypermedia and Adaptive Web-Based Systems* (pp. 41-52).

Brown, E., Brailsford, T., Fisher, T., & Moore, A. (2009). Evaluating learning style personalization in adaptive systems: Quantitative methods and approaches. *IEEE Transactions on Learning Technologies*, *2*(1), 10–22. doi:10.1109/TLT.2009.11

Brusilovsky, P., & Millan, E. (2007). User models for adaptive hypermedia and adaptive educational systems. In P. Brusilovsky, A. Kobsa, & W. Neidl (Eds.), *The Adaptive Web: Methods and Strategies of Web Personalization* (LNCS 4321, pp. 3-53). New York: Springer.

Carver, C. A., Howard, R. A., & Lane, W. D. (1999). Enhancing student learning through hypermedia courseware and incorporation of student learning styles. *IEEE Transactions on Education*, *42*, 33–38. doi:10.1109/13.746332

Cha, H. J., Kim, Y. S., Lee, J. H., & Yoon, T. B. (2006). An adaptive learning system with learning style diagnosis based on interface behaviors. In *Proceedings of Intl. Conf. E-learning and Games (Edutainment)*, Hangzhou, China.

Coffield, F., Moseley, D., Hall, E., & Ecclestone, K. (2004). *Learning styles and pedagogy in post-16 learning. A systematic and critical review.* London: Learning and Skills Research Centre.

DCMI. (2009). *Dublin Core Metadata Initiative.* Retrieved from http://dublincore.org

Felder, R. M., & Silverman, L. K. (1988). Learning and teaching styles in engineering education. *Engineering Education, 78*(7), 674-681. Retrieved from http://www4.ncsu.edu/unity/lockers/users/f/felder/public/Papers/LS-1988.pdf

Gardner, H. (1995). Reflections on multiple intelligences: Myths and messages. *Phi Delta Kappan, 77*(3), 200–209.

Graf, S., Lan, C. H., Liu, T. C., & Kinshuk. (2009). Investigations about the effects and effectiveness of adaptivity for students with different learning styles. In *Proceedings of ICALT 2009* (pp. 415-419). Washington, DC: IEEE Computer Society Press.

Honey, P., & Mumford, A. (2000). *The learning styles helper's guide.* Maidenhead, UK: Peter Honey Publications Ltd.

Lee, C. H. M., Cheng, Y. W., Rai, S., & Depickere, A. (2005). What Affect Student Cognitive Style in the Development of Hypermedia Learning System? *Computers & Education, 45*, 1–19. doi:10.1016/j.compedu.2004.04.006

Limongelli, C., Sciarrone, F., Temperini, M., & Vaste, G. (2009). Adaptive learning with the LS-Plan system: A field evaluation. *IEEE Transactions on Learning Technologies, 2*(3), 203–215. doi:10.1109/TLT.2009.25

Moodle. (2009). Retrieved from http://moodle.org

Papanikolaou, K. A., Grigoriadou, M., Kornilakis, H., & Magoulas, G. D. (2003). Personalizing the interaction in a Web-based educational hypermedia system: the case of INSPIRE. *User Modeling and User-Adapted Interaction, 13*, 213–267. doi:10.1023/A:1024746731130

Poole, D., Mackworth, A., & Goebel, R. (1998). *Computational Intelligence: A Logical Approach.* Oxford, UK: Oxford University Press.

Popescu, E. (2010a). A Unified Learning Style Model for Technology-Enhanced Learning: What, Why and How? *International Journal of Distance Education Technologies, 8*(3). (pp. 65-81)

Popescu, E. (2010b). Adaptation Provisioning with respect to Learning Styles in a Web-Based Educational System: An Experimental Study. *Journal of Computer Assisted Learning, 26*(4). (pp. 243-257) doi:10.1111/j.1365-2729.2010.00364.x

Popescu, E., & Badica, C. (2009). Providing personalized courses in a Web-supported learning environment. In *Proceedings of WI-IAT 2009, Workshop SPeL* (pp. 239-242). Washington, DC: IEEE Computer Society Press.

Popescu, E., Badica, C., & Moraret, L. (2009). WELSA: An intelligent and adaptive Web-based educational system. In *Proceedings of IDC 2009* (SCI 237, pp. 175-185). New York: Springer.

Popescu, E., Badica, C., & Trigano, P. (2008b). Learning objects' architecture and indexing in WELSA adaptive educational system. *Scalable Computing: Practice and Experience, 9*(1), 11–20.

Popescu, E., Trigano, P., Badica, C., Butoi, B., & Duica, M. (2008a). A course authoring tool for WELSA adaptive educational system. In *Proceedings of ICCC 2008* (pp. 531-534).

Sangineto, E., Capuano, N., Gaeta, M., & Micarelli, A. (2008). Adaptive course generation through learning styles representation. *Universal Access in the Information Society, 7*(1), 1–23. doi:10.1007/s10209-007-0101-0

Stash, N. (2007). *Incorporating cognitive/learning styles in a general-purpose adaptive hypermedia system.* Unpublished doctoral dissertation, Eindhoven University of Technology, The Netherlands.

Stash, N., Cristea, A., & De Bra, P. (2005). Explicit intelligence in adaptive hypermedia: Generic adaptation languages for learning preferences and styles. In *Proceedings of the Workshop CIAH2005, Combining Intelligent and Adaptive Hypermedia Methods/Techniques in Web Based Education Systems, in conjunction with HT'05* (pp. 75-84).

Triantafillou, E., Pomportsis, A., Demetriadis, S., & Georgiadou, E. (2004). The value of adaptivity based on cognitive style: an empirical study. *British Journal of Educational Technology, 35*(1), 95–106. doi:10.1111/j.1467-8535.2004.00371.x

Ullrich, C. (2005). The Learning-Resource-Type is dead, long live the Learning-Resource-Type! *Learning Objects and Learning Designs, 1,* 7–15.

Wang, T., Wang, K., & Huang, Y. (2008). Using a style-based ant colony system for adaptive learning. *Expert Systems with Applications, 34*(4), 2449–2464. doi:10.1016/j.eswa.2007.04.014

Witkin, H. A. (1962). *Psychological differentiation: studies of development.* New York: Wiley.

This work was previously published in the International Journal of Information Systems and Social Change, Volume 2, Issue 1, edited by John Wang, pp. 31-47, copyright 2011 by IGI Publishing (an imprint of IGI Global).

Chapter 4
Expertise Learning and Identification with Information Retrieval Frameworks

Neil Rubens
University of Electro-Communications, Japan

Dain Kaplan
Tokyo Institute of Technology, Japan

Toshio Okamoto
University of Electro-Communications, Japan

ABSTRACT

In today's knowledge-based economy, having proper expertise is crucial in resolving many tasks. Expertise Finding (EF) is the area of research concerned with matching available experts to given tasks. A standard approach is to input a task description/proposal/paper into an EF system and receive recommended experts as output. Mostly, EF systems operate either via a content-based approach, which uses the text of the input as well as the text of the available experts' profiles to determine a match, and structure-based approaches, which use the inherent relationship between experts, affiliations, papers, etc. The underlying data representation is fundamentally different, which makes the methods mutually incompatible. However, previous work (Watanabe et al., 2005a) achieved good results by converting content-based data to a structure-representation and using a structure-based approach. The authors posit that the reverse may also hold merit, namely, a content-based approach leveraging structure-based data converted to a content-based representation. This paper compares the authors' idea to a content only-based approach, demonstrating that their method yields substantially better performance, and thereby substantiating their claim.

DOI: 10.4018/978-1-4666-2922-6.ch004

INTRODUCTION

In today's knowledge-based economy, having the proper expertise is crucial to resolving many tasks. In the pedagogical world, such tasks range from educating others, to solving difficult problems, assessing/guiding the research directions of others, and reviewing the quality of conference/journal papers. The most traditional approach to expertise finding has been a burdensome process, involving manual referrals and direct contact. Luckily computers have mitigated this burden to a considerable degree. Several excellent surveys exist concerning this (e.g., Yimam-Seid & Kobsa, 2003; Maybury, 2006). As a result of the aid of computers, expert finding systems (EFS) have started to gain acceptance and are being deployed in a variety of areas. The Taiwanese National Science Council utilizes EFS to find reviewers for grant proposals (Yang et al., 2009); Australia's Department of Defense has deployed a prototype EFS to better utilize and manage its human resources (Prekop, 2007); ResearchScorecard Inc.'s EFS allows a user to find and rank scientists involved in biomedical research at Stanford University and at the University of California in San Francisco. There is also several expertise finding platforms that are applicable to wider domains and are utilized by an increasing number of companies (Maybury, 2006). Further, many methods have been developed to automate the task of expertise finding, including language and topic modeling (Yang et al., 2009), latent semantic indexing (LSI) (Lochbaum & Streeter, 1989), probabilistic modeling (Balog et al., 2006), and link analysis (Karimzadehgan et al., 2009; Halonen et al., 2010).

Essentially, the following two approaches are almost always employed in EFS. Typically, a description of a task is given, and the system aims at finding a person with the appropriate expertise to match it.

- Content-based approaches analyze the content (text) of both a given task's description and of candidates' papers (see Figure 1). For example, a candidate (researcher) may be knowledgeable about the given task (and therefore a good choice for the task) if the task description and the candidate's works share many of the same terms (keyword extraction).

- Structure-based approaches analyze the topographical structure of the search space by treating papers as nodes in a graph, interlinked by citations, authorship, affiliations, etc. (see Figure 2). For example, a researcher may be considered a match for a given task if both the task and researcher cite the same papers (showing familiarity with the subject area).

There are many pros and cons to each approach. For one, while the underlying data of structure-based approaches is usually rather precise, content-based approaches must grapple with the inherent ambiguity and complexity of the data expressed in natural language. One way to circumnavigate this problem is to reduce the number of possible dimensions by extracting salient keywords (Laender et al., 2002); these methods can perform quite well, but are nevertheless hindered by their ability to extract appropriate and completely representative keywords. To get beyond this, intricate knowledge of natural language processing (NLP) is required a field which often lacks usable off-the-shelf tools. On the other hand, many graph analysis tools exist (ideal for structure-based approaches) (Wikipedia, 2009a). Availability of frameworks is also important. Many powerful information retrieval (IR) frameworks exist that provide easy deployment and scale well (luc, lem, ter); this is something that structure-based approaches often lack.

The bottom line is that the majority of methods to date use one approach (content-based, "C") or the other (structure-based, "S"); though sometimes both approaches are used in tandem (C and S). Yet previous work (Watanabe et al., 2005b) has demon-

Figure 1. Traditional content based approach for expertise finding using keywords

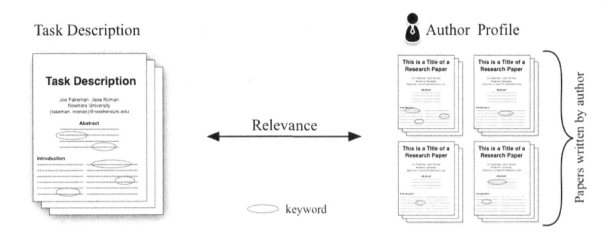

strated the merits of a synthesis-based approach (S using C), using a structure-based approach derived from content keywords. We posit that the reverse may also hold merit, in other words a structure derived synthesis (C using S). Our motivation is to use the precision of structure-based data while leveraging the advantages supplied by using content-based frameworks. We propose a way to transform the structure-based representation into one usable by content-based approaches, which further lets us integrate both content-based and structure-based analysis methods; both of which have been previously difficult to compare due to the intrinsic differences in their underlying data.

In our experiments we show that applying a content-based approach to graph data outperforms applying it to text data, the kind of data for which content-based approaches were originally designed. We demonstrate our approach by first creating a structure-based representation and then converting it to a content-based one (see Figure 3). We then compare it to a keyword-extraction

Figure 2. Traditional structure based approach for expertise finding using graph metrics

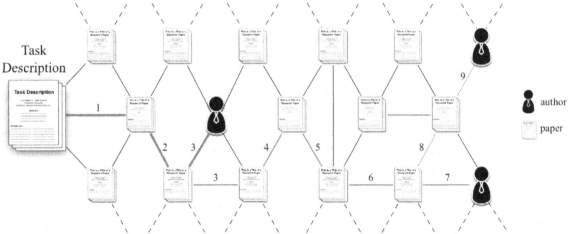

Figure 3. New approach for expertise finding leveraging a content based representation of network-based data (with max distance=2)

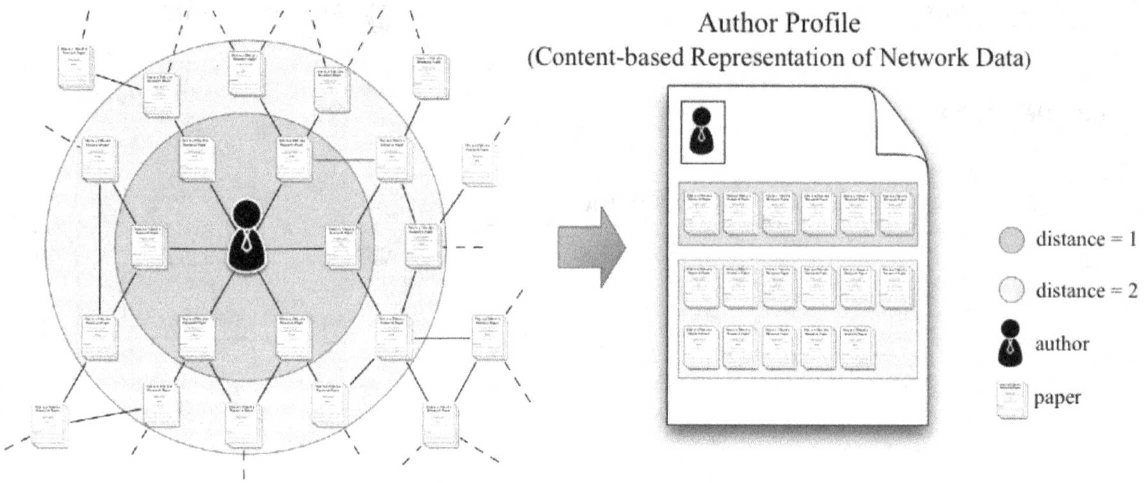

content-based approach that is quite popular, showing the potential merit of applying content-based approaches to network-based data in the realm of expertise finding.

RELATED WORK

In a practical setting, the most similar research to our experiments in Expertise Finding is the reviewer assignment problem (RAP), which attempts to assign suitable reviewers to submitted manuscripts (a challenging task for journal editors, conference program chairs, research councils, etc. (Wang et al., 2008)). It is clear that without proper matching of proposals (papers) to reviewers the integrity of the review process would crumble (Hettich & Pazzani, 2006), making it an important task in the EF field. Time constraints coupled with the labor intensive aspect of the reviewing process has led to research seeking ways to mitigate the problem since 1992 (Dumais & Nielsen, 1992).

The following section provides an overview of general ideas. For more specific implementation details, please see the corresponding sub sections in the Problem Background Section.

Content-Based Approach

Content-based approaches concentrate on the content of individual items, such as the words in a document. One approach to comparing related documents is through use of keywords. There are a number of drawbacks to this approach. For one, it has been noted in numerous studies that it is more the exception than the rule for keyword assigners to input the same terms (i.e., there is low agreement) when unrestricted text is used (Furnas et al., 1987). If one sets the taxonomy of keywords in advance, this problem is mitigated, but engenders others, namely, the difficulty of the scientific community in selecting a set of keywords, as often in one discipline finer or more course-grained labels are necessary than in another, and also that such a list of keywords is difficult to keep up to date (Hettich & Pazzani, 2006). It is argued, in fact, that such lists cannot keep speed with the fast pace of progressing technology (Wang et al., 2008). One means to combat this is to employ data-mining to automatically extract salient keywords via unsupervised clustering or supervised learning (Biswas & Hasan, 2007). The NSF Revaide system (Hettich & Pazzani, 2006) is an example of this, which employs a model using standard tf-idf, a term weight common in IR and text mining; the vector terms are taken from the space of all words in the document collection. Another approach is to use Latent Semantic Indexing (LSI), increasing being referred to as Latent Semantic Analysis (LSA), which determines the amount of overlap between two sets of content without a preconceived notion of relationships (Dumais & Nielsen, 1992) (Yarowsky & Florian, 1999).

Structure-Based Approach

Structure-based approaches, also called network- or graph-based, refer to the hierarchical relationship between papers, authors, affiliations, institutions, etc. The edges of the graph may or may not be weighted, and the nodes may represent different types of entities. In effect, such a structure represents the opinions of like-minded people (as they cite or do not cite based on merit

and opinion). Bibliographic coupling (Kessler, 1963) and Co-citation Indexing (Small, 1973) are perhaps the simplest in concept (see Figure 4) representing this idea. Co-citations occur at various granularities, such as at the sentence, paragraph, section, and article level (Elkiss et al., 2008), and their similarity has been shown to be proportional to their textual proximity to one another (Kessler, 1963; Small, 1973, Nanba & Okumura, 1999). These approaches allow one to find potentially similar works, but do not remedy the RAP or the more general EF problem. For this, (Wang et al., 2008) provides a survey of many structure-based approaches; some of these are summarized below.

One approach (Hartvigsen, 1999; Schirrer et al., 2007) is to use a network flow model that utilizes a bipartite graph, where the "flow" from one point (the task/paper/proposal) to the other (candidate reviewer/suitable researcher/etc.) is how much bandwidth would "flow" in that direc-

Figure 4. Citation indexing: bibliographic coupling (left), co-citation (right)

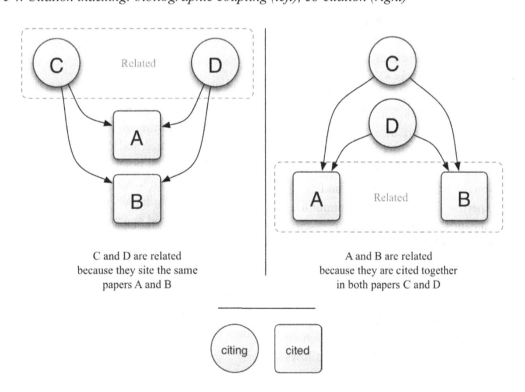

C and D are related because they site the same papers A and B

A and B are related because they are cited together in both papers C and D

citing   cited

tion across the network with weights assigned to edges to determine the strength of the connection.

It is also possible to recast a content-based representation to a structure-based representation (S using C) (Watanabe et al., 2005a); it creates a scale-free network with nodes representing keywords extracted from papers for both representing the paper as well as the author; these nodes are connected with a score approximating their similarity, which can then be used for EF by tracing a path between two nodes along adjoining edges.

A more technical explanation of structure-based approaches appears in the Problem Background section of this paper.

Expertise Finding Venues

There are several venues for the promotion of research in expertise finding.

The Expert Search task in the Enterprise track of the Text Retrieval Conference (TREC) has been running annually since 2005 (Balog et al., 2008). The task is to extract subject experts from a large corpa of data. Entrants from the 2008 task employed a variety of methods, ranging from special treatment of person occurrences (Shen et al., 2008; Yao et al., 2008; Jiang et al., 2008), citation network-based link analysis (Xue et al., 2008; Zhu, 2008), proximity-based techniques (Balog & Rijke, 2008; He et al., 2008; Zhu, 2008), to other text-based methods. A more detailed summary is available in (Balog et al., 2008).

The Expertise Retrieval Workshop at SIGIR 2008 also sought to assess progress and bring (literally) expertise from different communities pursing similar goals in EF together to discuss future research agenda.

Real-World Deployments

Very few expertise finding systems are available for public use. The Microsoft Academic Search system, which is perhaps the most relevant, allows

one to retrieve "related authors" with respect to entered keywords (Mic). Other systems that provide related functionality include ResearchScorecard Inc.'s EF system, which allows a user to find and rank scientists involved in biomedical research at Stanford University and at the University of California in San Francisco, and CiteSeerX and Google Scholar, which both provide a "related articles" feature. The latter two both provide a type of clustering feature, though differ in their implementations. CiteSeerX shows documents that cite a similar set of papers (i.e., bibliographic coupling), while Google Scholar uses a document similar metric and document relevance score to find related papers.

PROBLEM BACKGROUND

The objective of expertise finding is to find an expert suitable for a given task description (e.g., a reviewer for a given paper). We can formulate this problem as learning an expertise relevance function:

$$\hat{f}_{relevance}(q, d), \qquad (3.1)$$

where q corresponds to the task description and d is a candidate's profile. The value of $\hat{f}_{relevance}(q, d)$, should be high if the candidate possesses expertise required for the task and low otherwise. In content-based approaches it is assumed that a candidate profile is relevant to the given task description if both contain similar keywords in their contents. In structure-based approaches relevancy is often measured with regards to citation-based similarity, i.e., if entities cite similar papers.

We provide a formulation of the problem for both content-based and structure-based approaches so that our approach, which leverages their synergy, can be understood in the subsequent section.

Content-Based Approaches

We will discuss the handling of expertise finding in the framework of information retrieval (IR) since it is the framework most commonly used in practice for content-based representations (Balog et al., 2008). Let us briefly discuss some common terminology and methodology (Wikipedia, 2009c; Singhal, 2001). An information retrieval process begins when a user enters a query (a formal statement of an information need) into the system, e.g., "expertise finding". An IR system computes a relevance score (q, d) of how well each document in the database matches the query, then the top ranking documents are shown to the user. Query q is typically represented by a series of terms t, e.g. *expertise, finding* (from out sample user input given above); more formally, this becomes:

$$q = \left\langle t_{1,q}, t_{2,q}, ..., t_{n,q} \right\rangle. \tag{3.2}$$

A document is likewise represented by a list of terms that constitute it:

$$d = \left\langle t_{1,d}, t_{2,d}, ..., t_{m,d} \right\rangle. \tag{3.3}$$

In the EF setting, a query corresponds to a task description and documents correspond to candidate profiles. Since a query is represented by the task description/paper/proposal/etc., it potentially contains many terms. A document is represented by a candidate profile, made up of papers written by the candidate as well as all papers cited (up to a desired citation distance).

IR systems typically assign weights to each term to reflect a term's importance. The *tf—idf* weight (term frequency—inverse document frequency) is an often used weighting schema (Wikipedia, 2009c; Singhal, 2001).

The term count in the given document is the number of times a given term appears in the document. The more times the term appears in the document the more likely it is to be representative of the document. For example if the term *expertise* appears frequently in our paper, it could be indicative of the paper's topic. The term count is usually normalized to prevent a bias towards longer documents (which may have a higher term count simply due to the larger number of total terms). The term frequency *tf* is defined as follows:

$$\mathrm{tf}_{i,j} = \frac{n_{i,j}}{\sum_{k} n_{k,j}}, \tag{3.4}$$

where $n_{i,j}$ is the number of occurrences of the considered term t_i in document d_j. The denominator is a normalizing factor represented by the sum of the number of occurrences of all terms in document d_j.

The term frequency alone does not necessary make a term important; some terms may appear frequently in all of the documents, such as the article "the". There is therefore a need for a measure of the general importance of the term across the entire document set. This is achieved by means of an inverse document frequency *idf* which diminishes the weight of terms that occur very frequently in the overall collection and increases the weight of terms that occur rarely (Wikipedia, 2009c; Singhal, 2001). The *idf* is obtained by dividing the number of all documents by the number of documents containing the term, and then normalizing it by taking the logarithm of the quotient:

$$\mathrm{idf}_i = \log \frac{|D|}{1 + |\{d : t_i \in d\}|}, \tag{3.5}$$

where $|D|$ is the total number of documents in the corpus, $|\{\mathbf{d}: t_i \in \mathbf{d}\}|$ is the number of documents where the term t_i appears ($n_{i,j} \neq 0$). Since the absence of a particular term in the document set will lead to a division-by-zero, we prevent this by adding 1 to the quotient.

A term of high importance expressed by the *tf-idf* is reached by a high term frequency (in the given document) and a low document frequency of the term in the whole collection of documents:

$$(\text{tf} - \text{idf})_{i,j} = \text{tf}_{i,j} \times \text{idf}_i \qquad (3.6)$$

A relevance scoring of a document's relevance given a user query could then be computed by summing the tf-idf for each query term as it related to that document:

$$(\text{tf} - \text{idf})_{i,j} = \text{tf}_{i,j} \times \text{idf}_i \qquad (3.7)$$

Many relevance scoring functions are simply more sophisticated variants of the above model (Wikipedia, 2009c; Singhal, 2001).

Structure-Based Approaches

A paper contains both the content data (the paper's text) and structure data (citations, affiliations, authors, etc.). Structure-based approaches utilize the latter of these two data types. Further, structure-based data is often represented by a heterogeneous graph data-structure

$$\hat{f}_{relevance}(q,d) = \sum_{q \in q} \text{tf}_{q,d} \times \text{idf}_q \qquad (3.8)$$

where V is a set of vertices/nodes, and E is a set of edges/links. Node types are often paper, person, publication venue (e.g. conference, journal), or affiliation (e.g. university, company), and edge types (see Table1) wrote, cites, published in, or affiliation.

We are interested in the relations between the node corresponding to task description and the nodes corresponding to the candidate authors. The graph contains many nodes that are not related to the task description. We extract a subgraph by expanding the task description node which allows us to reduce the size of the graph and to speed up

analysis. As can be seen, the precision of the data types and their relations provides good data on how the data relates to one another.

Features

The relevance function is learned by utilizing machine learning methods based on the feature values from the subgraph that relate the task description with candidate profiles. Let us provide a brief description of some features utilized in the structure-based analysis.

Shortest path between the task description and a person may indicate that the person is familiar with the matters covered by the task, e.g., if both the task description and a person cite the same paper.

Using the shortest path alone may not be enough, since it is also conceivable that the shortest path could be due to coincidence, e.g., both papers citing the same funding source. The average path may provide a more complete idea of the relation between the nodes of interest.

The resistance distance is equal to the resistance between two nodes on an electrical network (Klein & Randić, 1993). The intuition behind this is that the denser the surrounding network is the larger the resistance distance.

The centrality of a node measures the relative importance of the node within the graph. We use the common measures of network centrality: degree centrality, betweenness, closeness, and eigenvector centrality (Wikipedia, 2009b). For example, a paper that cites many other papers may be less focused. On the other hand, a paper may be influential if it is cited by many other papers.

Graph strength could be used to compute partitions of sets of nodes and to detect zones of high concentration along edges. We use it as an indicator of strength of the relation between the task and the candidate members. As an alternative measure we can also use a clustering coefficient that measures the degree to which nodes tend to cluster together (T. Schank, 2005) and also vertex connectivity (White & Harary, 2001).

PROPOSED APPROACH

It was shown that potential merit exists in the conversion of data from one form of representation to be used in the other, such as converting keywords extracted from paper content into a network-based structure (Watanabe et al., 2005a). We hypothesize that the reverse may also be true, namely, that converting structure-based data to a content-based representation may also be worthwhile. We therefore propose to utilize structure-based data within a content-based framework. Simply passing a structure-based representation to content-based methods, however, is not possible, as the underlying representations are not compatible. Conversion from one underlying representation to the other is therefore necessary. During this conversation it is important to retain the key features of the structure-based representation or we will lose any advantage gained through such a conversion, while at the same time considering how such data will be leveraged by the content-based method. To test the merits of a content-based representation

of a network-based data approach, we consider only the data available in the original structure representation for our experiments.

Node Representation

As discussed in previously, there are nodes of different types: paper, person, publication venue (e.g., conference, journal), and affiliation (e.g., university, company). In this paper we primarily concentrate on citation analysis and therefore only keep the author node and paper nodes (including the papers that correspond to task descriptions). The precise textual representation of nodes is not important as long as it uniquely identifies the entry to which it refers. For readability, we have chosen to represent them by a bibliography type reference string, e.g., *Rubens* (2007a). As illustrated in Figure 5, the content-based representation of the task description could then simply be a document containing a list of reference-ids of cited papers. The candidate's profile could be represented by the list of reference-ids of written papers and papers

Figure 5. Structure based representation

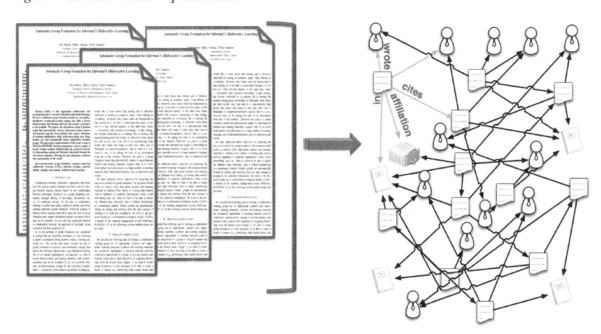

that are related to them through citations. In both cases we obtain a paper reference-id by performing a breadth-first traversal (up to a desired depth) starting from the nodes of interest (i.e., candidate researcher node or task description node).

Representation of Internode Relations

The above representation preserves only the nodes of the graph. This may result in each node having an equal contribution to the relevance of a document. However, the distance between the author node and other nodes can be considered representative of the author's expertise. That is, an author may be much more knowledgeable about

his own papers (one edge away) rather than the papers that are cited in that author's works (two edges away). Therefore, it may also be important to capture relations between the nodes (represented by the edges in a graph). One approach could be to assign a weight that corresponds to the influence of the node on the relevance score (see Figure 6). In our case, the closer the node the more important we consider it to be. We define a distance-based weight w_{distance} with regards to the minimum distance in the graph between the node of interest n_j and the author's node n_a

$$w_{\text{distance}}(n_j) = k^{-(\min(\text{distance}(n_j, n_a)))} \qquad (4.1)$$

Figure 6. Structure based approaches: a person may be considered an expert if he wrote and cited papers similar to those of task description

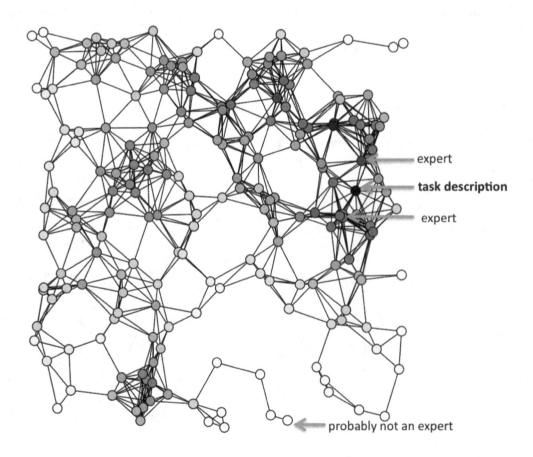

with the parameter k, in our case fixed: $k = 10$; i.e., the nodes that correspond to the papers that the author wrote will have 10 times larger weight than the papers that they cite. A common approach is to utilize machine learning to learn an appropriate weight. However, we believe that any reasonably defined weight should allow us to achieve a decent performance and therefore omit this step. In addition, text-based methods also assign weights to the nodes which we will briefly describe in the Content-based Approaches Section.

Next let us look at the ways to represent the concept of weight in a content-based representation. In a content-based representation, a concept of weight is often represented through the term frequency *tf* — terms that occur more often in a document are given a higher weight (Content-based Approaches Section), so one option would be to output the text representation of the node corresponding to the reference-id (Node Representation Section) with the frequency proportional to the node weight. However, this may not be the best option since it could not be easily changed, i.e., it would require changing the document content and re-indexing the document. Another drawback that changing the frequency of the terms may inadvertently affect other measures, e.g., document length, etc. Another option could be to define a field specific weight. Many IR frameworks allow specifying field-based weights, e.g., the word occurring in the title will have a higher weight than the word in the text. We therefore perform weight representation conversion in the following manner. Since we use distance-based weights (Representation of Internode Relations Section) for each distance l we define a field notated by *level$_l$* and populate each level with the textual representation of nodes at the corresponding distances. The weight of the field is then set to the same as the weight of the level in the graph; this is indicated by the different color circles in (Algorithm 1).

We have chosen to implement the field-based weight scheme since it allows us to change the

Algorithm 1. Pseudocode of the proposed representation conversion procedure (Section 4)

```
# converts structure-based representation of <node> to a
content-based one
function convert(node)

# create an empty document
doc = new Document()

# perform a bread-first traversal of the tree starting from the
<node>;
# nodes at the same level/depth at stored in the array
nodes[level]
nodes = bfs(node)
 for each level
 doc.field[level].terms = nodes[level]
 # weight is calculated by Eq.(9)
 doc.field[level].weight = weight(level)
 end

return doc
```

field easily without the need for changing content or re-indexing.

Another measure commonly used by content based methods is the inverse document frequency *idf* (Content-based Approaches Section), proportional to the inverse of the node degree in a structure-based representation. However, unlike the term frequency which is a document level measure (could be calculated given a document), *idf* is a corpus level measure (needs to be calculated with regards to all of the documents in a corpus). Incorporating *idf* measure on a document level would not be efficient, since if its value changes, the content-representation would need to be updated for all affected documents. Moreover, information retrieval frameworks already provide efficient ways to calculate *idf* measure from the document index. Information retrieval frameworks, may further change term weights.

Goal-Based Relations Representation

So far we have discussed measures in relation to the author's node. Our ultimate goal is to find an appropriate expert given a task description.

Network analysis tools usually accomplish this by considering graph measures in relation to the task description node and author nodes. However, our goal is to convert a network-based representation of both the author and the task description independently, and delegate the decision making to a content-based method. Therefore we do not calculate any of the measures between the task description node and the author nodes, but rather convert the task description node in the same manner as the author nodes and allow the content-based method decide relevancy.

EVALUATION

In this section we describe the implementation of our proposed method and evaluate its performance.

Data

Since our method is built around a structure-based representation, it is crucial that our dataset provide such structure. We therefore chose a database of citations, the CiteSeer dataset (cit, 2009); it is one of the most comprehensive datasets of academic publications openly available. The CiteSeer dataset contains data on 1.3×10^6 academic papers along with 26.5×10^6 corresponding citations that link the papers. For the content-based representation we use the 'description' field which often corresponds to a combination of the paper's abstract and keywords. For the structure-based method we use the paper reference-ids and citation links.

Evaluation Criteria

Our goal is to predict who has the needed expertise to accomplish a given task (represented by a task description) or evaluate a research paper. To simulate this setting, we try to find a person who is an expert on the matters covered by a selected paper. We assume that the first author of a paper has a sufficient level of expertise on the paper that

s/he wrote. This is not to say that the author has the most expertise on his own paper, e.g. an editorial committee could easily possess more extensive expertise on the paper's subject. However, only authorship properties are contained within our data (there is no information related to editorials). Our task is then, given a paper, to predict who the paper's first author is. We have chosen to evaluate based only on the first author, since usually their contribution can be more precisely identified (Hettich & Pazzani, 2006).

Typically the performance of information retrieval systems is evaluated by precision (fraction of the documents retrieved that are relevant to the user's information need), and recall (fraction of the documents that are relevant to the query that are successfully retrieved). However, in our settings these criteria are not applicable. According to our assumptions, we know that the first author is an expert on the paper's matters; however if the person is not an author it does not mean that the person does not have expertise on the paper's material. Therefore all we can accurately evaluate is the relevance score or ranking that the system assigns to the actual author (i.e., the actual should be highly ranked by the system). For such settings mean reciprocal rank (MRR) is often used (Wikipedia). Mean reciprocal rank is a statistic for evaluating a list of possible responses to a query, ordered by probability of correctness. The reciprocal rank of a query response is the multiplicative inverse of the rank of the first correct response. The mean reciprocal rank is the average of the reciprocal ranks of results for selected queries:

$$\text{MRR} = \frac{1}{|Q|} \sum_{i=1}^{Q} \frac{1}{\text{rank}_i}. \tag{5.1}$$

Implementation

Changing the representation of the data could be a time consuming process, which includes data parsing, tree traversal etc. Initially we load the data into MySQL database as indicated below. Then, by using the database we obtain the information on selected tasks and authors, transform its representation from a graph-based one to a content-based one by following the proposed procedure (Proposed Approach Section). Finally, we output a content-based representation of each entity as a text file.

At this stage, the information retrieval system loads and indexes the content-based representation created in the previous step. We have chosen a free open source search library, Apache Lucene. For the deployment option we use Apache Solr, a high performance search server that provides integration with Lucene and provides caching and replication layers.

Experiment

First we construct a set of experts; for this we use the first author that appears on a work with the assumption that s/he is an expert on his/her own paper. We do this by first selecting a paper (hereafter "query-paper") at random from the dataset, and subsequently its first author (i.e, the expert). The data on the author is then extracted from the database, with data related to the query-paper omitted; in other words, we remove any record that the author wrote that paper from their generated profile. Based on this data we construct both a structure-based representation and a content-based representation so we can compare the two (Problem Background Section). We then apply the proposed procedure (Proposed Approach Section) to convert the structure-based representation into a content-based representation of the structure-data. We repeat this procedure to create 100 query-papers and the profiles for their corresponding experts.

Next, we create a set of non-experts by randomly selecting 1,000 authors and creating their profiles by the above mentioned method.

At the evaluation stage, we pass the query-paper to the information retrieval system, and the system is expected to return a ranked list of candidate experts from the set of 1,000 + 1. The task is not finding a needle in a haystack, but is nonetheless difficult; to be successful the expert should receive a high ranking out of 1,000 candidates. We then record the resultant rank of the paper's author and use it to calculate mean reciprocal rank (MRR) as described in Evaluation Criteria Section.

Results and Discussion

The system is evaluated by using both a content-based representation, structure-based representation and the proposed structure-data to content-based representation. The mean reciprocal rank was almost three times higher 0.692 (higher means better) for the proposed input format than for the pure content-based input 0.253. The information retrieval approach is designed to deal with content-based data. However a better accuracy was achieved for the data provided by the proposed approach, which was converted from a structured-based representation into a content-based one. We believe that this improvement is achieved due to the reduction in the ambiguity of the data.

Subgraph Considerations

All of the measures we have mentioned are node-based, that is, weights assigned to nodes. However, some of the characteristics are based on the graph or subgraphs (Structure-based Approaches Section). For example, a clustering coefficient measures the degree to which nodes tend to cluster together (Schank, 2005), where nodes in a cluster may correspond to papers in the same subfield. One way in which this relation could be captured is by ensuring that the nodes that are in the same cluster belong to the same paragraph (or field if more flexibility is necessary) of the content-based representation. Since we are trying to keep things relatively simple we do not currently consider subgraph features.

Other Considerations

It may not be possible to represent all of the features of the graph in a content-based representation. Some graph information may therefore be inadvertently lost during the conversion process. If the unaltered graph representation is important, then it may be beneficial to use a hybrid approach of combining both the content-based methods and structure-based methods.

CONCLUSION

Performance of content-based methods in expertise finding is often hindered by the inherent ambiguity of the input data, expressed in natural language. Structure-based representations are much less ambiguous. A previous attempt (Watanabe et al., 2005a) showed that potential merits exist in the conversion of data from one form of representation to be used in the other. As this method used content-data to network-based representation approach, we posited that similarly the reverse, a network-data to content-based representation approach may also yield promising results. Our results show that this is indeed true, with nearly three times the MRR of the comparable content only-based system. Future work includes finding a means to encode more robust structure-based features into a form usable by content-models.

REFERENCES

Balog, K., Azzopardi, L., & de Rijke, M. (2006). Formal models for expert finding in enterprise corpora. In *Proceedings of the 29th annual international ACM SIGIR conference on Research and development in information retrieval* (p. 50). New York: ACM.

Balog, K., & Rijke, M. D. (2008). Combining candidate and document models for expert search. In *Proceedings of the Seventeenth Text REtrieval Conference (TREC 2008)*.

Balog, K., Soboroff, I., Thomas, P., Csiro, A., & Craswell, N. (2008). *Overview of the TREC 2008 Enterprise Track*.

Biswas, H., & Hasan, M. (2007). Using Publications and Domain Knowledge to Build Research Profiles. In *Information and Communication Technology (ICICT'07)* (pp. 82-86).

Centrality. (n.d.). In *Wikipedia, The Free Encyclopedia*.

Dumai, S. T., & Nielsen, J. (1992). Automating the assignment of submitted manuscripts to reviewers. In *Proceedings of the Annual ACM Conference on Research and Development in Information Retrieval*.

Elkiss, A., Shen, S., Fader, A., States, D., & Radev, D. (2008). Blind men and elephants: what do citation summaries tell us about a research article. *Journal of the American Society for Information Science and Technology, 59*.

Furnas, G., Landauer, T., Gomez, L., & Dumais, S. (1987). The vocabulary problem in human-system communication. *Communications of the ACM, 30*(11), 971. doi:10.1145/32206.32212

Halonen, R., Thomander, H., & Laukkanen, E. (2010). DeLone & McLean IS Success Model in Evaluating Knowledge Transfer in a Virtual Learning Environment. [IJISSC]. *International Journal of Information Systems and Social Change, 1*(2).

Hartvigsen, D. (1999). The Conference Paper-Reviewer Assignment Problem. *Decision Sciences, 30*(3), 865–876. doi:10.1111/j.1540-5915.1999.tb00910.x

He, B., Macdonald, C., Ounis, I., Peng, J., & Santos, R. (2008). University of Glasgow at TREC 2008: Experiments in blog, enterprise, and relevance feedback tracks with terrier. In *Proceedings of the 17th text retrieval conference, Gaithersburg,* MD.

Hettich, S., & Pazzani, M. (2006). Mining for proposal reviewers: lessons learned at the national science foundation. In *Proceedings of the 12th ACM SIGKDD international conference on Knowledge discovery and data mining* (pp. 862-871). New York: ACM.

Information Retrieval (*n.d.*). Wikipedia, The Free Encyclopedia.

Jiang, J., Lu, W., & Zhao, H. (2008). CSIR at TREC 2008 Expert Search Task: Modeling Expert Evidence in Expert Search. In *Proceedings of the 2008 Text REtrieval Conference (TREC 2008)*, Gaithersburg, MD.

Karimzadehgan, M., White, R., & Richardson, M. (2009). Enhancing expert finding using organizational hierarchies. In *Proceedings of the 31th European Conference on IR Research on Advances in Information Retrieval* (pp. 177-188). New York: Springer.

Kessler, M. M. (1963). Bibliographic coupling between scientific papers. *American Documentation, 14*(1), 10–25. doi:10.1002/asi.5090140103

Klein, D., & Randić, M. (1993). Resistance distance. *Journal of Mathematical Chemistry, 12*(1), 81–95. doi:10.1007/BF01164627

Laender, A. H. F., Ribeiro-Neto, B. A., da Silva, A. S., & Teixeira, J. S. (2002). A brief survey of web data extraction tools. *SIGMOD Record, 31*(2), 84–93. doi:10.1145/565117.565137

Lochbaum, K., & Streeter, L. (1989). Comparing and combining the effectiveness of latent semantic indexing and the ordinary vector space model for information retrieval. *Information Processing & Management, 25*(6), 665–676. doi:10.1016/0306-4573(89)90100-3

Maybury, M. (2006). *Expert finding systems (Tech. Rep.)*. MITRE Corporation.

Mean Reciprocal Rank (*n.d.*). Wikipedia, The Free Encyclopedia.

Nanba, H., & Okumura, M. (1999). Towards multi-paper summarization using reference information. In *Proceedings of IJCAI* (pp. 926-931).

Prekop, P. (2007). *Supporting Knowledge and Expertise Finding within Australia's Defence Science and Technology Organisation*. HICSS.

Schank, D. W. T. (2005). Approximating clustering coefficient and transitivity. *Journal of Graph Algorithms and Applications, 9*(2).

Schirrer, A., Doerner, K., & Hartl, R. (2007). Reviewer assignment for scientific articles using memetic algorithms. *Metaheuristics, progress in complex optimization systems, 39*, 113-134.

Shen, H., Wang, L., Bi, W., Liu, Y., & Cheng, X. (2008). Research on Enterprise track of TREC 2008. In *Proceedings of the 2008 Text REtrieval Conference (TREC 2008)*.

Singhal, A. (2001). Modern information retrieval: A brief overview. *A Quarterly Bulletin of the Computer Society of the IEEE Technical Committee on Data Engineering, 24*(4), 35–43.

Small, H. (1973). Co-citation in the scientific literature: A new measure of the relationship between two documents. *JASIS, 24*, 265–269. doi:10.1002/asi.4630240406

Social network Analysis Software (*n.d.*). Wikipedia, The Free Encyclopedia.

Wang, F., Chen, B., & Miao, Z. (2008). *A survey on reviewer assignment problem (LNCS 5027)*. New York: Springer.

Watanabe, S., Ito, T., Ozono, T., & Shintani, T. (2005). *A Paper Recommendation Mechanism for the Research Support System Papits*. Washington, DC: IEEE.

Watanabe, S., Ito, T., Ozono, T., & Shintani, T. (2005). *A Paper Recommendation Mechanism for the Research Support System Papits*. Washington, DC: IEEE.

White, D., & Harary, F. (2001). The cohesiveness of blocks in social networks: Node connectivity and conditional density. *Sociological Methodology*, 305–359. doi:10.1111/0081-1750.00098

Xue, Y., Zhu, T., Hua, G., Zhang, M., Liu, Y., & Ma, S. (2008). THUIR at TREC2008: Enterprise track. trec.nist.gov.

Yang, K.-H., Kuo, T.-L., Lee, H.-M., & Ho, J. M. (2009). A reviewer recommendation system based on collaborative intelligence. In *Web Intelligence and Intelligent Agent Technology, IEEE/WIC/ACM International Conference on* (Vol. 1, pp. 564-567).

Yao, J., Xu, J., & Niu, J. (2008). Using Role Determination and Expert Mining in the Enterprise Environment. In *Proceedings of the 2008 Text REtrieval Conference (TREC 2008)*, Gaithersburg, MD.

Yarowsky, D., & Florian, R. (1999). Taking the load of the conference chairs: towards a digital paper-routing assistant. In *Proceedings of the 1999 Joint SIGDAT Conference on Empirical Methods in NLP and Very-Large Corpora*.

Yimam-Seid, D., & Kobsa, A. (2003). Expert finding systems for organizations: Problem and domain analysis and the demoir approach. In *Sharing Expertise: Beyond Knowledge Management*. Cambridge, MA: MIT Press.

Zhu, J. (2008). The University College London at TREC 2008 enterprise track. In *Proceedings of the 2008 Text REtrieval Conference (TREC 2008)*.

APPENDIX

Table 1. Edge types

Edge Type	Node Types	Directed	Semantics
wrote	person, paper	no	paper's author
cite	paper	yes	a paper cites another paper
published in	paper, publication venue	no	paper's publication venue
affiliation	person, affiliation	no	person's affiliation

Table 2. Result obtained by using different input formats

Input Format	Mean Reciprocal Rank
	(higher is better)
content-based	0.253
structure-based	0.507
proposed	0.692*

This work was previously published in the International Journal of Information Systems and Social Change, Volume 2, Issue 1, edited by John Wang, pp. 48-63, copyright 2011 by IGI Publishing (an imprint of IGI Global).

Chapter 5
SocialX:
An Improved Reputation Based Support to Social Collaborative Learning through Exercise Sharing and Project Teamwork

Andrea Sterbini
Sapienza University of Roma, Italy

Marco Temperini
Sapienza University of Roma, Italy

ABSTRACT

SOCIALX is a web application supporting social-collaborative and cooperative aspects of e-learning, such as sharing and reuse of (solutions to) single exercises, and development of projects by group-work and social exchange. Such aspects are supported in the framework of a reputation system, in which learners participate. We describe design and motivational issues of the system, show implementation details and describe a small-scale experimentation that helped evaluating the effectiveness of the system as well as the best lines for further development. In the system, learner's reputation is computed, presented and maintained during her/his interactions with the system. The algorithm to compute reputation can be configured by the teacher, by tuning weights associated to various aspects of the interactions. To enhance collaboration on exercises, we support contextual (to the exercise) micro-forum and FAQs, together with a currency-based concretization of the perceived usefulness of questions/answers. Group responsibilities, peer-assessment and self-evaluation are supported by group-based projects with self/peer-evaluated phases: Different stages of a project are assigned to different groups; a stage-deliverable is both self-evaluated (at submission) and peer-evaluated (by the group receiving it for the next stage). This paper is an extension of the original version, published on the International Journal of Information Systems and Social Change: mainly two sections were added, describing the latest improvements and some experimentation.

DOI: 10.4018/978-1-4666-2922-6.ch005

INTRODUCTION

Web-based cooperative and collaborative learning can improve e-teaching and e-learning considerably. Traditionally the terms *Cooperative* and *Collaborative* bear different meaning; the former appears to be more related to interactions in a well-structured framework, mainly with the aim of producing a deliverable, possibly through plain and planned division of the work activity; the latter is preferably related to more loosely structured interactions, in which possibly clear roles and responsibilities are not directly pre organized (Panitz, 1997; Slavin, 1990). Yet commonalities among the mentioned concepts are extensive: for instance, the learner and the teacher are involved in active and sharing experiences, the teacher is in a role of facilitator, the learners co-works in group activities. For a more comprehensive set of commonalities cf. Kirschner (2001) and Kreijns et al. (2003). We think that, in the web-application we are going to present, those concepts are implied distinctly in some modules and concurrently in some others; so in the following we shall not make distinctions, and use only the term *collaborative*.

In this paper we describe the design and the present implementation of a system, SOCIALX, supporting e-learning in a reputation-based social network environment; in such framework, learners participate by interaction, exchange of information, and collaboration over common problems (e.g. mandatory exercises in a subject matter and/ or group projects).

In SOCIALX we try and integrate the social dimension of e-learning (Cheng & Ku, 2009; Kirschner, 2001; Weller, 2007), with the more experienced group dimension (Panitz, 1997; Slavin, 1990). In particular we support the "social dimension" through management of reputation. So, in the next two subsections, we discuss collaborative e-learning and reputation in social-based e-learning, trying to point out the contribution that we hope SOCIALX can bring into the picture.

Collaborative Learning

Collaborative learning is considered a strong methodology to allow the development of critical thinking in learners, and the acquisition of new knowledge. Moreover it is seen as a way to support the retention of knowledge and its deepening in time (Kreijns et al., 2003). In a collaborative environment, learners can be supported in sharing training experience, combining their skills, and eventually preparing for team-based working activity (Cheng & Ku, 2009). Whereas collaborative learning is usually discussed and applied on small groups, a further aspect of interest is, then, in the vision of e-learning in a social dimension, as a community in which social activities take place and social interaction skills are developed by the *participants* (Wenger, 1998).

Moreover, the methodologies and technologies, developed to support collaboration environments, are starting to produce effects on the design, and extension, of present e-learning standards (Yu & Chen, 2007), namely in the IMS Learning Design (IMS Learning Design, 2009). In this respect we may note that, as well as the internet has been growing around the key technological factors of openness, robustness and decentralization, e-learning is likely to develop around the same factors (Weller, 2007).

Regarding the model of group-work applied in SOCIALX, it is intended as to allow that the relational and technical improvements, gained by the participants through usual group-work, can be augmented through social exchange (namely, in the development of a project work, intermediate products are exchanged and peer-evaluated through groups).

Reputation Systems and e-Learning

A reputation system captures (and makes evident to the learner) the contributions the learner is giving to the group, to the class and to the course. A reputation system is both a motivational tool and a way to evaluate and understand learner's psychological preferences, relations with others, and ability to analyze/judge others' work (thus conceptual competences).

Being, conceivably, applicable to "the measuring and representing of persons", the concept of reputation shows both appealing and worrisome aspects (Doctorow, 2003). As a matter of facts, it is present in several areas of social web applications, and it is also used with educational purposes. With respect to e-learning, reputation based techniques are used to increase the relevance of the learning material that each individual student is exposed to, and provide quicker system reaction and feedback.

Wei et al. (2007) try and catch *credibility ranks* of participating users in an e-learning system called Learning Village. Basically, citation relations among contributions (*articles*) are managed. Analysis of such relations can unveil dependency relations among authors (and the related credibility). Simulated experiments confirm that "top users" are mostly classified in the "top rank" area of the users list.

In Jin et al. (2008) a user reputation model is applied, in the framework of the *Digital Library Digital Education Learning 2.0* community. The aim is to promote qualified participation by learners, through activities such as uploading learning resources, rating them, challenging them, discussing them in blogs. Points, gained, or lost, during activities, make up the individual reputation: it is computed as a function of a *systemScore* and a *collaborationScore;* the latter *score* measures activities among users, and comprises marks coming from collaboration, popularity (*encouragementScore*) and kind-of "usefulness to the others" (*transferScore*). Experiments showed that the activities in the system affect learning, in that the reputation scores tend (in time) to be sufficiently discriminated, so to witness skill differences among the learners.

In Yang et al. (2009) a *page-rank* model is applied to learners' contributions (in a similar way, page-rank weighting mechanisms are used,

to compute the degree of relevance of a web-page during a search). Encouragement is managed, so to reward those learners that help the others, by contributing *knowledge items*, more often and more to the point.

In SOCIALX we use a reputation based approach, initiated in Sterbini & Temperini (2007), as a basic didactic tool for social collaborative learning. The reputation of a participant in the system is calculated basing on several aspects of the person / system interaction. Examples of such aspects are the submission of contributions, the availability to judge others' contributions, the usefulness of one's contributions for the others' work, the self-judgment capabilities. The original reputation model has been enhanced in several respects; moreover it has been made "configurable by the teacher": A teacher, responsible for a given course, is allowed to interactively revising and redesigning the reputation algorithm, limited to that course. In this way the weights associated to each component of the reputation can be tuned, according to aspects, of the student activity, that the teacher does actually intend to encourage. For instance, a student could be encouraged to produce numerous contributions in a course, rather than, more frequently, to reuse contributions coming from others; or the encouragement could be the other way around: reusing (and ameliorate) more frequently than producing new contributions; or the learner could be stimulated to try and assess more frequently the work of others, rather than reuse it. All the components of the reputation model can be configured, so to let the reputation itself to measure certain (combinations of) qualities rather than others.

This paper is an extension of (Sterbini & Temperini, 2011), in which we mainly add two sections, describing the latest improvements in the implementation of the SOCIALX system and a small-scale experimentation we conducted in late 2011 and beginning of 2012. In the rest of this paper we firstly and briefly describe the originating version of the SOCIALX system, that allowed the basic management of social exchange and reuse of exercises; then we give an account of the design of the new version of the system, comprising extensions to the management of micro-forums for exercises and a social-collaborative approach to group working. In Section 4 we report in greater detail about the implementation of the reputation system. Then we describe the experimentation, while the final Section tries to draw some conclusion remarks about the presented work and some lines for future developments.

OLD VERSION OF SOCIALX: WHAT IS PRESERVED

A previous version of our system (Sterbini & Temperini, 2008) paired a reputation system together with an exercise-sharing web tool. It aimed at:

- Increasing the motivation of students in doing homework.
- Increasing / encouraging higher level cognitive learning activities (cf. the taxonomy of cognitive levels in Bloom, 1964), both by rewarding the student grading others' solutions, and by rewarding the reuse and correction of others' solutions.

To obtain this we built a reputation-based system, within which a student is able to do homework, share exercises solutions, and judge and reuse others' solutions.

A reputation system is normally used to motivate interaction and to elicit good behaviors by awarding credits to the user's actions that are deemed more useful to the community. In this earlier version of the system, the student's reputation is a blend of five facets that describe how well s/he is working within the class: *involvement*, *usefulness*, *competence*, *judgment*, and *critical thinking* (see later for details on the meaning of such topics, which are preserved in the new version of the system). Reports of the student's reputations

can be shown both at the course and topic level, with details displaying all the facets to allow the student to improve his/her reputation by focusing on the type of social activity s/he likes more.

THE NEW SOCIALX

With respect to the previous version of the system (see Figure 1 in the Appendix) SocialX is extended by:

- Increasing collaboration and peer-based help by introducing contextual micro-forums, in which "direct rewards" (tokens) are used to explicitly capture the perceived usefulness of others' help.
- Introducing a better perception of personal skills and problems, via self-evaluation and judgment on others' contributions.
- Managing group-based projects, also in a social dimension, by allowing a project to be the result of work performed by different groups.
- Supporting the teacher (a very valuable, yet limited resource), by exploiting the recorded social interactions among the students; we also support direct teacher interventions, on the very definition of the reputation (limited to the portion of system – the *course* – in which the teacher is responsible).

In the following we define the structure of the reputation that is managed in the system, its components and their basic meaning.

Definition 1 (SocialX Reputation): The reputation of a learner is an overall representation of certain learner's qualities as they come out from his/her interaction with the SocialX system. It can be calculated at different levels of detail in the system: course topic, whole course, and whole system (encompassing several courses).

There are six basic aspects that are taken care of in the system:

- **Involvement:** The degree of active participation in the system, measurable by the amount of work that the learner has been available to submit, also in terms of participation (such as the number of solutions submitted, questions proposed and grades given, as well as the propriety and extension of judgments).
- **Usefulness:** How the learner's work is beneficial to others in the system (such as the reuse of learner's solutions, and the appreciation of her/his questions).
- **Competence:** An appraisal of the skills shown by the learner (deriving from the grades and judgments coming from peer students and from the teacher.
- **Judgment:** How well the student has evaluated others' solutions, questions, answers and products (with respect to the teacher's grades and evaluations).
- **Self-judgment:** How well the student has evaluated her/his own answers and products (with respect to the teacher's grades and evaluations).
- **Critical appraisal:** A measure of the conceptual work issued to understand and critically appraise others' work, in order to modify, reuse, and start from such work (such as when a solution is the first produced for a problem, or is the correction of another).
- **Group-reputation:** This item represents the reputation of the group (or the average reputation of the groups, if more than one) in which the learner takes part. It is actually representative only of those aspects of group participation, which cannot be clearly distinguished among the group members. E.g., the judgment or self-judgment capabilities of a learner don't contribute to the group reputation, even if they are dis-

played by the learner during group-related activities; in this case they are used only to compute learner's personal reputation. In some sense, the undistinguishable part of the group reputation is related only to the products delivered by the group as a whole, and to their evaluation from peers and teacher.

Figures 2 and 3 (in the Appendix) show parts of the SOCIALX system, namely two screenshots of the student interface.

Figure 2 shows the homepage: two *courses* are visible (LWEB and TDP - names are irrelevant here) with some of the available exercises. Each exercise is related to a certain topic ("tematica", first column). Topics are used to classify exercises into courses; they are course-specific. In figure, exercises are labeled also by a short description ("commento"), the teacher that proposed it ("docente"), an optional deadline ("scadenza") and creation date. The menu in the page redirect to other parts of the system, of which the most relevant are the group-work area ("Gruppi"), and the interface to consult personal reputation and interactions with the system ("Dati").

Figure 3 shows the page where a student ("marte") can see records of an exercise ("Test a Scatola Nera" – blackbox testing in computer programming). The solution proposed by another student ("martest") is visible; "marte" can download and reuse the solution, in order to propose another solution ("estendi"). In that case "marte" will have to declare the degree of reuse of the downloaded solution, and this implicitly will affect "martest" reputation. Moreover, "marte" can assess the solution proposed by the colleague ("vota!"). The assessment is done with respect to, possibly, several criteria ("completeness", "correctness" and "documentation" in figure). The criteria were defined by the teacher proposing the exercises. Each criterion has a verbal description, provided by the teacher in order to explain its application. If an exercise has more criteria applicable, all of them can be applied in the assessment, yet one was defined as primary with respect to the ("parametro di giudizio primario" in figure); this just means that the teacher considers that criteria as paramount in the assessment of the solution.

Now, through the following subsections, we discuss some aspects of the interaction supported by the system. Then, in the next section, it will be possible to describe the implementation of the reputation and the parameters and weights that are actually used in its computation.

Increasing Collaboration

We introduce both contextual micro-forums attached to each exercise, so that students help each other by asking/answering questions, and FAQs to collect the most interesting discussions. The students' exchanges are moderated by the teacher, that can "promote" the discussion threads by re-factoring the most interesting pairs of questions/answers to the exercise's FAQ. When a discussion is re-factored as a new FAQ entry, the students involved in the originating discussion are rewarded by increasing their usefulness and competence reputation levels.

To make the best use of the teacher's time, we highlight the Q/A exchanges that the students have already implicitly selected as the most interesting / appropriate (as an effect of exchanging "tokens of perceived usefulness").

Tokens of Perceived Usefulness

To enhance the motivation of the students in helping each other, and to make explicit the perceived usefulness of others, we apply the classical currency-based approach (token exchange) so that students can use tokens to acquire services by other students or by the teacher. Each student is awarded an initial number of tokens; one's tokens can be conceded in exchange for good answers to one's questions. Whenever the learner needs help s/he can issue a question (consuming one token)

and later reward the best answer received (with the consummated token). Tokens are a limited resource, and thus a student needing more answers should first "work" for the community to collect the tokens needed to ask more questions. The total number of tokens received this way is a direct indication both of the usefulness of the student in the community and to his/her ability to solve the question issued, thus it contributes to the usefulness and to the competence dimensions of his/her reputation. The number of tokens spent, instead, counts how many times the student has asked questions to the community, and thus it contributes to the involvement factor of his/her reputation.

To avoid students cheating the system (e.g. by exchanging useless questions/answers) we mildly discourage "off-topic" and "dummy" discussions. The teacher or tutor flags this kind of useless exchanges so that they contribute zero to the reputation, while the token employed remains spent (and so it is lost for both the students involved). *Discouraged exchanges* may affect the reputation of both parties involved.

Therefore the participation of the students at the contextual micro-forums produces reputation through the rules:

- The answers given to others (even if not awarded with the token) contribute to the student's *involvement* factor.
- The tokens received show how much a student has been useful to the others, increasing her/his *usefulness* factor.
- The tokens spent to propose questions show how much the student has participated, and contribute to his/her *involvement* factor.
- A Q/A pair promoted to FAQ shows that that contribution is important, thus contributing to the answering student's *competence* factor.

- "Dummy" and "off-topic" discussions are completely ignored, while the corresponding token is lost.

To make the best use of the teacher's time we highlight the token exchanges to help her/him to evaluate faster the dummy and FAQ candidates.

The Teacher is the Bottleneck

The teacher's work in the system is a crucial, limited resource. S/he should correct solutions, moderate answers, promote good Q/As to the FAQs, manage the group projects. We must make the best use of the teacher's expertise, even if s/he would be able to check/test/correct just a small part of the solutions submitted. To this aim we exploit the network of social exchanges between the students to guide the teacher by selecting the most interesting items to be evaluated. Then, the social network is used to propagate the evaluation results to the neighbor items to adjust the authors' reputations accordingly.

In particular, the tokens exchanged in the contextual micro-forums are used (also) to highlight the best answers (elected as such by the questions authors, and so entitled for a token). This can reduce the teacher workload, while analyzing the questions and the answers in the system to find good candidates for FAQ promotion. On the other hand, unanswered questions have nothing to highlight: after a time span, they are managed by the teacher, so to provide an answer, and, if it is the case, promoted into the FAQ list

Regarding the evaluation of the solutions to exercises, the teacher is guided by the judgments expressed, by the agreement/disagreement on them, and by the reputation of the intervening students.

Social Collaborative Projects

As we have seen so far, SocialX allows the use of a reputation system in an e-learning environment, supporting the development of collaborative-social exercising activities within a potentially large group of students. So far, such "exercising activities" have been made out of single exercises, freely reusable by each learner. So, in the context of SocialX the learning activity is a trade-off between individual work and social exchange, where the individual work is in the selection and comprehension of others' contributions, in the reuse and adaptation of such contributions, and in the development of new solutions / contributions.

However in certain courses the development of projects is a relevant part of learning, in both cases of an activity performed by an individual learner or a collaborative work carried on by a small group of learners. So we extended SocialX to embed also the support to a *partially social collaborative approach* to the development of projects. The meaning of that "partially" being in that, while a small group (possibly singleton) of learners is still the basic operating unit, the products of such units are submitted to social exchange with the other units (to be reused and assessed).

A project is usually a prolonged and organized activity, made of a sequence of tasks, each one depending on the previous one and depended upon by the following ones. Usually a project is entrusted to a small group of learners, and collaborative work among them is instructed and supported by the teacher, in order to produce the deliverable for the whole project. We add to SocialX the support to a partially social collaborative approach to the development of projects. Instead of having a small group working on the various steps of a single project, the idea is to have the group working on different steps of different projects: all the projects share a similar structure, made of a sequence of tasks (the steps); the n-th task of a project is expected to be "similar" to the n-th task of another (with respect to the general learning goals related to the project development methodology); so the group would be assigned a path of tasks, each one possibly involving a step in a different project.

In this framework, at each step the group as a whole is expected to deliver a product; moreover, the learners in the group provide evaluations of the product(s) received from earlier step(s) in the same project (from which the group should start to work on its task) and of the deliverable released by the group (to show self-evaluation skills).

We define a social collaborative project (SCP), on a given course-topic T, as a set of tasks $P^T = \{t_i\}_{i \in (1,...,n_T)}$. Each task is assigned to a group of learners $(g = \{l_i\}_{i \in (1,...,n_g)})$, that will do the corresponding learning activity (such as the construction of a deliverable product). Moreover, the sequence of tasks in a SCP provides a complete span of learning activities about the related project methodology. (Henceforth, where possible we shall assume that projects are all on the same topic T and avoid the related indexes).

In the following definition, an SCP-path is a sequence of tasks, selected from different projects in such a way to provide the aforementioned *complete span of learning activities.*

Definition 2 (work-field - WF - and SCP-path): A work-field WF is a set of projects on the same topic T

$$\{P_j\}_{j \in (1,...,n_{WF})}$$

A SCP-path for a group g in a work-field WF is a set of tasks, selected from the projects in WF, such that two adjacent tasks are never in the same project:

$$\text{SCP-path (g)} = \{t_{i,j}\}_{i \in (1,...,n_T), j \in (1,...,n_{WF})}$$

where $t_{i,j}$ is the i-th task in the j-th project of WF, and for all $p \in (1,...,n_T - 1)$ if $\{t_{p,q}\}$ and $\{t_{p+1,q'}\}$ are adjacent tasks in SCP-path (g), then $q \neq q'$

In a work-field, the projects are supposed to share a common structure, meaning that the number of steps and their logical sequence are homogeneous, so that it is acceptable that a SCP-path

provides group learners with a reasonably standard and complete project activity in the course topic. Once a suitable work-field is defined, SCP-paths can be assigned to groups. The following is an example of path assigned to a group *g,* where it is supposed as granted that adjacent tasks are never in the same project, and for each $h \in (1,...,n_T)$ it is $1 \leq k_h^g \leq n_{WF}$ (the learning path in the work-field is in the sequence of tasks to be fulfilled by the group, so we render it as that sequence):

$$\text{SCP-}path\ (g) = t_{1,k_1^g} \rightarrow t_{2,k_2^g} \rightarrow ... \rightarrow t_{n_T,k_{n_T}^g}$$

In particular, the path is made by n_T tasks (to make the overall activity complete according to the course topic definition); every *h-th* task is the *h-th* task in one of the projects of the work-field (the *k-th,* with *k* established as the order number in the set of available projects). We assume that a generic task in a project is likely to be 1) depending on the fulfillment of the previous task in the sequence, and 2) depended upon by the following task. We also assume that for almost each task undertaken by a group, along its SCP-path, the group is going to depend on the work done by other groups and will as well produce material for other groups to use. This gives the social dimension to the activities in a SCP-path, and gives also the opportunity to add feedbacks over the reputation of learners, beyond the evaluation of their technical skills related to project deliverables.

Let us have:

- A work-field $\{P_j\}_{j \in (1,...,n_T)}$,

- And, *for a fixed group* $\overline{g}$ a *SCP-path* $(\overline{g})$

$$\{t_{1,k_1^{\overline{g}}}, t_{2,k_2^{\overline{g}}}, ..., t_{n_T,k_{n_T}^{\overline{g}}}\},$$

and let us call *g* the generic group, and $p(g, t_{i,j})$ the product delivered by the generic group *g* after fulfillment of the task $t_{i,j}$. Then, the expanse of the tasks available in the *WF* (with $n_{WF} = 3$) is shown in Figure 4 (in Appendix). In figure we draw a possible interpretation for the *SCP-path* $(\bar{g})$, also pointing out the tasks to be received and delivered in correspondence to each task in the path. The abovementioned social dimension of learners' activities in a SCP-path, is captured by the finalization of such activities in the system. Such finalization is given in the next Definition 3, as the fulfillment of a task by a group. (In considering the dependencies of a task from others, we limit the scope to those immediately preceding and succeeding the task; this is done in order to simplify a bit the notation, with no prejudice for the general discussion).

In particular, figure 4 shows a three-projects WF, with a *SCP-path* for group $\bar{g}$. For each one of the tasks to be fulfilled

$$(t_{1,1}, t_{2,2}, t_{3,3}, ..., t_{n_T-1,1}, t_{n_T,2})$$

the products delivered by the group are pointed out by an upper pointed triangle and the products received by others are indicated by a bottom oriented triangle. The product received before of the first task is supposed to be a set of specifications and directions; the product delivered after last step is the final fulfillment for the project.

Definition 3 (fulfillment of a task by a group): Given a task $t_{i,j}$ assigned to group $g = \{l_p\}_{p \in L_g}$ and assuming that the previous and successive tasks in the same project P_j, i.e. the tasks $t_{i-1,j}$ and $t_{i+1,j}$, are resp. assigned to the groups g' and $\bar{g}$, then we say that g fulfills the task $t_{i,j}$ when it provides the system with

- a product $p(g, t_{i,j})$;
- a set of evaluations $\{\text{VAL}(l_p, g', t_{i-1,j})\}_{p \in L_g}$ over the product received from the previ-

ous task in the project (one explicit evaluation for each member of the group);

- A set of self-evaluations $\{\text{AVAL}(l_p, g, t_{i,j})\}_{p \in L_g}$ over the product released by the group itself (one explicit evaluation for each member of the group.

In particular, a further set of evaluations will be available on the product $p(g, t_{i,j})$ (and, indirectly, on the group g and its members) when the group $\bar{g} = \{l_q\}_{q \in L_{\bar{g}}}$ will endeavor the fulfillment of its task $t_{i+1,j}$: $\{\text{VAL}(l_q, g, t_{i,j})\}_{p \in L_{\bar{g}}}$.

So, from the work of a group of learners, g, in a SCP-work-field, and from Definition 3, many items may produce a feedback over the reputation of the group members:

- For each task $t_{i,j}$ of the SCP-path assigned to g we have a set of evaluations of the product $p(g, t_{i,j})$ issued by the members of group $\bar{g}$ that followed g in the same project P_j;
- Each grade, as well as the principal one given by teachers, is imparted to the whole group and can easily be spread, mediated by the teachers' judgment, to feedback over *usefulness* and *competence* of each $L_p \in g$.
- For each task, $t_{i,j}$, we also have the evaluations issued by group members about the product $p(g, t_{i-1,j})$ inherited from the previous task in the same project: those are single learner's evaluations, that can be compared with teachers', affecting both learner's *competence* and *judgment*.
- The various evaluations mentioned at point 1) are also to be taken into account to measure the ability of group g to build a good product, basing on the one they received from previous task: the relationship between the grades of the former, $p(g, t_{i,j})$,

and those of the latter, $p(g, t_{i-1,j})$ can provide a feedback over the *active critical thinking* component in the reputation of the members of group g. Of course, as it is apparent that the evaluations over the previous product, $\{\text{VAL}\,(l_p, g, t_{i-1,j})\,\}_{p \in L_g}$ are coming from g's members, on its account only the teachers' grades will be taken into account.

- Finally, for each $t_{i,j}$ a task assigned to g, we also have the evaluations issued by group members about their own product $p(g, t_{i,j})$: those are single learner's self-evaluations, that can be compared with teachers' evaluations, affecting each learner's *self-judgment*.

IMPLEMENTATION OF THE SOCIALX SYSTEM

The Making of a Reputation: the Implementation of the Reputation System

As earlier mentioned, the reputation of a learner is a representation of 7 dimensions (*indexes*), of the interaction in the system. It is computed by the following expression, where l is the learner and the p_i are integer weights associated to the indexes. See Box 1.

Each index is an integer number, which is in turn computed and maintained, during the participation of the learner in the system, basing on sub-indexes that give numerical interpretations of the various levels and quality of the relevant interactions. Each one of such sub-indexes is in turn weighted by an integer value. In the following we give an account of such interpretations:

- Involvement of a learner is proportional to:
  - The number of solutions to exercises posted by the learner;
  - The number of solutions and tasks that s(he) marked;
  - The number of questions posted (or, of tokens spent);
  - The number of answers posted.
- Usefulness is proportional to:
  - Number of reuses of the learner's solutions by others, at a light level (declared by the user as simple inspiration);
  - Same as above, but at a higher level of reuse;
  - Number of times a solution has been extensively reused;
  - How many questions originated from this learner and has been included in a micro-forum FAQ;
  - How many best answers to micro-forum questions from this learner (corresponding to tokens got);
  - How many answers went in a micro-forum FAQ.
- Competence is determined after:

Box 1.

$$
reputation(l) =
$$
$$
\begin{aligned}
(\text{involvement} \cdot p_1 &+ \text{usefulness} \cdot p_2 + \\
+\text{competence} \cdot p_3 &+ \text{judgment} \cdot p_4 + \\
+\text{critical\_appraisal} \cdot p_5 &+ \text{self\_judgment} \cdot p_6 + \\
+\text{group\_reputation} \cdot p_7 &)
\end{aligned}
$$
$$
\overline{}
$$
$$
7
$$

- The average marks obtained by the learner's solutions and tasks (meaning tasks delivered by groups in which the learner participated); these are teacher's marks.
- The same as above, but marks are those assigned by peers;
- Bonus credits obtained after several occurrences, such as a particularly good or negative evaluation from the teacher, the same from peers (yet with a lower weight than the teacher's), first solution submitted for an exercise, and declared correct;
- Number of best answers given, questions and answers eventually sent to FAQ.

- Judgment is determined after:
  - Occurrence of marks given by the learner that match with teacher's mark (over solutions submitted by others and groups not participated by the learner); *matching*, here, is intended as mark being precisely as the teacher's;
  - As above, with some more tolerance in the matching (mark is *around* the teacher's);
  - Number of tokens assigned to best answers (confirmed as "best" by the teacher).
- Critical_appraisal is meant at representing the capability to select the objectives of an activity, so the factors considered are:
  - The number of solutions submitted by others, that have been reused by the learner while they were not yet accepted by the teacher (so it was not sure they were at least basically correct and reuse was more risky);
  - As above, but limited to the solutions that have been accepted and not yet marked by the teacher (so their selec-

tion cannot be done on the basis of a positive sanction by the teacher.
- Self_judgment is quantified similarly to judgment, except for only solutions and products originated and evaluated by the learner are considered:
  - Occurrence of marks, self-given by the learner, that match exactly with teacher's mark;
  - As above, with some more tolerance in the matching.
- Group_reputation is measured after the sub-indexes listed below; its computation is limited to those aspects that cannot be associated directly to other indexes of the individual reputation of the learners participating in the group. (The interactions conducted by the learner, during group-related activities, that can be related to such "other indexes" do actually influence directly the individual reputation, according to their definitions). The sub-indexes are:
  - Group_involvement, which in turn is computed on the basis of the number of products delivered by the group and the number of marks given by the members of the group to products received from other groups;
  - Group_competence, which in turn is taken from the average teacher's marks on group products.

Reputation's Personalization and Reputation History

The structure we presented for the reputation algorithm is completely parameterized: each index is weighted, in the final computation, by the associated p_i; each index is in turn computed basing on the values of sub-indexes that are individually weighted as well.

So the reputation of a learner is depending on indexes and sub-indexes chosen by the designer (us) to represent the learner's interaction in the system, and it is also depending on the weights that indexes and sub-indexes are associated with. And SocialX supports full configurability, for all those weights, by the teacher.

Such parameters can be changed and tuned by the teacher, in order to conform the interaction environment to privilege (push participants toward) certain activities rather than others. In particular, the system allows the construction of different sets of parameters, stored in the teacher's profile. When a teacher is granted responsibility for a given course, (s)he can choose which of those sets to apply in the course depending. This selection is done to reflect the particular teaching aims stated by the teacher, that is the specific trade-off between social activities and exercise work (s)he wants to focus the students on.

The history of students' interactions in the system is maintained and stored, so to support the teacher with the analysis of the reputation factors, as they evolve in time for each learner. The teacher can tune the reputation weights even when the course is ongoing; however that is not advisable, as it could determine learners' disorientation, and should be limited to cases such as simulation, or the need for correction of an initial mistake in the definition of reputation. After such a change, all the reputation levels are recomputed (with the new weights) basing on the stored logs.

The SocialX Architecture

The general software architecture of SocialX is shown in Figure 5 (in Appendix); the system is built as a PHP application interfacing a Mysql database, using the PHP development framework Symfony (Potencier, 2008).

Simfony is an MVC (Model View Controller) framework in which it is relatively easy to build Web-applications. It supports the construction of a set of PHP classes (the object-model) wrapping the Mysql database tables and implementing all the business-objects manipulated by the application. Similarly, PHP classes are built to generate

the application's web pages and to react to the user's actions.

In particular, SocialX is built around a set of objects implementing:

- The actors (student, teacher, administrator);
- The exercises and their related data (course, topic, exercise statement, student's solution, student's and teacher's marks);
- The groups and projects (sequence of stages/tasks, group of students);
- The discussions related to an exercise (question, answer, best answer, off-topic);
- The computation of the student's reputation (weights of the reputation components, computations formulas);
- The daily log of the reputations (both at topic/keyword and course level).

Further Notes about the Implementation and Latest Improvements

During the final part of 2011 and the beginning of 2012 the web application SocialX has incurred a set of improvements and corrections. A set of unit tests have been developed, and some bugs and deficiencies have been fixed. The SocialX system has been improved in three ways: in the metadata associated to each exercise, in the implementation of its reputation system, and in its usability.

Regarding Metadata, all exercises, now, contain a description of the prerequisites (RK, for Required Knowledge) and of the post-requisites (AK, for Acquired Knowledge) of the exercise. The RK and AK metadata are described as sets of keywords (that we called *kitem*, for Knowledge Items). This metadata allows us to properly sequence the SocialX activities respect to the prerequisite relation, so that the student moves within the exercises in a regular structured way. In particular, the metadata allow us to connect the SocialX activities to our Lecomps system (Sterbini & Temperini, 2009a), a personalized e-learning environment that produces and delivers learning paths, tailored both to the students' current knowledge and to their learning styles. By building a bridge between SocialX and Lecomps we plan to merge the social-collaborative activities of SocialX with the personalized learning paths of Lecomps, to deliver social personalized activities appropriately chosen in the Zone of Proximal Development (Vygotskij, 1978) of the student or of the group (De Marsico, Sterbini & Temperini, 2010 and 2011).

With respect to the peer assessment and to the computation of the reputation, the following changes have been made:

- The peer assessment of others' solutions can be expressed as a set of different facets/criteria (e.g. Correctness, clarity of exposition, thoroughness, etc.). Earlier, when the students' votes were compared to the teacher's one to compute the "judgement" reputation index, only the "main" criteria was compared. Now weights can be associated to the assessment criteria, so that the students know how much each facet is important for the teacher. Similarly, the "judgment" reputation index has been updated to consider the weighted difference of the grading criteria values.

- The peer-assessment workflow has been slightly changed, requiring the students to express their assessment of a solution before being able to "reuse" it.

- Teachers, now, can assess the "reuse" factor declared by the students, and thus the reputation, now, is penalized when the alleged "reuse" factor is wrong.

- To motivate students to work to improve their ability to judge, now, a penalization is given to the "judgement" and "self-judgement" factors when a peer assessment is very far from the teacher's one (unacceptable).

Then, after a series of interviews with users, we have updated the SocialX system to improve its usability, mainly respect to the following issues:

- The teachers, now, can attach an explanation to their grades, for each of the grading criteria/facets.

- The level of agreement between the students' assessments and the teachers' ones is now shown to the students to help them understand where and why the grades they gave are excellent/good/acceptable/bad.

- The overall SocialX layout and style has been completely revamped (both by revising its templates and with an extensive use of AJAX) to improve the workflow, the information displayed, the navigation menu bar, the notifications and the error messages.

- Field validation and file upload checks have been added to reduce errors during data submission.

- Student's reputation levels now are shown in a "Hall of Fame" page, to improve student motivation.

The workflow the teacher should follow, in order to define a new exercise, has been streamlined to make the task clearer and simpler.

A SMALL-SCALE EXPERIMENTATION

The first experimentation on SocialX was conducted during the period "half Nov 2011" – "end Jan 2012" with the students of the course "Linguaggi per il Web" (literally "Languages for the Web") given in the third year of the bachelor degree in Computer Engineering at Sapienza University. In the system we defined 64 exercises (tasks) distributed over the topic areas of the course (they were XHTML, CSS, CGI, PHP, MySQL, XML, Javascript). In the task definition we included the specification of the appropriate RK and AK, as requested by the recent developments described in the last sub-section: to this aim we defined 135 kitem (an example of kitem is server_post_request-use-3, meaning the knowledge of the "use", at cognitive level 3 in the Bloom's taxonomy, of the clause "post" in the request sent to a web server). Moreover we defined five assessment criteria: 1) presentation (e.g. comments and clarity of the code), 2) correctness, 3) well-formed and valid (referred to the web pages produced), 4) interface effectiveness, 5) data management (related to the data defined in the MySQL databases and XML archives). The criteria were useful to give guidance in the evaluation of tasks execution, to both teacher and student. A presentation of the system features was given to the participants, on the system, as a pdf file, and by the teacher. The course had 12 students, of which 10 started their participation. No particular inducement was established for the participation in the experiment; in particular no automatism was ensured on how the reputation would correspond in the end to marks in the exam (or to a part of marks in the exam). The participating students submitted 28 solutions, with 24 judgments and only one micro-forum discussion. Half of the people actually stopped quite early (basically, and sadly, after a few inconclusive sessions). During the early stage of the experiment the system incurred in a series of changes in its interface, aimed at making the usability of the system better; the effective switch from an interface to the other happened in the second half of December 2011. In order to gather users' reactions and evaluations we administered two questionnaires, one dedicated to the early quitters and another one for the participants that did stay active longer (possibly just using the system at a minimal level, yet still logging in every now and then). Here, for the sake of space, we don't get into details on the questionnaires and give only a description and discussion of the answers.

As of the students that left the system soon (first questionnaire), we saw that basically they got a clear idea about the services offered in SOCIALX, yet they were not attracted by the idea of participating in a reputation system. They mostly didn't, of course, notice the change in the interface happened in December; those that noticed it, though, appreciated it as an important amelioration.

The students that remained in contact with the system more (second questionnaire) were somewhat attracted by the idea of participating in a reputation system, and in a social collaborative learning experience. They had a clear idea of the services offered by the system, praised them infact (see below), and appreciated very much the changes in the interface; in particular they were assessing the system usability as barely sufficient prior of the changes, and quite satisfactory after the improvements. On the other hand, unfortunately, they didn't see clear effects on their learning experience. The answers to open questions in this questionnaire were useful to get some further insight and directions: in particular we were hinted to extend the system with some asynchronous messaging services, to improve the possibility of interpersonal dialogue between students. Another request was to add a notification mechanism, to raise a clear alert in the student's homepage whenever new answers to questions in a relevant micro-forum are sent. We were also suggested to add a view of the exercises showing how many assessment and solution were posted for each of them: the motivation of this request is to provide an help in the selection of the next exercise to work on.

Finally we like to report on two further observations we got from the questionnaires.

The first is basically reproaching us, for having placed exceedingly too many exercises in the system: the idea is that if in each topic area there were just "2 or 3" tasks, the student could have just felt the actual possibility to "complete the path". Of course the set of tasks was meant (by us) to give a wide variety of possible exercises, for the student to pick up some of them (by no means a checklist to be exhausted); yet we will have to take care, in future, of the "sense of accomplishment" demanded by this notice.

The second observation amounted to the request to allow posting "barefaced wrong or incomplete" solutions, so to allow for a discussion and collaborative correction of the work. This is intended to happen, in the suggestion given, outside of the reputation system, to avoid damaging who is trying to solve the problem. We would rather think that this could be done "in" the reputation system, applying the countermeasure of the case, to turn it into a valuable learning experience.

Although the experimentation is very limited, we gained a lot of suggestions on how to improve the system.

CONCLUSION AND FUTURE WORK

We have presented the ongoing development of the SOCIALX system, which introduces collaborative group projects, and contextual micro-forums, with rewards for best answers, within a reputation based e-learning environment. In such framework we tried and integrate the social and collaborative/ group dimensions of e-learning. The resulting environment comprises the support to several instructional activities, and, in our opinion, affects learning in various educational respects, of which we discuss three significant examples in the following items:

- The execution of exercises by an individual learner, and the related feedback, has direct implications with learning, being it the practical/applicative counterpart of the theoretical studying.
- The group-work, conducted over tasks of projects (possibly different, yet commonly shaped), allows to develop team-work capabilities, that are of growing importance in both technical/scientific and humanistic education. In this module of the system,

moreover, the need to use products delivered by other groups, and the requirement of assessment and self-assessment for such products, can allow the learner to deepen the practical experience, anticipating some additional aspects of real work environments.

- The social dimension of the system, operating as a common background in different modules, allows individuals and groups to select/exploit/assess the work done by others (exercises, and project-tasks deliverables). This, in our opinion, affects the comprehension and refinement (the learning) of certain skills that are referred, at higher cognitive levels, in the Bloom's Taxonomy (Bloom, 1964), and are valuable outcomes of a learning process.

While we think that the participation process supported by the system does affect truly the learning experience of a student, we must also admit that we have not yet presented enough new fresh empirical data about the overall effectiveness of this particular process. This is due to the present lack of extensive experimentation of the system, which we think could be partially claimed to be a consequence of the hard work that was needed to design and develop the system till a functional stage. We valued the functional feasibility of the system as a preliminary requirement of crucial importance. Being such a requirement reached, we are presently planning a comprehensive experimentation. So here we presented the design issues of a system incorporating concepts and features from different dimensions of e-learning, and from the Web2.0 area of the applications for social activities, and offered the implementation of SOCIALX as a proof-of-concept for such design.

Among the system features, we discussed the *configurability of reputation*, which is the possibility for a teacher, responsible for a given course, to redesigning the reputation algorithm, limited to that course. This is done by tuning the weights associated to each component of the reputation, so to encourage certain activities (and learners' behaviors) more than others. This would be done by the teacher, according to her/his teaching aims in the course. We think this gives a touch of *teacher personalization* to the system, which could be a profitable issue, once it is used consciously. For instance, it is conceivable that different teachers do pursue different teaching strategies, so requiring to *stress differently* the diverse learning activities, in the reputation algorithm. Moreover, different courses may really profit of different reputation approaches. The danger, here, is to incur in a kind of "fragmentation of assessment principles" for students participating in different courses (with or without different teachers). We think that such a bad outcome does not actually depend directly on the sole availability of the configuration mechanism, and would rather be the result of misuse. So we proposed it as a tool to enrich the learning and teaching service of the system, granted that its use needs surveillance.

Regarding the concept of reputation, we have discussed its use in some e-learning initiative, and we have shown the numerous aspects on which it is computed in SOCIALX. One interesting application area of reputation is the web-based personalized and adaptive e-learning (Brusilowsky & Vassileva, 2003; Limongelli et al., 2011). In this field, learner personalization of the learning activity is managed, usually, through maintenance and update of a *student model*. As learner's reputation can be a valid enhancement to the structure of a student model, we foresee a fruitful application of a reputation system in the above mentioned area.

In terms of future work, related to the development of the system, we think to concentrate on some aspects in particular, which we discuss in the following two subsections.

Improving the Reputation Model

The current reputation system tries and captures the usefulness of the student's work (for both her/

himself and the class. What we are not capturing, instead, is the cost to make the system run (in terms of how much effort and work each activity does actually cost to the teacher and to the student). To increase the cost-efficiency of the system (that is to make more *valued* the valuable work of the actors) we plan to:

- Instrument the system to collect data on the time spent doing each activity
- Analyze the collected data to rate each activity's effectiveness
- Model (e.g. through agent-based social simulations) the system interaction with a population of students
- Adapt the reputation parameters to obtain the best student results from the available resources

Student's Fairness

Among the students there are those who prefer to try and cheat the system, to gain high reputation without working too much, just similarly to what Tom Sawyer does to get success in the Bible Competition (Twain, 1876). As we have seen with the *discouraged exchanges*, to keep a high level of quality we discourage misbehaviors. The penalization should be done very mildly to avoid discouraging less evil participation; thus we currently just make all misbehaviors void. We would like to introduce a *fairness* factor to capture how much the student agrees with the "didactic pact", i.e. with the proper behavior rules in the course. This factor is probably meaningful only for the teacher/tutor, and is updated whenever the student misbehaves within the system, either by annoying others or by trying to cheat the system.

AKNOWLEDGEMENT

The authors feel quite indebted with the anonymous reviewers, for their willingness and professional work, which made it possible to improve significantly the paper. The authors also thank very much the guest editors of the journal issue, for their valuable and continuous commitment on the endeavor.

REFERENCES

Bloom, B. S. (1964). *Taxonomy of educational objectives*. New York, NY, USA: David McKay Company Inc.

Brusilowsky, P. (2001). Adaptive hypermedia. *Journal of User Modeling and User-Adapted Interaction, 11*, 87–110. doi:10.1023/A:1011143116306

Brusilowsky, P., & Vassileva, J. (2003). Course sequencing techniques for large-scale web-based education. *Journal of Continuing Engineering Education and Life-long Learning, 13*, 75–94.

Cheng, Y., & Ku, H. (2009). An investigation of the effects of reciprocal peer tutoring. *Computers in Human Behavior, 25*.

De Marsico, M., Sterbini, A., & Temperini, M. (2010). Tunnelling Model between Adaptive e-Learning and Reputation-based Social Activities. Proc *21st International Conference on Database and Expert Systems Applications*, DEXA 2010, *3rd International Workshop on Social and Personal Computing for Web-Supported Learning Communities*, SPeL 2010, Aug.30th-Sept.3 2010, Bilbao, Spain, A Min Tjoa, R. Wagner (Eds.), IEEE Computer Society.

De Marsico, M., Sterbini, A., & Temperini, M. (2011). The Definition of a Tunneling Strategy between Adaptive Learning and Reputation-based Group Activities. In Proc. *11th IEEE International Conference on Advanced Learning Technologies* (ICALT 2011), July 6-8, 2011, Athens, Georgia, USA. IEEE Computer Society, Los Alamitos.

Doctorow, C. (2003) *Down and out in the magic kingdom*. Tor Books. ISBN 0765304368. Under licence Creative Commons, Attribution-NonCommercial-ShareAlike, 2004. Retrieved March, 5th 2010, from http://www/craphound.com

Fernandez, G., Sterbini, A., & Temperini, M. (2007). On the specification of learning objectives for course configuration. In *Proc. Int. Conf. on Web-Based Education (WBE07)*.

Jin, F., Niu, Z., Zhang, Q., Lang, H., & Qin, K. (2008). A user reputation model for DLDE learning 2.0 community. In Buchanan, G., Masoodian, M. and Cunningham, S. J. (Eds) *Proc. 11th Int. Conf. on Asian Digital Libraries: Universal and Ubiquitous Access To information*, Dec. 02 - 05, 2008, Bali, Indonesia. Lecture Notes In Computer Science, 5362, 61-70. Berlin, Heidelberg, Springer-Verlag.

Kirschner, P. A. (2001). Using integrated electronic environments for collaborative teaching/learning. *Research Dialogue in Learning and Instruction, 2*(1), 1–10. doi:10.1016/S0959-4752(00)00021-9

Kreijns, K., Kirschner, P. A., & Jochems, W. (2003). Identifying the pitfalls for social interaction in computer-supported collaborative learning environments: a review of the research. *Computers in Human Behavior, 19*, 335–353. doi:10.1016/S0747-5632(02)00057-2

Learning Design, I. M. S. (2009). *IMS Learning Design best practice guide* and *information binding* and *information model*. IMSGLOBAL publication, retrieved November 13th 2009 from website http://www.imsglobal.org/learningdesign/.

Limongelli, C., Sciarrone, F., Temperini, M., & Vaste, G. (2011). The Lecomps5 framework for personalized web-based learning: a teacher's satisfaction perspective. *Computers in Human Behavior* 27:4, Elsevier, pp.1285-1466, ISSN 0747-5632

Lin, B., & Hsieh, C. (2001). Web-based teaching and learner control: a research review. *Computers & Education*, 37.

Panitz T. (1997). Collaborative versus cooperative learning: comparing the two definitions helps understand the nature of interactive learning. *Cooperative Learning and College Teaching*, 8(2).

Potencier, F. (2008). The symfony 1.3 & 1.4 reference guide. Sensio Labs Books, ISBN: 9782918390145. License Creative Commons Attribution-Share Alike 3.0 Unported. Retrieved March 10th 2010 from http://www.symfony-project.org.

Slavin, R. (1990). *Cooperative learning: theory, research, and practice*. Prentice Hall.

Sterbini, A., & Temperini, M. (2007). Good students help each other: improving knowledge sharing through reputation systems. In *Proc. 8th IEEE Conference on Information Technology Based Higher Education and Training (ITHET07)*.

Sterbini, A., & Temperini, M. (2008). Learning from peers: motivating students through reputation systems. In *Proc. Int. Symp. on Applications and the Internet, Social and Personal Computing for Web-Supported Learning Communities. Turku, Finland*.

Sterbini, A., & Temperini, M. (2009a). Adaptive construction and delivery of web-based learning paths. *39th ASEE/IEEE Frontiers in Education Conference, Oct.18-21 2009, San Antonio, TX, USA*.

Sterbini, A., & Temperini, M. (2009b). Collaborative projects and self evaluation within a social reputation-based exercise-sharing system. *IEEE/WIC/ACM Int. Conf. on Web Intelligence (WI'09) Intelligent Agent Technology (IAT'09), Workshop on Social and Personal Computing for Web-Supported Learning Communities (SPeL). Milano, Italy*.

Sterbini, A., & Temperini, M. (2011). SOCIALX: reputation based support to social collaborative learning through exercise sharing and project teamwork. *Journal of Information Systems and Social Change, IGI Global, 1*(2), 64–79. doi:10.4018/jissc.2011010105

Twain, M. (1876). *The adventures of Tom Sawyer* (2003rd ed.). Bath, UK: Paragon.

Vygotskij, L. S. (1978). Mind in society: the development of higher psychological processes. (M. Cole, V. John-Steiner, S. Scribner, & E. Souberman, Eds. and Trans.) Cambridge MA: Harvard University Press. (Original material published between 1930 and 1935).

Wei, W., Lee, J., & King, I. (2007). Measuring credibility of users in an e-learning environment. In *Proc.16th Int. Conf. on World Wide Web*, 1279-1280, New York, NY, USA, ACM.

Weller, M. (2007). The distance from isolation: Why communities are the logical conclusion in e-learning. *Computers & Education*, 49.

Wenger, E. (1998). *Communities of practice: Learning, meaning, and identity*. Cambridge Un. Press.

Yang, S., Zhao, J., Zhang, X., & Zhao, L. (2009). Application of pagerank technique in collaborative learning. In Leung, E. W., Wang, F. L., Miao, L., Zhao, J. and He, J. (Eds.) *Revised Selected Papers from Advances in Blended Learning: Second Workshop on Blended Learning*, Jinhua, China, August 20-22, 2008. Lecture Notes in Computer Science 5328 (102-109). Berlin, Heidelberg, Springer-Verlag.

Yu, D., & Chen, X. (2007). Supporting collaborative learning activities with IMS LD. In Proc. *International Conference on Advanced Communication Technology* (ICACT2007).

APPENDIX

Figure 1. Evolution from previous version of SOCIALX. The main "operating" functions are shown (we left implicit the support to management of all interactions records).

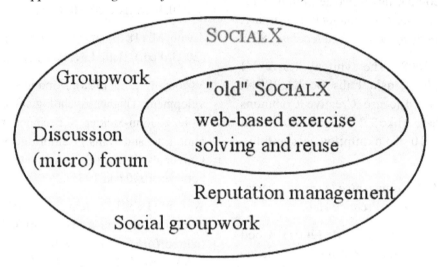

Figure 2. The student interface in SOCIALX. Homepage

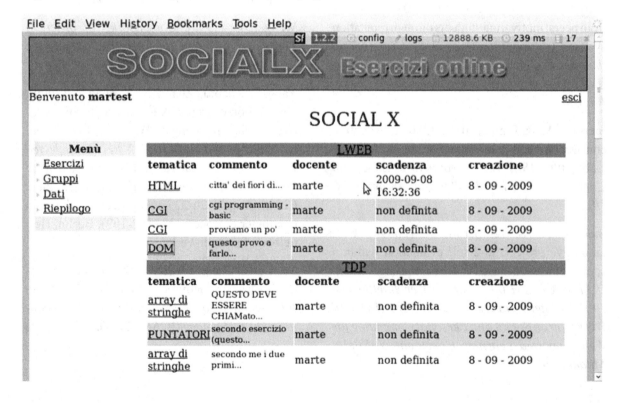

Figure 3. The student interface in SOCIALX. A page related to the exercise "Test a Scatola Nera" - Blackbox testing in computer programming). The student "marte" can see, reuse ("estendi") and assess ("vota!") the solution proposed by another student ("martest")

esercizio

Test A Scatola Nera		
autore:	marte	Blackbox testing, over a simple main function
creato il:	13/09/2009	
scadenza:	non definita	allegato aggiungi soluzione

	from martest to blackbox	
autore:		
data:	martest	solution from martest
esteso:	13/09/2009	
	0 volte	allegato

estendi

completeness of the solution: 3   1 ⬦ vota!

correctness: 3   1 ⬦ vota!

documentation: 10   1 ⬦ vota!

Parametro di giudizio primario:
documentation

Figure 4. A three-projects WF, with a SCP-path for group $\overline{g}$

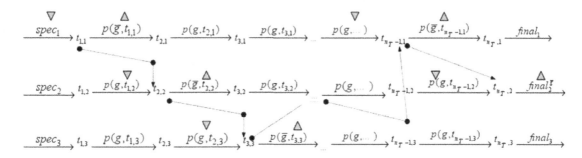

Figure 5: SOCIALX software architecture

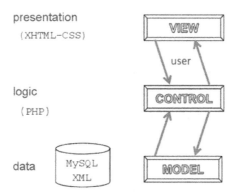

This work was previously published in the International Journal of Information Systems and Social Change, Volume 2, Issue 1, edited by John Wang, pp. 64-79, copyright 2011 by IGI Publishing (an imprint of IGI Global).

Section 2
Information Systems and Information Technologies Issues in Nonprofits

Chapter 6
Exploring Barriers to Coordination between Humanitarian NGOs:
A Comparative Case Study of Two NGO's Information Technology Coordination Bodies

Louis-Marie Ngamassi Tchouakeu
Prairie View A&M University, USA

Harold Robinson
The Pennsylvania State University, USA

Edgar Maldonado
The Pennsylvania State University, USA

Carleen Maitland
The Pennsylvania State University, USA

Kang Zhao
Iowa State University, USA

Andrea H. Tapia
The Pennsylvania State University, USA

ABSTRACT

Humanitarian nongovernmental organizations (NGOs) are increasingly collaborating through inter-organizational structures such as coalitions, alliances, partnerships, and coordination bodies. NGO's information technology coordination bodies are groups of NGOs aimed at improving the efficiency of ICT use in humanitarian assistance through greater coordination. Despite their popularity, little is known about these coordination bodies, specifically the extent to which they address inter-organizational coordination problems. This paper examines coordination problems within two humanitarian NGO's information technology coordination bodies. Based on data collected through interviews, observation, and document analysis, despite positive attitudes toward coordination by members, seven of eight widely accepted barriers to coordination still exist among members of these coordination bodies. Further, in a comparison of mandate-oriented, structural and behavioral coordination barriers, research finds mandate issues were most significant and structural factors were found in the greatest numbers. Findings suggest that effective humanitarian NGO's information technology coordination bodies must pay attention to both organizational design and management issues, although the former are likely to have a greater impact on coordination.

DOI: 10.4018/978-1-4666-2922-6.ch006

INTRODUCTION

In recent years, as the number of man-made and natural disasters has risen, so has the number of non-governmental organizations (NGOs) engaged in international humanitarian relief and development (UNDP, 2002). This growth has in part contributed to their increasing importance in the humanitarian field but at the same time has increased the range of challenges they face. One of these challenges is inter-organizational coordination around information technology (Saab et al., 2008).

In an attempt to mitigate this challenge, humanitarian NGOs are forming structures such as coalitions, alliances, partnerships, and coordination bodies (Guo & Acar, 2005; Zhao et al., 2008). NGOs' coordination bodies are groups of NGOs brought together with the purpose to improve coordination of their activities. Coordination efforts among NGOs members of a coordination body are thought to function as a solution to the duplication of efforts, poor planning and implementation of relief efforts, and a lack of knowledge among humanitarian organizations on the developing situation. This NGOs' coordination entails developing strategies, determining objectives, planning, sharing information, dividing roles and responsibilities, and mobilizing resources. Coordination among NGOs is also concerned with synchronizing the mandates, roles and activities of the stakeholders and actors at higher organizational levels. NGOs coordination ensures that priorities are clearly defined, resources are efficiently utilized and duplication of effort minimized in order to provide coherent, effective and timely assistance to those in need (Harpviken et al., 2001).

The issues involved in forming and maintaining these entities, as well as inter-organizational relationships more broadly, have been the subject of some studies (Bennett, 1995; Donini, 1996; Harpviken et al., 2001). These studies find that while coordination bodies share a limited number

of common traits, they vary in several dimensions. Common features include (1) independence from government; (2) existence of a semi-permanent secretariat; and (3) a variety of participants sharing common ideology (Bennett, 1994).

Within the frame of these common elements, coordination bodies have been found to vary in their structure, size, formality and duration. Structural variations are observed in their variety of missions, organizational forms, and decision making processes. Size variations are reflected in coordination bodies that attempt to coordinate intensely among a small subset of NGOs, or target larger memberships and less complex interactions. Variation in the level of formality and authority depends on who has taken the initiative to set up the coordination entity, and which agencies are involved (Harpviken et al., 2001). Moreover, coordination bodies may be temporary initiatives, ongoing inter-agency bodies or permanent incorporated nonprofit organizations (Zhao et al., 2008).

A number of coordination bodies focus exclusively on information technology and management (IT/IM) related issues. We refer to them in this paper as information technology coordination bodies. These coordination bodies aim at reducing redundancies and pooling limited IT resources, while at the same time promoting inter-organizational information sharing to improve humanitarian relief and disaster response. They help to ensure that organizations that are members have access to the best information and communication technology and practices when assisting during or after disasters. The two cases investigated in this paper are examples of such coordination bodies. HumaniNet is a salient example of an information technology coordination body. HumaniNet consists of over a hundred organizations. HumaniNet provides its members with practical assistance in global information and communication technologies, especially in remote areas.

Despite their popularity, the existing scholarship on humanitarian NGOs has yet to investigate the impact of humanitarian NGO's information technology coordination bodies. The literature is especially silent on the extent to which obstacles obstruct an effective inter-organizational information technology coordination under the umbrella of a coordination body. In response to this limitation, our research explores the issues that humanitarian NGO's information technology coordination bodies face when carrying out their activities.

Adopting the label of 'coordination body', this research addresses the question: "What barriers face NGO's information technology coordination bodies in the humanitarian relief field?" Using a comparative case study design, this research investigates coordination problems within two humanitarian NGO's information technology coordination bodies. The two coordination bodies[1], ReliefTechNet International and Information Technology for Emergency Alliance (ITEA), have respectively twenty-two (22) and seven (7) organizational members. Our unit of analysis is the coordination body, not the member NGOs which comprise the bodies in question. The study introduces an analytic framework that divides coordination barriers into mandate, behavioral, and structural categories and finds that the coordination bodies studied here differentially influence these areas. Taking into account past literature, the study finds that from the eight identified coordination barriers, the coordination bodies seem to be able to overcome only one barrier, namely competition for resources among members. In addition, descriptions provided by the subjects elaborated on the nature of the obstacles helps to add detail to the framework introduced in the first part of this study.

When approaching this research, we identified the eight barriers to inter-organizational coordination as said earlier, but did not anticipate each of the eight to receive equal weight from our study participants. Both the special context of our research, humanitarian relief, and the special type of coordination bodies, those focused on IT, signaled to us that the weighting of these barriers would be differently distributed. The context of humanitarian relief led us to believe that the barriers involving resources and costs would be paramount. From the literature on these large-scale humanitarian relief organizations we knew that finances, resources and personnel are always stretched thin. We anticipated that competition and the perceived increasing costs of coordination would prevent some organizations from entering into coordination body project activities. Conversely, we assumed that barriers involving conflicting goals and values would receive little weight since most organizations shared the larger mission of humanitarian relief.

The fact that both coordinating bodies in question were also special, focused on technology issues, also led us to anticipate an unequal weighting of these barriers. We assumed that since the body was focused on more technical, rather than organizational joint projects and activities, barriers involving information and communication issues and divergent goals would matter less to decisions to coordinate. The members of the coordinating bodies who sat around the table shared an interest and expertise in IT. This led us to believe that those problems that could be classified as technical problems would be treated as such and dealt with easily by the body. Those that were more organizational may have been seen as out of their scope of expertise and may have presented more of a barrier to coordinate.

The paper is structured as follows: the next section provides a background on coordination barriers, which is followed by the third section that introduces the analytic framework. In section four, the research methods are discussed and thereafter, the two coordination body case studies are presented. In section six, the research findings are articulated followed by the discussion and conclusions in section seven.

BACKGROUND

The Need for NGO's Coordination

Humanitarian non-governmental organizations provide assistance to people who have been struck by natural or man-made disasters, through disaster relief and subsequently development projects. Relief activities, which are typically short-term, focus on providing goods and services to minimize immediate risks to human health and survival. Alternatively, development activities are longer-term assistance, focusing on community self-sufficiency and sustainability. These activities include establishing permanent and reliable transportation, health care, housing, and food.

While growth in the international humanitarian sector is widely recognized (UNDP, 2002), the global nature of this growth is less so. Thus, whereas the decade of the 1980s international NGOs grew to 2,500 in number, within the developing world, the number of local NGOs with a relief and development focus is now approximately 30,000 (UNDP, 2002). Naturally, these increases generate further challenges for inter-organizational coordination.

Understanding NGO Inter-Organizational Coordination

NGO Coordination: Despite the variety of academic perspectives from which research on coordination and inter-organizational coordination is approached (Comfort & Kapucu, 2006; Crowston, 1994; Grandori, 1997; Lewis & Talalayevsky, 2004; Mulford & Rogers, 1982; Mulford, 1984; Thompson, 1967; Van de Ven et al., 1976; Whetten & Rogers, 1982), a common theme across all of them is that coordination requires the sharing of information, resources and responsibilities to achieve a common goal.

In the particular realm of NGO coordination, initiatives are seen as a solution to duplication of efforts in assistance projects, badly planned

and implemented relief efforts, and the lack of knowledge among humanitarian organizations on the actual situation in which they operate. These initiatives entail developing strategies, determining objectives, planning, sharing information, the division of roles and responsibilities, and mobilizing resources. They are also concerned with synchronizing the mandates, roles and activities of the various stakeholders and actors at higher organizational levels. In a nutshell, NGO coordination is intended to ensure that priorities are clearly defined, resources more efficiently utilized and duplication of effort minimized; the ultimate goal being to provide coherent, effective and timely assistance to those in need (Harpviken et al., 2001).

Coordination among NGOs, as well as between NGOs and other humanitarian actors, takes place at different levels. Harpviken et al. (2001) identify these levels as international, national, regional and local. At the international level, the formulation of policy, general guiding principles and strategies are of concern. At the national level, coordination typically revolves around program development and policy articulation. At this level, local groups are typically less involved, while United Nations agencies, government departments and NGOs representatives assume a central role. Coordination at the local level usually takes place between representatives from NGOs, United Nations agencies, and local communities. It is at the local level where humanitarian priorities can be most readily identified and articulated. Figure 1 depicts these different levels of coordination, within which inter-organizational relationships may vary, depending on the level of coordination pursued. Our study focuses on coordination at the international level.

Inter-organizational Coordination Forms: Identifying and classifying the various forms of inter-organizational coordination has been a subject of research in both the for- and non-profit domains. Research on for-profit organizations has identified two general structures of coordination

Figure 1. Level of NGOs coordination; source: Author adaptation from Harpviken et al. (2001)

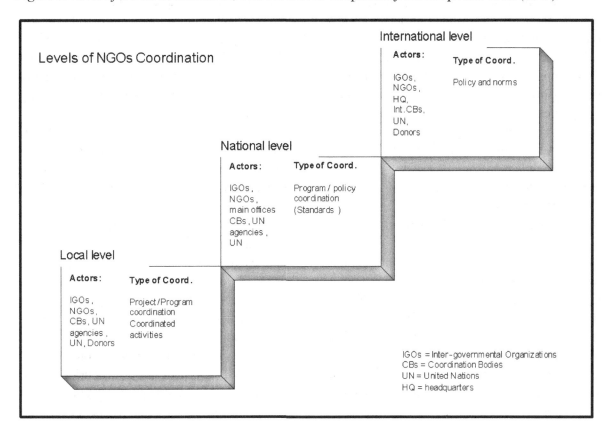

(Malone, 1987; Thompson et al., 1991). The first is a hierarchical coordination structure, characterized by long-lasting relationships with fixed rules of behavior and clear authoritative relationships. Put simply, one organization has control over the other(s). The second is a "market" coordination structure, in which all organizations are fully autonomous and make decisions in their own interest.

In the non-profit domain, research has similarly identified multiple structures (Donini & Niland, 1999). The first is "coordination by command," in which the lead NGO has authority to pursue coordination through the use of carrots or sticks and possesses strong leadership abilities. In such a situation, a central authority has the power to define the agenda, instigate preferences and enforce sanctions. Power can come in the form of control of information or resources, but also

the institutionalized legal means, through which preferences might be implemented. The second form is "coordination by consensus". In this form, organizations develop agreed-upon guidelines and standards to achieve similar goals, and there is no authority to enforce compliance. The last form, "coordination by default" describes ad-hoc coordination in which a division of labor is generally the only exchange of information among actors. Obstacles to inter-organizational coordination may vary depending on these various forms of coordination.

Alternatively, research on coordination structures in the humanitarian sector finds that structure within NGOs themselves. Enjorlas (2008) argues that collectively NGOs on their own serve as coordination structures. Due to the nature of their individual governance structures, they reinforce the norm of reciprocity; making possible the pooling of resources and, because of these features, thereby facilitate collective action oriented toward public or mutual interest as well as advocacy. Moreover, this nonprofit governance structure is also compatible with other types of coordination mechanisms, and thus NGOs are able to operate in complex environments, mobilizing resources from market operations, governmental subsidies, or from reciprocity (Enjorlas, 2008).

Information Technology Coordination in the Humanitarian Relief Sector

Around the world, the adoption of information technology for disaster relief is increasing among humanitarian organizations including NGOs (Quarantelli, 1997). For these NGOs, information technology plays a vital role in disasters relief. The sooner humanitarian organizations are able to collect, analyze and disseminate critical information, the more effective the response becomes and the more lives are potentially saved. Studying inter-organizational disaster response, many researchers have looked at the use of IT as a coordination tool. A rich body of literature points to the critical role IT plays in complex inter-organizational disaster response plans (Comfort, 1993; Comfort et al., 2001; Moss & Townsend, 2006). Wentz (2006) presents current knowledge and best practices in creating a collaborative, civil-military, information environment to support data collection, communications, collaboration, and information-sharing needs in disaster situations and complex emergencies. Comfort (1993) identifies three main roles of information technology in managing humanitarian disaster including. According to the author, information technology enables disaster managers to create an interactive network that facilitates communication and focuses attention on the same problem at the same time. The second role identified by Comfort (1993) is that information technology allows the representation of information in graphic form, thus simplifying

complex data and increasing the speed and accuracy of communication. Thirdly, information technology enables and facilitates the development of a database for a given community which stores relevant information about the community and its population and assists managers in quickly formulating alternative solutions for assistance.

Inter-organizational coordination of humanitarian NGO around information technology however gives rise to many challenges. They originate not only from the general organizational characteristics but also from those of IT. These challenges include issues related to the inter-organizational context, to the nonprofit sector and to the emergency response context. Researchers have explored coordination related issues in humanitarian NGOs IT coordination bodies. Saab et al. (2008) investigates the extent to which organizational characteristics such as structure, number of members and funding influence outcomes as well as what they see as the critical priorities for facilitating coordination. Van Gorp et al. (2008) investigate how and in which situation coordination does occur within a humanitarian coordination body. The study also explores the benefits and constraints for coordination of VSAT deployment for development and relief purposes. Maitland et al. (2009) identify similarities and differences between information management and information technology challenges to inter-organizational coordination. They also identify requirements for resolving these challenges.

In our study, we are not directly concerned with the development and use of information technologies and information systems within individual NGOs. Rather, we investigate coordination problems within humanitarian NGO's information technology coordination bodies seeking those that these entities help to address. We also seek, when appropriate, to make recommendation on how to address other barriers. To this end, we introduce an analytic framework that helps us to group well-known inter-organizational coordination barriers into mandate, behavioral, and structural categories.

We focus on coordination at the international level first because in the two case studies investigated in the paper, coordination is performed at the international level. Humanitarian relief, which often implies services provided in low income countries by organizations from high income countries, is inherently international. National level coordination challenges do exist but these are the domain largely of domestic oriented NGOs which are not included in our study. Another reason is that some of our previous work explored other levels of coordination (Tapia et al., 2010).

NGOs' Inter-Organizational Coordination Problems

Research on barriers to inter-organizational co-ordination has been undertaken in both general organizational contexts (Burbridge & Nightingale, 1989; Comfort, 1990; Comfort & Kapucu, 2006; Crowston, 1997; De Bruijn, 2006; Faraj & Xiao, 2006; Quarantelli, 1982; Thompson, 1967), as well as among NGOs specifically (Bennett, 1995; Bui et al., 2000; Foster-Fishman et al., 2001; Saab et al., 2008; Uvin, 1999; Van Brabant, 1999). After an analysis of the literature, we found a fairly consistent set of eight coordination barriers. They include (1) bureaucratic and turf-protection, (2) divergent goals and conflicting interests, (3) resource dependency, (4) coordination cost, (5) information and communication issues, (6) assessing and planning joint activities, (vii) competition for resources, and (7) emergency response time.

Bureaucratic barriers and turf-protection refer to the desire to maintain autonomy and thus avoid having individuals in other organizations interfere within one's own organization. Burbridge and Nightingale (1989) note a common fear among organizations is that coordination may somehow result in a take-over or a loss of decision-making autonomy. Furthermore, the discipline of coordination can limit maneuverability, and hence poses a major challenge (Uvin, 1999). Coordination may be perceived as increasing bureaucracy, generating institutional resistance among bureaucratically burdened NGOs (Van Brabant, 1999).

A common problem in inter-organizational coordination is that divergent goals or an over-emphasis on individual organizational goals as opposed to those of beneficiaries may lead to conflicting interests (Bennett, 1995; Bui et al., 2000; Quarantelli, 1982; Saab et al., 2008; Van Brabant, 1999). Goal conflicts occur when a party seeks divergent or incompatible ends. Further, divergent goals may also lead to an exacerbation of turf issues or other coordination problems (Bui et al., 2000).

Resource dependency is both a motivation for and barrier to coordination (Crowston, 1997; Dawes et al., 2004; Thompson, 1967). Interdependencies, whether of the pooled, sequential or reciprocal type, require coordination (Thompson, 1967). However, at the same time they can create problems for coordination and constrain the efficiency of task performance (Crowston, 1997). One of these problems is the associated cost of coordination, as to be effective it is time and staff intensive and the benefits must outweigh these costs (Aldrich, 1972; Bennett, 1995; Van Brabant, 1999).

Coordination cost is yet another barrier that hampers coordination among organizations. Inter-organizational coordination is believed to limit an organization because scarce resources and energy have to be invested in the maintenance of relationships with other organizations. Negotiation of resources allocation can lead to difficult bargaining among parties engaged in coordinated activities. Usually, organizations find it difficult to allocate scare resources (Bui et al., 2000). Aldrich (1972) argued that it is costly for organizations to initiate and/or maintain linkages with other organizations. For example, the costs can be seen as in term of additional staff-time necessary to attend a joint board of directors' meeting; or the additional funds necessary to participate in joint database. According to Uvin (1999), the high cost in time and money that effective co-ordination entails constitute one of the major barriers to inter-organization coordination.

Another frequently encountered barrier is related to the availability and the quality of information. This is usually due to the inconsistency in data collection and management across organizations and to the mismatch between the informational demands and supplies (De Bruijn, 2006; Fisher & Kingma, 2001). According to Bui, et al. (2000), there are varying levels of mistrust, misrepresentation of facts, and incomplete information exchange among organizations. Further, the high level of uncertainty in humanitarian operations likely requires greater amounts of information to be processed between decision makers (Galbraith, 1977).

General assessment and planning of joint activities can lead to disagreement about the means and the ends of a coordinated activity (Bui, et al., 2000). Situations tend to worsen when organizations are unsure of their role, and act independently, without consulting or coordinating with others. Joint activities must also confront problems of understanding, which emanate from the fact that participants in inter-organizational relationships are accustomed to different structures, cultures, functional capabilities, cognitive frames, terminologies, and management styles and philosophies (Vlaar et al., 2006).

In addition to the resources related to coordination itself, competition for scarce resources in general may inhibit the initiation of inter-organizational coordination generally (Uvin, 1999; Van Brabant, 1999). Given the increasing numbers of NGOs, combined with decreasing overseas development assistance budgets, competition for funding between organizations is heating up (Salm, 1999; Van Brabant, 1999).

Finally, response time is considered yet another obstacle to coordination among organization. Coordination is often perceived as increasing response time especially in case of emergency. According to Van Brabant (1999), there is the fear that the coordination effort will cause delays

in providing relief. Comfort (1990) observed that coordination activities generated delays in response in the four events she analyzed.

Thus, inter-organizational coordination between international humanitarian NGOs will seek to share information, resources and responsibilities that through more efficient use of resources and minimization of duplicate activities will provide effective and timely assistance to those in need (Harpviken et al., 2001). This coordination can occur at multiple levels and may be carried out through one of several forms, including command, consensus or default (see Table 1). Whatever the form, it must contend with a wide range of challenges.

ANALYTICAL FRAMEWORK

To date, research on NGO inter-organizational coordination generally, as well as that specifically related to coordination bodies, lacks a coherent framework for analysis. In particular, coordination barriers are considered as separate constructs without being categorized into useful higher order concepts. These higher order constructs should be useful both in terms of theory building as well as generating actionable insights for nonprofit organizations. In this section we present an analytic framework that can be applied to analyze coordination bodies and systematically analyze barriers they confront. The analytic framework presented here borrows from one presented in a New Zealand State Services Commission (2008) report. This report was designed for analyzing factors for successful coordination among government agencies. Although, there is an organizational difference between the coordination of institutions under the government of a country, and the coordination among independent NGOs, the framework is useful pointing the three critical aspects for the coordination process. The original framework distinguishes three broad factors for successful inter-organizational coordination, including the

Table 1. Summary of eight coordination barriers

Barriers	Issues	Authors
• Bureaucratic and turf protection	• Desire to maintain autonomy and thus avoid having individuals in other organizations interfere within one's own organization	Burbridge and Nightingale (1989) (Uvin, 1999). (Van Brabant, 1999).
• Divergent goals and Conflicting interests	• Divergent goals or an over-emphasis on individual organizational goals	Bennett 1995; Bui et al, 2000; Quarantelli, 1982; Saab et al, 2008; Van Brabant, 1999.
• Resource dependency	• Interdependencies require coordination but at the same time they can create problems for coordination and hamper performance.	Crowston, 1997; Dawes et al., 2004; Thompson 1967). Aldrich 1972; Bennett, 1995; Van Brabant 1999
• Coordination cost	• Scarce resources have to be invested in the maintenance of relationships with other organizations.	Bui et al, 2000; Aldrich,1972; Uvin, 1999
• Information and communication issues,	• Information availability and accessibility, • Information quality, • Information Sharing • Information system quality, • Standards and interoperability • Systems integration	De Bruijn, 2006; Fisher & Kingma, 2001; Bui, et al 2000; Galbraith, 1977.
• Assessing and planning joint activities	• Disagreement about the means and the ends of a coordinated activity	Bui, et al, 2000; Vlaar et al., 2006
• Competition for resources	• Competition for scare resources may inhibit the initiation of inter-organizational coordination	Uvin, 1999; Van Brabant, 1999; Salm, 1999.
• Emergency response time	• Coordination is often perceived as increasing response time especially in case of emergency	Van Brabant, 1999; Comfort, 1990.

organizational mandate, system and behavior. Here we substitute 'structure' for 'system' as we perceive it to be more reflective of the crucial issues, as well as for its connection to the coordination literature. Further, this framework creates an ordering of the well-known coordination barriers discussed above, as each one can be associated with one of the three factors (Figure 2). Consequently, delineating coordination issues into those related to mandate, structure and behavior separates those more closely aligned with organizational design (mandate, structure) from those associated with coordination body management (behavior). To provide a better sense of the meaning of each of the three areas, they are discussed in turn below.

For successful coordination, the mandate category recognizes that each member organization of the coordination body must emphasize the importance of effective coordination and commit to making it work by prioritizing the coordinated activities. This requires leadership and clear goals. Coordination is best achieved when senior leaders have invested significant time and energy modeling and supporting this way of working (Gratton & Erickson, 2007). Particular behaviors required of leaders include: ensuring that resources and time are available for the team and managing external pressures so that coordination can occur. Further, clearly-defined and mutually-agreed upon joint outcomes are critical for successful coordination. If objectives are unclear or not shared, participants may work towards different, incompatible goals and fail to achieve desired outcomes.

In terms of structure, successful coordination requires appropriate governance and accountability frameworks, as well as adequate resources. The roles, responsibilities and contributions of each organization must be made clear. Further, governance frameworks will influence the way interactions among organizations develop over

Figure 2. Humanitarian NGOs coordination bodies' coordination barriers framework; source: Author adaptation from State Services Commission (2008)

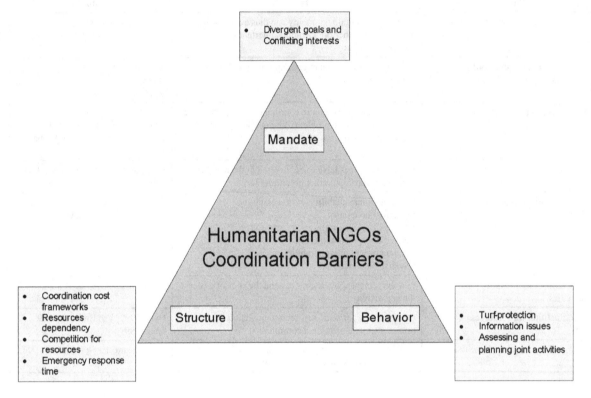

time and must be designed to sustain them. Governance frameworks must also specify appropriate resource allocations. The main resource requirements are a dedicated budget, a working pace that can sustain progress without overwhelming the group and, sufficient time to establish working relationships, achieve outcomes, and nurture the required behaviors.

With regards to behavior, successful coordination requires organizations be represented by people with the appropriate authority, and the right skills and competencies to work collaboratively. There must be clear leadership among the group. Participants in a coordination initiative should represent a cross-section of agencies whose involvement is necessary for the coordinated initiative to succeed. Representatives need the ability to negotiate, sense when to compromise and have the patience to allow the relevant parts of their

agency time to act deliberately and thoughtfully to reach decisions.

Further, each organization's culture must support coordination so that, over time, people involved in the coordinated activity come to share common culture, language and values. Shared culture is important if members are to develop a sense of joint ownership of the way the group works and of the results it produces. This is easier to achieve when agencies have a prior history of working well together, primarily because the issue of shared culture has already been partially resolved.

We feel that this higher order analytical structure that organizes these barriers into three larger categories is appropriate to both the context of humanitarian relief as well as to coordinating bodies focused on IT. In the case of the first factor, mandate, we feel that this may be of particular interest in this special case. We perceive each member of the coordinating body to be operating under the onus of three distinct mandates—first to the large goals of humanitarian relief and saving lives, second, to the particular goals of IT and building more effective, robust, secure and efficient systems, and lastly the mandate to represent the interests of their home organization as a whole within the coordinating body. This juxtaposition and potential conflicts therein, may prove to be scientifically interesting.

Secondly, we feel that the higher order classification of structure also has potential in this special case. The coordinating body members are coming from unique, established, large and hierarchical relief organizations to form a coordinating body between themselves which will not mirror any of their home organizations, yet must account for each in some fashion. The authority and governance structures that might be created between these member organizations which allows IT resources and distributed decision-rights to control them to flow from the home organizations may take on very interesting forms.

Lastly, in terms of the higher order factor of behavior and culture, we believe that unlike other arenas of coordination, those in this special case already share the overarching goal of humanitarian relief and the instrumental goal of building and maintaining better IT infrastructure. We believe that this common ground may go a long way in smoothing the ability of the coordinating body to successfully coordinate across organizational boundaries.

RESEARCH METHODS

This study employs a comparative case study research design to capture holistic detail in natural settings (Creswell, 1998). This method is particularly well-suited to studying phenomena that cannot easily be distinguished from their context and provides insight into contemporary phenomena within real life settings (Yin, 2003).

As will be discussed in more detail below, the two cases, The Information Technology for Emergency Alliance (ITEA) and ReliefTechNet, exhibit a variety of similarities and differences. They both seek to foster coordination between information technology managers at the headquarters level of large international humanitarian NGOs. Generally, NGO information technology units are considered part of overhead and hence budgets are limited. The two bodies are ideal for comparison in that they differ on a variety of organizational characteristics, yet have some overlap of membership. This overlap enabled in-depth understanding, developed through observation, of both cases. Furthermore, one coordination body manager familiar with both bodies was able to discuss with us perceived similarities and differences. However, to control for potentially confounding effects, organizations with dual membership were omitted from interviews related to the larger organization. Our unit of analysis is the coordination body, not the member NGOs which comprise the bodies in question.

As is common in the case study method, multiple data collection methods were employed (Yin, 2003). Data for the two cases were collected over a period of 15 months (October 2006 through December 2007). The data sources included semi-structured interviews, direct observation, and document analysis. The specific data collection activities for each case are outlined in Table 2.

The data collection emphasized semi-structured interviews as they allow interviewees to convey their experiences and assumptions in a way that is not permitted by completely structured questions (see interview guide in appendix A). The interviews, which were used to follow-up on questions arising from the archival, documentary, and observational data, were guided by the researchers to cover specific topics, but were flexible enough to pursue avenues of inquiry as they arise during the interview process (Berg, 1989).

Table 2. Case Study Data Collection Activities

Case Study	Interviews	Other
ITEA	12 in-person and through telephone	Background documentation; access to conference calls; observations at a meeting
ReliefTechNet	19 through telephone	Background documentation; access to project conference calls; limited field office survey and observations at meetings

We conducted nineteen (19) interviews with ReliefTechNet staff and representatives of member organizations. All the organizations member of ReliefTechNet did not take part to the interview. Out of the nineteen (19) interviews, sixteen (16) were conducted among representatives from eight organizations. The rest (3) interviews were conducted with ReliefTechNet management staff. We also conducted two in-person observations at ReliefTechNet meetings in 2006 and 2007. For the ITEA case, we conducted twelve (12) interviews with representatives, as well as participated in conference calls. Each interview lasted between forty-five (45) to seventy-five (75) minutes, was recorded and transcribed. All the 12 interviews were conducted through phone and involved four out of the seven ITEA members. Two ITEA interviews were conducted with consultants. In either of the two cases, we believe the views of the organizations represented in the interviews are representative of the whole coordination body.

For this study, we used a mixture of deductive and inductive coding (Epstein & Martin, 2005). Deductive codes were developed based on our research questions. In this way, open and selective coding was carried out for each interview, so that themes and categories could be developed. As such, we were able to compare these themes (about assumptions and interpretations) across

interviewees as well as against the research questions and the theoretical framework. During the coding process we also let some codes emerge from the data. The inductive approach reflects frequently reported patterns used in qualitative data analysis. The process of coding was iterative and cyclical based on the framework developed by Seidel (1998).

This study is part of a larger research agenda that seeks to understand how aid agencies can organize themselves to promote higher levels of coordination. For the particular research question of this study, the initial coding of interview's transcription focused on identifying the problems within the coordination bodies. Although there was a particular question to inquiry about this aspect (Could you mention the issues or problems that you have faced while working with the coordination body?), subjects mentioned issues and problems while describing coordination bodies' activities. Those answers were also included in the coding process.

The interview results were complemented by direct observation and document analyses. We conducted two multi-day, in-person observations at ReliefTechNet meetings in 2006 and 2007. We also reviewed and analyzed organizational documents including meeting minutes, annual reports and organization publications. The ITEA Initiative provided documentation to establish context and background to ITEA project outcomes.

GENERAL CASE DESCRIPTIONS

This section presents general descriptions of the two coordination bodies, providing background and contextual information. It is followed in Section 6 by an analysis of interview data.

ReliefTechNet is a coordination body of humanitarian NGOs. With help from initiators, the organization sought to pool NGOs' demand for IT donations, but quickly took on a range of other activities including coordinating information and communication technologies (ICTs), both during disaster response and development activities. ReliefTechNet membership grew from 7 organizations in 2001 to 22 in 2008. The organization's administration and projects are funded through a combination of grants and membership dues. ReliefTechNet is wholly autonomous, having established itself as a non-profit organization. ReliefTechNet has three major stakeholder groups including (1) ReliefTechNet members, (2) ReliefTechNet management, and (3) ReliefTechNet supporters. There exists significant interplay among these three groups. ReliefTechNet has a board as well as a project committee that approves project ideas from the membership. ReliefTechNet's activities initially focused on the headquarters level of its member organizations, which allowed for collective bargaining with vendors, to provide ICT services such as satellite telecommunications, coordination of ICT policies and practices, and more. Within ReliefTechNet, project involvement is voluntary and funded by participating organizations. While some member organizations are larger, having more resources to contribute to particular projects, these larger organizations do not appear to have disproportionate control over the decision-making process, despite their financial leverage. The consensus surrounding projects has been fairly easily achieved because participation is voluntary and thus those uninterested are unlikely to stand in the way of others for whom the projects are a priority.

With regards to activities, ReliefTechNet develops and implements tools (e.g. NetReliefKit) which provides data and voice connectivity in a small, transportable suitcase allowing its members to quickly establish a short-term communications solution in the event of a disaster or emergency. ReliefTechNet tests and manages the deployment of communications infrastructure to provide its members with access to the Internet at remote project sites where relief and development operations are carried out. ReliefTechNet provides forums for member organizations to document

and share their field experiences regarding the effectiveness of their telecommunication technology and to suggest ways to improve future delivery of services. ReliefTechNet provides its members with ICT skills capacity building to improve emergency response.

The Information Technology for Emergency Alliance (ITEA) is a coordination body consisting of seven agencies and was funded by a large foundation. Its goal was to improve preparedness for relief efforts of NGOs over a two-year period. In particular, it focused on four specific areas: Staff Capacity Development (Initiative 1); Accountability and Impact Measurement (Initiative 2); Disaster Risk Reduction (Initiative 3); and Information and Technology Requirements (Initiative 4). ITEA had a decentralized project management structure that coordinated the implementation of its activities for its planned two-year program. ITEA4, the last initiative of ITEA focusing specifically on ICTs, is the one discussed in this paper. The main activity of the ITEA4 was to conduct an assessment of how information is managed in emergency response and what tools and resources are available for these activities.

Similarities and Differences

Besides their obvious common interest in facilitating coordination, ReliefTechNet International and ITEA share much in common. In addition to their members being engaged in humanitarian assistance and international development, all members of ITEA are also members of ReliefTechNet International. Furthermore, coordination in both bodies is at the international level and is by consensus (Donini & Niland, 1999).

With regard to differences, ReliefTechNet International and ITEA differ in size, their primary focus, their funding mechanisms and their duration. With regards to the size, ITEA is a smaller coordination body with seven (7) members as compared to twenty-two (22). ReliefTechNet International's focus is primarily on technology change, while ITEA's is primarily organizational change. With regards to funding mechanisms, ITEA is funded by one donor while ReliefTechNet International is funded through a combination of private sector support and membership fees. Finally, with regards to their duration, ITEA is a fixed term (2 years) initiative while ReliefTechNet International is an ongoing initiative. Table 3 summarizes the demographics of the two organizations.

CASE DATA

Through systematic coding of our data, we identified fifteen different types of coordination barriers. In the coding process we noted each time a type of barrier was mentioned and aggregated these occurrences as presented in Table 3. Subsequently we ranked and performed basic statistics on the occurrences of each issue. While many barriers are similar to the general and well-known ones, others appear to be specific to the functional domain of the coordination bodies, namely information technology. Further, whereas 6 of the 15 barriers were mentioned in both cases, the other 9 were identified in only one case or the other. We elaborate on these barriers in Table 4.

Barriers in both Coordination Bodies

One of the most frequently cited barriers to coordination in both cases (19% in aggregate) was the conflict between the goals of the member's home organization and those of the coordination body. This was also expressed as a conflict of goals, a conflict of interests and competing interests. To a large extent, members of the coordination body have individual goals they tend to prioritize, overlooking the general interest of the group. For one of the subjects, the span of attention to coordination activities last as long as the meetings of the body.

I think the main issue could be that a lot of the people, once they leave the meeting and go back, are more focused on their organization rather than ReliefTechNet.

The second most cited barrier (14% in aggregate) was a lack of resources, providing yet more evidence for what is known to be a significant problem. However, here while it was common to both groups, it was more significant in the ITEA case, in which the coordination body had more ambitious goals but also had external funding. More specifically, the difference of resources and capacities among organizations was mentioned by one of the subjects as an important factor.

...it is a challenge for everybody to be able to do that when they are not funded or skilled or staffed equally. It is bad for those that are lagging behind as well as those that are leading.

Table 3. ITEA and ReliefTechNet demographics

	ITEA	ReliefTechNet
Number of members	7 Agencies	22 agencies with varying numbers of representatives
Open/Closed Membership	Closed	Closed; by invitation only
Funding Sources	Private foundation	Yearly membership dues
Mission Focus	Preparedness, Relief	Preparedness, Relief, moving towards Development
Degree of Autonomy	High	High
Organizational Level Focus	All Levels	Executive, field
IT Centricity	Low to Moderate	High - entirely devoted to ICT utilization and enhancement
Governance model	Consensus	Consensus with opt-in/opt-out of specific projects

Table 4. Aggregated responses to coordination barriers

	ReliefTechNet		ITEA		Total	
	Occ.	%	Occ.	%	Occ.	%
Conflict of goals or interests	5	20.83	3	16.67	8	19.05
Lack of resources	2	8.33	4	22.22	6	14.29
Problems of standards	3	12.50	2	11.11	5	11.90
Institutional or bureaucratic	4	16.67	1	5.56	5	11.90
Lack of incentives	2	8.33	1	5.56	3	7.14
Lack of technical skills	0	0.00	3	16.67	3	7.14
Lack of tools for collaboration	1	4.17	1	5.56	2	4.76
Lack of time and timing	2	8.33	0	0.00	2	4.76
Geographical distance	2	8.33	0	0.00	2	4.76
Lack of trust /sharing spirit	1	4.17	0	0.00	1	2.38
Speed for emergency	0	0.00	1	5.56	1	2.38
Staff turn-over	0	0.00	1	5.56	1	2.38
Communications / language	1	4.17	0	0.00	1	2.38
Membership/ size of organization	1	4.17	0	0.00	1	2.38
Different organization structure	0	0.00	1	5.56	1	2.38
	24	100.00	18	100.00	42	100.00

The third most frequently cited problem (in aggregate nearly 12%) is the issue of standards. The issue was mentioned nearly equally in both cases. One respondent noted:

So now I think that's the biggest obstacle. To get the standards, you have to get everybody together, key people, enough key people, to reach agreement. Once you reach agreement, building it out, once you get that, the technology is there.

The fourth mostly commonly cited barrier (in aggregate nearly 12%) is institutional and/or bureaucratic issues. Interviewees expressed their reluctance to pursue coordination, perceiving it as bringing about more bureaucratic and institutional constraints. This issue was more significant among members of the ongoing ReliefTechNet than those of the fixed-duration ITEA.

Although less frequently mentioned, other barriers cited by members of both coordination bodies include a lack of incentives and a lack of collaborative tools. Representing in aggregate roughly 7% of responses, one of the subjects described the lack of incentives as follows:

There is no resources allocated, it is pretty much on a voluntary basis. There is no pressure to do it... a lack of incentives.

A similar level of concern (roughly 5% in aggregate) was shown in both cases towards a lack of collaborative tools. A subject described this problem both within and beyond the humanitarian NGO community.

For all of us in the for-profit, or not-for-profit, tools for collaboration are a real challenge.

Exclusively ITEA Barriers

As described above, ITEA is a coordination body that seeks organizational change among IT units in humanitarian organizations with a particular focus on relief operations, and as such has a relatively ambitious goal. Accordingly it also has fewer members. Issues found only among members of this coordination body are a lack of technical skills, different organizational structures, speed for emergency assistance, and staff turn-over.

Lack of technical skills represented more than sixteen percent (16.67%) of the responses in this case, whereas different organization structures speed for emergency response and staff turnover represented a little more than 5% each. As regards organization structures, one member observes

One of the biggest barriers is we are structure[d] differently, internally to our organizations. You know some people have their field people coordinated one way, some people have their systems one way, and some people have it another. Sometimes I think that is a big barrier because we have different ways of getting stuff done in the field.

As regards the issue of speed for emergency assistance, one member observes:

I guess the biggest imperative these ones, and that causes a lack of coordination is the really the imperative to respond to the emergency as quickly as possible

As regards the issue of staff turn-over, a subject indicates:

obviously one of the big things, the big problems when you do disaster relief is that there is a relatively high turnover of staff so there is not a lot of institutional knowledge. Organizations are aware of that and seem to be trying very hard to change that.

Exclusively ReliefTechNet Issues

As compared with ITEA, ReliefTechNet is a larger coordination body, with less ambitious goals and of an on-going duration. Barriers exclusively mentioned by its partners include a lack of time for and/or timing for coordination, geographical distance, communications and/or language, membership size and lack of trust.

Among these time and distance represented each roughly 8% of responses. Successful coordination requires time for appropriate planning and gathering information. As said by one of the subjects, the fact that all organizations are independent makes harder to spend enough time all together working in common issues.

We all work for independent organizations. I think probably time, you know, time to work on things together. Time to work things out, timing, is the other.

As concerns geographical distance, a subject expresses its importance in the following terms:

Take these conference calls for example, if you have people in the eastern US, western US, Australia, India, the UK, all of those people, there are people in all of those time zones represented, how do you have a meeting?

Of less concern, representing roughly 4% of responses each are communications, size and trust. Of the first, one member observes:

Of course you add in the language problems and the misunderstandings and misconceptions that can be found when one person will say a lot of things and it can be misconstrued by a second person whose native language is not the language that the first is using.

These issues of communications exist concurrently with issues that arise due to differing sizes (and hence resources, structures, etc.) of member organizations as well as with those of doubting whether fellow members will honor their commitments. This latter issue has particular implications for information sharing, a requirement for effective coordination.

Summary

Overall, out of the fifteen different types of inter-organizational issues, six were identified in both cases, although perhaps receiving different levels of emphasis in the two cases. Problems caused by (i) conflict of interests or goals, (ii) institutional and/or bureaucratic constraints (iii) lack of and/or conflicting standards and (iv) lack of resources appear to be the major barriers for coordination. These four types of coordination problems registered in total more than half of all the responses of interviewees to coordination problems (see Figure 3).

ANALYSIS

Given the above findings related to coordination barriers, we now analyze their implications through the lens of our analytic framework. The framework views coordination challenges as aligning with one of three components: mandate, structure or behavior. Structural barriers arise when appropriate governance and accountability frameworks are lacking, as well as adequate resources. Mandate barriers arise when coordination body member organizations are not committed to effective coordination and do not prioritize the coordinated activities. Finally behavioral barriers result when organizations are represented by people without the appropriate authority, culture, skills and competencies to work collaboratively. Taking the above mentioned coordination barriers and aligning them to these three components (see Table 4) we observe that whereas the distribution of the 8 commonly known coordination barriers created a somewhat balanced triangle, once applied to our two cases the triangle becomes significantly skewed. Only one barrier is placed in the mandate corner, but it receives significant

Figure 3. Frequency of barriers to coordination

Frequency (in%) of Barriers to Coordination

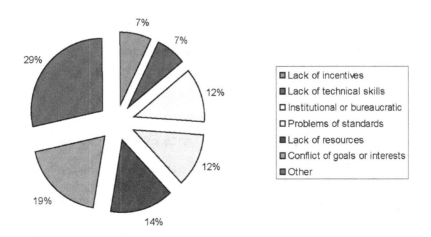

attention from the participants. This suggests the barrier concerning conflicting goals and interests is universal and powerful. Four barriers were placed in the behavior corner, yet this corner received little attention. Nine barriers are placed in the structure corner (see Table 5), showing that great attention was paid to structural barriers, but that the forms of those barriers were diverse.

From this we can claim that for these two NGO coordination bodies, mandate and structural barriers were more important than behavioral barriers in undermining coordination. We can also state that while structural barriers were important, they were also diffuse in comparison with mandate barriers. Below, these findings will be discussed in greater detail.

Conflict of goals and or interests is a single, but very significant, inter-organizational coordination issue that represents the "mandate" category. This finding suggests that, irrespective of differences such as organizational structure, mission focus, and sources of funding, if members of coordination bodies do not clearly commit to joint activities and give them priority within their individual organizations, chances are high that coordination would fail. This finding highlights the implications for coordination of the circumstance that nonprofits serve a multitude of stakeholders whose goals and needs are often very heterogeneous (Beamon & Balcik, 2008). This would explain why conflicting

Table 5. Cross study coordination barriers per category

Framework Category	Barriers identified
Mandate	Conflict of goals or interests
Behavior	Lack of trust /sharing spirit Speed for emergency Staff turn-over Lack of incentives
Structure	Communications / language Membership/ size of organization Different organization structure Lack of tools for collaboration Lack of time and timing Geographical distance Lack of technical skills Institutional or bureaucratic Problems of standards Lack of resources

goals and interests are perceived as major barrier to coordination among humanitarian NGOs. As shown in Figure 4, mandate-related barriers accounted for twenty percent (20%) of the occurrences of coordination issues.

Inter-organizational coordination barriers in the "behavior" category were identified in both cases. This finding suggests that since the humanitarian relief sector is a relatively new and growing field, organizations in this field have not yet developed and matured a shared culture and work practices that would favor coordination. This is particularly the case in the field of hu-

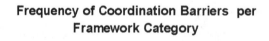

Figure 4. Frequency of coordination barriers per framework category

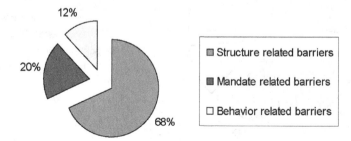

manitarian information management and technology functional areas within humanitarian NGOs. With approximately twelve percent (12%) of occurrences, behavioral related issues were the least identified in the data.

As pictured in Figure 4, the majority (sixty eight percent) of the aggregated responses of coordination issues identified in the study, fall in the "structure" category of the framework. This finding suggests that problems of governance, of responsibilities and contributions for joint activities, are those that most undermine coordination in a coordination body. This observation also highlights the fact, as mentioned earlier, that the information technology function within the humanitarian NGO sector is still young. It is not surprising to find that issues, such as a lack of standards, are identified as major coordination problems. Moreover, as seen in the review of relevant literature, appropriate and sufficient resources are necessary to successfully coordinate activities. The fact that NGOs rely primarily on donations as their source of funding, emphasizes the issue of resources. However, we did not identify competition for resources, a generally well-known barrier, as an issue in these cases.

The fact that some coordination barriers were discussed only in one of the two cases draws attention to specific characteristics of these bodies. For example, members of ReliefTechNet, a larger organization with greater variance among its members in terms of location and size, unsurprisingly mentioned structural barriers associated with membership size and geographical distance, as well as the possibly related issue of communications and/or language and a lack of time for and/or timing for coordination. ReliefTechNet members also identified a lack of trust as a behavioral barrier. The structural coordination barriers unique to ITEA include a lack of technical skills and different organizational structures, whereas behavioral barriers include speed for emergency assistance and staff turn-over. While these findings do not clearly associate behavioral or structural barriers as being

more significant to one or the other, clearly there exist a set of common barriers, as well as those that are somewhat idiosyncratic to the mission of the coordination body. For example, relief response time and staff turn-over, both of which emerged from inductive coding, likely arise from ITEA's primarily relief orientation, as opposed to relief and development in ReliefTechNet, and ITEA's relatively short duration.

Furthermore, in addition to being influenced in part by idiosyncrasies of the coordination body, coordination barriers may also be functionally determined, in this case related to IT. The above analysis shows that members of these two coordination bodies experience both common and unique IT-related coordination barriers. Both groups experienced frustration over the lack of standards. By standards, we mean accepted and common data formats, transmission protocols, and hardware which support information sharing. A lack of standards can affect the quality and timeliness of information, which is so important to inter-organizational coordination. Conversely, only ITEA experienced the coordination barrier of a lack of IT skills. This could potentially be associated with the smaller size of the coordination body, which would result in a smaller pool of expertise from which to draw.

Structural Coordination Barriers: Internal and External

The initial analysis from the interviews revealed that most of the barriers described by the subjects especially all those related to information technology and information management can be classified under the Structure category. When analyzing these barriers, it is clear that some of the problems are inherent to the individual structure and operation of each NGO. On the other hand, other factors are related to the infrastructure and logistic needed for communications links to take place among the organizations.

For the purpose of this paper, we have categorized the structural barriers in two: internal and external. The structural internal problems are those that also affect the normal working of the NGOs and that can be addressed individually by each organization. Structural internal barriers include the lack of time and timing, the lack of technical skills, and bureaucracy. The structural external barriers are those that should be tackled taking a team effort. Structural external barriers consist of communication and language issues, lack of tools for coordination/collaboration, lack of standards, and lack of resources.

The analytical framework used as guide for this study was built assuming that the barriers that faced the coordination bodies were specific to the coordination issue. After the categorization of internal and external barriers, we found that NGOs still have to solve internal problems that may affect an effective coordination (see Figure 5).

DISCUSSION AND CONCLUSION

The purpose of the study was to investigate the effectiveness of humanitarian NGOs' information technology coordination bodies in addressing inter-organizational coordination problems. To guide the study, we employed an analytic framework that enabled organizing the myriad of well-known inter-organizational coordination barriers into three categories, which are recognized as factors for successful coordination among organizations in coordination bodies. The analytic framework was applied to data from two coordination body case studies that revealed fifteen different barriers to coordination among humanitarian NGOs. We present below a discussion of these findings.

Our first observation is that, in general, our findings corroborate previous research that has explored inter-organizational coordination problems in the specific context of humanitarian NGOs research (Bennett, 1995; Bui et al., 2000; Uvin, 1999; Van Brabant, 1999). For example, some of the major inter-organizational coordination problems identified by Bennett (1995) include

Figure 5. Structural coordination barriers

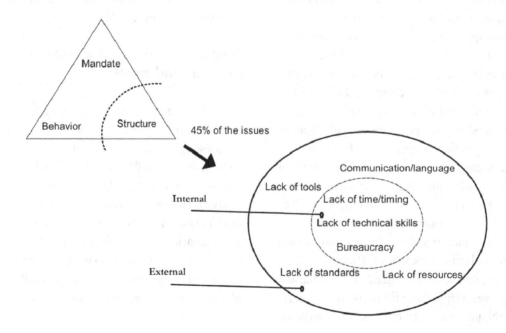

(1) conflicting interests, (2) coordination cost in terms of resource inputs, especially staff-time. Van Brabant (1999) also identifies and discusses obstacles to coordination in the humanitarian sector. Our findings are also consistent with all the issues discussed in that paper.

Our second observation is that members of these coordination bodies continue to face seven out the eight major coordination problems as identified in the literature. The one barrier they failed to mention in either case and, hence do not face, is competition for resources. None of the 31 interviewees discussed this issue as a coordination problem within the coordination body. This would suggest that humanitarian NGO coordination bodies are valuable for addressing this type of coordination problem and this insight constitutes a significant contribution to the literature on humanitarian NGOs coordination bodies.

Our third observation is that coordination barriers categorized in our study as "internal barriers" play an important role in the coordination initiatives. Consequently, coordination bodies should tackle first those issues when trying to foster coordination and collaboration.

Our fourth observation concerns information technology and information systems. In our study, they were not mentioned as an issue, by members of the coordination bodies. Here, we make two key assertions,

First, IT/IS collaboration is often the first form of collaboration entered into by NGOs. Organizational coordination between NGOs is often perceived as difficult, if not impossible, especially when NGOs must change some of their basic operations, procedures or come to significantly depend on others for key elements of their operations. IT/IS is different. From our research, IT/IS collaboration is perceived as easier to accomplish, less risky, and poised for success. In addition, donors also support these collaborative IT/IS efforts in that they often have the goals of increased accountability, visibility, and efficiency. Whether many of these IT/IS joint system develop-

ments actually result in successful collaboration is beside the point (most fail). The NGOs, and their donors, strongly believe that the first step in collaboration is through IT/IS.

Second, in traditional IS/IT research, collaborations are often contractual networks of dependent firms, interlocked into supply chains. These contractual relations are often of mutual benefit, but also, often coercive. Our findings suggest that the IT/IS collaborations among NGOs are entered into voluntarily and operate under the assumption of consensus as the decision-making parameter. While there may be some impetus from outside donor agencies to collaboration on IT/IS, the pressure to collaborate never is exerted between partners. This unique flat, yet pluralistic, space in which information systems are developed across organizations is a valuable contribution to IT/IS literature.

Finally, our study has introduced an analytic framework that divides coordination barriers according to their relationship to mandate, structure and behavior. This framework can serve as a basis for further theory development that views coordination barriers associated with structure and mandate as ex-ante organizational design issues and those associated with behavior as ex-post management issues.

Earlier in this paper we explained that drawing from the literature on coordination barriers in the humanitarian relief context, we anticipated finding that a lack of resources, and competition for those resources, would play a significant role in making coordination more difficult. This did not prove to be the case.

Our single greatest contribution from this research is that the strong value of creating coordinating bodies in the humanitarian relief sector that are focused on IT-issues is that it reduces or eliminates the barrier to coordinate around resources. We believe that the coordinating body created a structure and mechanism for the home organizations and outside donors to channel funding, staff and supplies to create collaborative IT projects that may have been impossible within any single NGO. Given the other (seven) significant barriers that still hold true in this sector with IT coordinating bodies, it is significant when we can see the diminished effects of one barrier.

This study has several limitations that prevent us from claiming that all coordinating bodies or even all IT-focused coordinating bodies held to resolve resource barriers to coordination. However, the implications are that perhaps with a well-structured coordinating body with the appropriate mandates and culture might facilitate coordination around IT issues across organizations, at least in the area of resources.

Future research needs to validate the usefulness of the framework beyond the consensus coordination structures considered here, to those that exhibit coordination by command, in which mandate and structural barriers may be less. One other major limitation of this framework is that it was developed to assess successful coordination in public sector coordination bodies. In our study we apply it to a context it was not originally intended for.

In addition, future research is needed to overcome other limitations of this study. Findings from this research cannot be generalized to all NGO coordination bodies. Generalizing from two case studies would be epistemologically problematic and would run the risk of being easily falsified by a single counterexample (Benbasat et al., 1987). However, this risk can be partially overcome by conducting several similar case studies. Moreover, as our study is conducted at the international headquarters level, is biased toward those *least* likely to face resource challenges (although they still do), future research should examine NGO coordination bodies at either the local or national level to validate whether they also are able to overcome the 'competition for resources' barrier for their members.

ACKNOWLEDGMENT

This work was partly supported by National Science Foundation grant number CMMI-0624219.

REFERENCES

Aldrich, H. (1972). *An organization-environment perspective on cooperation and conflict between organizations in the manpower training system*. Kent, OH: The Kent State University Press.

Beamon, B., & Balcik, B. (2008). Performance measurement in humanitarian relief chains. *International Journal of Public Sector Management*, *21*(1), 4–25. doi:10.1108/09513550810846087

Benbasat, I., Goldstein, D., & Mead, M. (1987). The case research strategy in studies of information systems source. *MIS Quarterly*, *11*(3), 369-386.

Bennett, J. (1994). *NGO coordination at field level: A handbook*. Oxford, UK: ICVA.

Bennett, J. (1995). *Meeting needs: NGO coordination in practice*. London, UK: Earthscan.

Berg, B. (1989). *Qualitative research methods for social sciences*. Boston, MA: Allyn & Bacon.

Bui, T., Cho, S., Sankaran, S., & Sovereign, M. (2000). A framework for designing a global information network for multinational humanitarian assistance/disaster relief. *Information Systems Frontiers*, *1*(4), 427–442. doi:10.1023/A:1010074210709

Burbridge, L. C., & Nightingale, D. S. (1989). *Local coordination of employment and training services to welfare recipients*. Washington, DC: The Urban Institute.

Comfort, L. K. (1990). Turning conflict into cooperation: organizational designs for community response in disasters. *International Journal of Mental Health*, *19*(1), 89–108.

Comfort, L. K. (1993). Integrating information technology into international crisis management and policy. *Journal of Contingencies and Crisis Management*, *1*(1), 15–26. doi:10.1111/j.1468-5973.1993.tb00003.x

Comfort, L. K., & Kapucu, N. (2006). Inter-organizational coordination in extreme events: The World Trade Center attacks, September 11, 2001. *Natural Hazards*, *39*(2), 309–327. doi:10.1007/s11069-006-0030-x

Comfort, L. K., Sungu, Y., Johnson, D., & Dunn, M. (2001). Complex systems in crisis: Anticipation and resilience in dynamic environments. *Journal of Contingencies and Crisis Management*, *9*(3), 144–158. doi:10.1111/1468-5973.00164

Creswell, W. J. (1998). *Quality inquiry and research design*. Thousand Oaks, CA: Sage.

Crowston, K. (1994). *A taxonomy of organizational dependencies and coordination mechanisms*. Cambridge, MA: MIT Center for Coordination Science.

Crowston, K. (1997). A coordination theory approach to organizational process design. *Organization Science*, *8*(2), 157–175. doi:10.1287/orsc.8.2.157

Dawes, S., Cresswel, A., & Cahan, B. (2004). Learning from crisis: Lessons in human and information infrastructure from the world trade center response. *Social Science Computer Review*, *22*(1), 52–66. doi:10.1177/0894439303259887

De Bruijn, H. (2006). One fight, one team: The 9/11 commission report on intelligence, fragmentation and information. *Public Administration*, *84*(2), 267–287. doi:10.1111/j.1467-9299.2006.00002.x

Donini, A. (1996). The Bureaucracy and the free spirits: Stagnation and innovation in the relationship between the UN and NGOs. In Weiss, T. G., & Gordenker, L. (Eds.), *NGOs, the UN and Global Governance*. London, UK: Lynne Rienner Publishers.

Donini, A., & Niland, N. (1999). *Rwanda, lessons learned: A report on the coordination of humanitarian activities*. Retrieved from http://www.grandslacs.net/doc/2404.pdf

Enjolras, B. (2008). A governance-structure approach to voluntary organizations. *Nonprofit and Voluntary Sector Quarterly*, *20*(10).

Epstein, L., & Martin, A. (2005). *Coding variables*. London, UK: Academic Press.

Faraj, S., & Xiao, Y. (2006). Coordination in fast-response organizations. *Management Science*, *52*(8), 155–189. doi:10.1287/mnsc.1060.0526

Fisher, C. W., & Kingma, D. R. (2001). Criticality of data quality as exemplified in two disasters. *Information & Management*, *39*(2), 109–116. doi:10.1016/S0378-7206(01)00083-0

Foster-Fishman, P. G., Salem, D. A., & Allen, N. A. (2001). Facilitating inter-organization collaboration: the contribution of inter-organizational alliances. *American Journal of Community Psychology*, *29*(6), 875–905. doi:10.1023/A:1012915631956

Galbraith, J. R. (1977). *Organization design*. Reading, MA: Addison-Wesley.

Grandori, A. (1997). An organizational assessment of inter-firm coordination modes. *Organization Studies*, *18*(6), 897–925. doi:10.1177/017084069701800601

Gratton, L., & Erickson, T. (2007). Eight ways to build collaborative teams. *Harvard Business Review*, 100–109.

Guo, C., & Acar, M. (2005). Understanding collaboration among nonprofit organizations: Combining resource dependency, institutional, and network perspectives. *Nonprofit and Voluntary Sector Quarterly*, *34*(3), 340–361. doi:10.1177/0899764005275411

Harpviken, K. B., Millard, A. S., Kjellman, K. E., & Strand, A. (2001). *Sida's contributions to humanitarian mine action: Final report* (Tech. Rep. No. 01/06). Stockholm, Sweden: Swedish International Development Cooperation System.

Lewis, I., & Talalayevsky, A. (2004). Improving inter-organizational supply chain through optimizing of information flows. *Journal of Enterprise Information Management, 17*(3), 229–237. doi:10.1108/17410390410531470

Maitland, C., Ngamassi, L., & Tapia, A. (2009, May). *Information management and technology issues addressed by humanitarian relief coordination bodies.* Paper presented at the 6th International ISCRAM Conference, Göteborg, Sweden.

Malone, T. (1987). Modeling coordination in organizations and markets. *Management Science, 33*(10), 1317–1332. doi:10.1287/mnsc.33.10.1317

Moss, M., & Townsend, A. (2006, May). Disaster forensics: Leveraging crisis information systems for social science. In F. B. Van de Walle & M. Turoff (Eds.), *Proceedings of the 3rd International ISCRAM Conference*, Newark, NJ.

Mulford, C. L. (1984). *Inter-organizational relations: Implication for community development.* New York, NY: Human Science Press.

Mulford, C. L., & Rogers, D. L. (1982). *Definitions and models.* Ames, IA: Iowa State University Press.

Quarantelli, E. L. (1982). Social and organizational problems in a major emergency. *Emergency Planning Digest, 9*, 7–10.

Quarantelli, E. L. (1997). Problematical aspects of the information/communication revolution for disaster planning and research: Ten non-technical issues and questions. *Disaster Prevention and Management, 6*(2), 94–106. doi:10.1108/09653569710164053

Saab, D., Maldonado, E., Orendovici, R., Ngamassi, L., Gorp, A., Zhao, K., et al. (2008). Building global bridges: Coordination bodies for improved information sharing among humanitarian relief agencies. In F. Fiedrich & B. Van de Walle (Eds.), *Proceedings of the 5th International ISCRAM Conference*, Washington, DC (pp. 471-483).

Salm, J. (1999). Coping with globalization: A profile of the northern NGO sector. *Nonprofit and Voluntary Sector Quarterly, 28*(4s), 87. doi:10.1177/089976499773746447

Seidel, J. (1998). *Qualitative data analysis.* Retrieved from http://www.scribd.com/doc/7129360/Seidel-1998-Qualitative-Data-Analysis

State Services Commission. (2008). *Factors for successful coordination: A framework to help state agencies coordinate effectively.* Wellington, New Zealand: State Services Commission. Retrieved from http://www.ssc.govt.nz/upload/downloadable_files/successful-coordination-framework.pdf

Tapia, A., Maitland, C., Maldonado, E., & Ngamassi, L. (2010, August 12-15). *Crossing borders, organizations, hierarchies and sectors: IT collaboration in international humanitarian and disaster relief.* Paper presented at the 16th Americas Conference on Information Systems, Lima, Peru.

Thompson, D. (1967). *Organizations in action.* New York, NY: McGraw-Hill.

Thompson, F. J., Frances, J., & Mitchell, J. (1991). *Markets, hierarchy and networks: The coordination of social life.* London, UK: Sage.

UNDP. (2002). *Human development report 2002: Deepening democracy in a fragmented world.* New York, NY: Oxford University Press.

Uvin, P. (1999). *The influence of aid in situations of violent conflict: A synthesis and commentary on the lessons learned from case studies on the limit and scope of the use of development assistance incentives and disincentives for influencing conflict situations.* Paris, France: OECD. Retrieved from http://www.ndu.edu/itea/storage/610/Impact%20 of%20Aid%20Uvin.pdf

Van Brabant, K. (1999). *Opening the black box: An outline of a framework to understand, promote and evaluate humanitarian coordination.* London, UK: Humanitarian Policy Group.

Van De Ven, A. H., Delbecq, A. L., & Koenig, R. Jr. (1976). Determinate of coordination modes within organizations. *American Sociological Review, 41*(2), 322–338. doi:10.2307/2094477

Van Gorp, A., Ngamassi, L., Maitland, C., Saab, D., Tapia, A., Maldonado, A., et al. (2008, June 24-27). *VSAT deployment for post-disaster relief and development: Opportunities and constraints for inter-organizational coordination among international NGOs.* Paper presented at the 17th Biennial Conference of the International Telecommunications Society Montreal, QC, Canada.

Vlaar, P., Van den Bosch, F., & Volberda, H. (2006). Coping with problems of understanding in inter-organizational relationships: Using formalization as a means to make sense. *Organization Studies, 27*(11), 1617–1638. doi:10.1177/0170840606068338

Wentz, L. (2006). *An ICT primer: Information and communication technologies for civil-military coordination in disaster relief and stabilization and reconstruction (Tech. Rep. No. OMB 0704-0188).* Washington, DC: National Defense University.

Whetten, R. A., & Rogers, D. L. (1982). *Inter-organizational coordination: Theory, research and implementation* (1st ed.). Ames, IA: Iowa State University Press.

Yin, R. K. (2003). *Case study research: Design and methods.* Thousand Oaks, CA: Sage.

Zhao, K., Maitland, C., Ngamassi, L., Orendovici, R., Tapia, A., & Yen, J. (2008, July 14-17). *Emergence of collaborative projects and coalitions: A framework for coordination in humanitarian relief.* Paper presented at the 2nd World Congress Conference on Social Simulation, Washington, DC.

ENDNOTES

[1] ReliefTechNet International and ITEA are pseudonym used to protect the confidentiality of the coordination bodies

APPENDIX

General Interview Guide

1. Who are you? Describe your role in the coordinating body.
2. Describe the structure of the coordinating body.
    Formal? Specialized/General?
    Centralized/Decentralized?
    Adaptable? Responsive to environment? To members' needs?
    Formal Meetings? Minutes? Agendas?
3. Describe the relationship between your home organization and the coordinating body.
    Kind of decisions able to make? Resources? Time?
    Level of power granted by home organization/position within organization? Need to check back
        home? In what circumstances?
    Reports back to home organizations?
    Coordination Body's impact on home organization?
4. Describe barriers to coordination?
    Failed process? Conflict? Consequences? Lack of follow through?
    Major challenges? What would you change in how this coordinating body runs?
5. Describe the decision making process the coordinating body has gone through leading to project
        X.
    Communication between members leading to decision?
    Offline? Online? Email discussions? Tele-meetings?
    Subgroup? Whole group?
    Roberts Rules? Voting? Consensus?
    Disagreement? Leadership? Persuasion?
    Documentation? History? Repository?
    Evaluation of projects? Monitoring? Maintenance?
    Enticements?
    Other projects considered? Criteria for decision?
6. Coordination in this environment? How do these make it easier/harder to coordinate?
    NGO? Relief? IT work?
    Inter-organizational relations?
    Policy/national government/ international government?
7. General Impacts of Coordination Body
    Coordinating Body's impact on disasters?
    Impact on field?

This work was previously published in the International Journal of Information Systems and Social Change, Volume 2, Issue 2, edited by John Wang, pp. 1-25, copyright 2011 by IGI Publishing (an imprint of IGI Global).

Chapter 7
Technology Acceptance:
Are NFPs or their Workers Different?

Geoffrey Greenfield
University of Queensland, Australia

Fiona H. Rohde
University of Queensland, Australia

ABSTRACT

During the past decade there has been an increasing interest in research within Not-for-Profit (NFP) organisations. Research has indicated that there are a number of characteristics that make NFPs different from other organisations. This paper considers whether workers within the NFP sector have different attitudes to technology and whether such differences affect the measures used within technology acceptance models. An exploratory study of workers within two NFPs indicated that workers within the service delivery functions of NFPs have different attitudes to technology than workers within the standard business functions of a NFP organisation e.g., marketing. These attitudes affected their perceptions of the use of and ease of technology.

INTRODUCTION

All businesses, including not-for-profit businesses, have had to adapt and become more efficient and effective to survive. Differences exist between for-profit and Not-for-Profit (NFP) organisations. Exemplifying firm type differences within the two sectors are environmental demands, managerial roles, managerial perceptions of external control,

and work-related attitudes among employees (Damanpour, 1991; Cunningham et al., 2004; Saidel & Cour, 2003).

The ability of workers within businesses to accept technology is still an important issue, with technology acceptance critical to the ability of organisations to successfully implement technology. With this in mind, a person's positive or negative attitude towards technology, formed by prior experience with technology, may influence their preferred career choice.

DOI: 10.4018/978-1-4666-2922-6.ch007

The Technology Acceptance Model (TAM) (Davis, 1989: Davis et al., 1989) is well accepted as a model to predict acceptance of new technology by individuals (Benamati & Rajkumar, 2008; Khalifa & Ning Shen, 2008). TAM argues that a person's belief about the perceived usefulness of a technological artifact and perceived ease of use of that artifact determine an individual's behavioral intention to use the artifact. Although TAM has been and still is extensively used, its universality to predict across all situations has recently been called into question (McCoy et al., 2007).

The purpose of this paper is to investigate whether two groups of people who have entered different careers, have *ex ante* different attitudes to technology that in turn could affect the variables contained in technology acceptance models. In particular we explore the relationship between attitude, perceived usefulness, and perceived ease of use in relation to technology for workers within NFP organisations. This research examines whether employees in a NFP organisation and working in service provider roles differ in their attitude, perceived usefulness, and perceived ease of use in relation to technology from workers within NFP organisations that are working in a traditional business area, e.g., marketing.

The results reveal differences between the groups in relation to the technology acceptance variables. The results may impact the manner in which firms need to handle different users when implementing new technologies.

This paper is structured as follows. Section 1 discusses the theory behind technology acceptance, before linking technology acceptance to career choice. The next sections describe the research method, results and implications. Finally, the conclusions are presented together with the research limitations and directions for future research.

BACKGROUND LITERATURE

TAM: A Model of Technology Acceptance

The majority of models investigating technology acceptance focus on an individual's ability to accept new technology within specific circumstances (Fishbein & Ajzen, 1975; Davis, 1989; Thompson & Higgins, 1991; Davis et al., 1992). The most popular and well-supported of these models is the Technology Acceptance Model (TAM) (Davis et al., 1989)[1]

The basic premise of TAM is that an individual's reaction to using technology, in terms of their perception of ease of use and usefulness, affects their intention to use technology, and ultimately their actual usage. Researchers have extensively tested and found support for TAM's ability to predict acceptance of new technology by an individual employee (Venkatesh & Davis, 2000).

Researchers have, however, begun questioning the generalisability of TAM. TAM does not always hold across all cultural groups (McCoy et al., 2007). More recently, Greenfield and Rohde (2009) found that social sciences students' attitudes to technology had an effect on both their perceptions of the usefulness of and the ease of use of technology, when compared to business students. Social sciences students were also less likely to act in accordance with their attitudes than business students and consider technology more of a necessary evil. The findings of this more recent TAM research calls for an examination of the model across a wider variety of situations.

This lack of variety is further supported when considering TAM research published since 1989 as 63% of the studies were conducted within the USA (Yousafzai et al., 2007) and focused on the business or for-profit sector using participants from within traditional business functions (Venkatesh et al., 2003). There is, however, another business sector, the not-for-profit (NFP) sector.

Differences exist between for-profit and NFP organisations. Exemplifying firm type differences within the two sectors are environmental demands, managerial roles, managerial perceptions of external control, and work-related attitudes among employees (Damanpour, 1991; Cunningham et al., 2004; Saidel & Cour, 2003). In general, NFP organisations are incorporated entities relying on multiple sources of funds, often the government, to operate. Recent reports reveal that, within the NFP sector, real funding by government has decreased, forcing organisations to reduce services. In addition, demand for NFP services is stretching their resources (Saidel & Cour, 2003). Thus, allocating funds to technology requires greater acceptance of the artifact because of the increased funding pressures.

TAM and Career

TAM normally focuses on an individual's acceptance of technology within specific circumstances. The current research follows this view of TAM, investigating perceived ease of use and perceived usefulness, in relation to a specific artifact at the time of deployment.

During the development of TAM, Davis et al. (1989) found that attitude was not significant. An attitude is defined as "a learned predisposition to respond to an object ... in a consistently favorable or unfavorable way" (Kothandapani, 1971, p321). A person's attitude to IS is important as it affects an individual's behavior and social influence - attitude has a social function. Attitude is contagious; people express their own and listen to others in the work environment (Yang & Yoo, 2004). People though may not act in a manner that is consistent with their attitude (Winter et al., 1998). This ability to act in a manner consistent with our attitude is termed volitional control.

Career choice may be made based on the decision to work in either the for-profit or NFP sector. Prior research has indicated that students studying social sciences have different perceptions of technology and also different attitudes towards technology (Greenfield & Rohde, 2009). These are the students who are most likely to enter into the fields of social work and associated business functions. Workers in the social sciences discipline area have a specific attitude to their work (Saidel & Cour, 2003). They are committed to the public good of an organisation's mission, seek more work-related challenges, look for job and task variety, autonomy and collegiality, and place a high value on non-monetary compensation. Thus, staff are often working for non-profit-motivated social reasons, and may be less likely to see the advantages of technology within their chosen work field leading to different attitudes towards technology. This difference in attitude towards technology will affect their perceptions about the usefulness of technology and the ease of use of technology (see Figure 1).

RESEARCH METHODOLOGY

This research examines the TAM constructs and attitude via a questionnaire administered to workers within two NFP organisations. Within one NFP, the questionnaire was distributed to only workers involved in the day-to-day service delivery. In the other, the questionnaire was distributed to different sections. One section consisted of workers involved in the day-to-day service delivery and the second section consisted of workers within the marketing division.

Participants

The study used participants from two organisations. Organisation A provides community support in the inner-city area of a state capital, organisation B is state-based and supports persons with a specific disability. The two case study organisations were selected by purposeful sampling as, at the time of the research, a new IS/IT artifact needed to be in the process of being implemented within

Figure 1. Research model

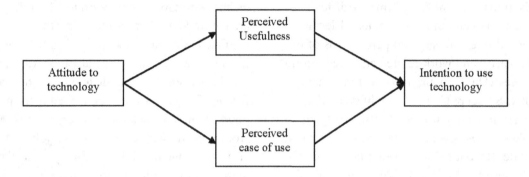

the firm. The systems were database driven Client Management Systems (CMS) shifting data collection from paper-based client record systems.

The participants were surveyed at the time that the new system was about to be implemented within their organisation. In each organisation the questionnaire was administered at staff meetings to all staff involved with the new system being invited and agreeing to participate.

A total of 17 responses were received for Organisation A and a total of 48 responses from Organisation B. Within Organisation B, 22 responses were from persons in the division involved in service delivery and 26 responses were from persons in the marketing division.

Measures

Respondents were required to complete two parts to the survey. The first section of the survey required respondents to provide basic demographic information about their use of computers including the amount of computer usage. The second section of the questionnaire required respondents to indicate their agreement or disagreement with a series of statements concerning computer usage based on their work environment. This section measured the constructs of perceived usefulness, perceived ease of use, attitude, and the control measure volitional control.

Perceived usefulness (Pu) is defined by Venkatesh and Davis (2000, p187) as the "extent to which a person believes that using the system will enhance his or her job performance" and was measured based on prior research (Igbaria et al., 1995). Each respondent was asked to indicate their agreement or disagreement with five statements (see Appendix A for questions) using a seven point Likert-type scale.

Perceived ease of use (Peu) refers to the "extent to which a person believes that using the system will be free of effort" (Venkatesh & Davis, 2000, p187). This was measured using construct based on prior research by Igbaria et al. (1995). To measure this construct the respondent was asked to indicate their agreement or disagreement with four statements (see Appendix A for questions) using a seven point Likert-type scale.

Attitude (Att) is defined as 'a learned predisposition to respond to an object ... in a consistently favorable or unfavorable way' (Kothandapani, 1971, p321). Four statements, based on prior research (Winter et al., 1998), (see Appendix A for questions) assessed attitude using a seven point Likert-type scale indicating agreement or disagreement with the statements.

Volitional control (Vc) is the ability of people to act in a manner that is consistent with their attitudes (Winter et al., 1998). This was measured with two statements (see Appendix A for questions) using a seven point Likert-type scale by indicating agreement or disagreement. This measure was included for completeness and as a control to ensure that the two groups of participants act in a manner consistent with their attitude.

RESULTS

Demographics

Of the 65 respondents, the average age of participants was 38 in Organisation A and 40 in Organisation B. The average age of the 39 participants directly involved in service delivery was 39 and in marketing 41 (see Table 1). Furthermore, 76% of the Organisation A respondents were female and 87% of Organisation B's were female with service delivery and marketing being fairly equally represented. These proportions are consistent with the proportions in the Australian workforce

for NFP organisations (Australian Bureau of Statistics, 2003; Australian Institute of Health and Welfare, 2003).

Of the participants almost half of the service delivery workers (combined) have achieved a tertiary qualification (Organisation A, 45%; Organisation B, 50%) compared with none of the marketing workers. However, 65% of the marketing workers achieved a Diploma/Certificate compared to approximately half the service delivery workers (combined) in Organisation A at 55% and 41% for Organisation B.

Details of the participants' computer usage revealed that marketing workers used their computer for more than three hours per day whereas the service delivery workers used their computer between 30 minutes and three hours per day. 96% of the marketing workers spend more than three hours using a computer compared with 42% of service delivery workers.

Table 1. Participant demographics

	Organisation A	Organisation B	
	Service Delivery	Service Delivery	Marketing
Gender			
Male	24%	14%	13%
Female	76%	86%	87%
Age			
Average Male	37yrs	39 yrs	30 yrs
Average Female	39 yrs	40 yrs	44 yrs
Average Age	38 yrs	40 yrs	41 yrs
Highest education level achieved			
Tertiary degree	45%	50%	0%
Diploma/Certificate	55%	41%	65%
Other or not stated	0%	9%	35%
Average computer use per day			
Almost never	0%	0%	0%
Less than ½ hour	24%	0%	0%
From ½ hour to 1 hour	6%	5%	4%
1-2 hours	24%	38%	0%
2-3 hours	6%	14%	0%
More than 3 hours	40%	43%	96%

Statistical Results

The remainder of the analysis is directed at looking at differences between the NFP workers within service provider roles (pooled responses from Organisations A and B) and the NFP workers within more traditional roles.

A one-way between-groups multivariate analysis of variance (MANOVA) was performed to investigate group differences. The four dependent variables were used: perceived usefulness (Pu), perceived ease of use (Peu), attitude (Att), and volitional control (Vc). Summated scores were used to allow for generalization of results. The independent variable was area of work e.g. marketing or service provider.

Comparing the service provider group with the marketing group, MANOVA results (see Table 2), indicate the following: Perceived usefulness of technology was significantly associated with the area of work ($F_{(1,61)} = 8.55$, $p = 0.005$); Perceived ease of use of technology was significantly associated with the area of work ($F_{(1,61)} = 12.11$, $p = 0.001$); Attitude to technology was significantly associated with the area of work ($F_{(1,61)} = 80.62$, $p = 0.000$); and volitional control was significantly associated with the area of work ($F_{(1,61)} = 30.90$, $p = 0.000$).

The means for both groups indicate that marketing workers perceived technology as more useful and easier to use, and that their attitude to technology was more positive than the workers within the service provider roles. When looking at the control measure volitional control, the ability of people to act in a manner that is consistent with their attitudes, the results indicate one subset of workers tends to act in alignment with their attitude and the other does not.

Post Hoc Analysis of Differences between Service Providers and Marketing Workers

Prior research that examined a similar research question into the relationship between technology-related factors and a student's chosen area of study (Greenfield & Rohde, 2009) found that social sciences students (and thus those more likely to enter careers within NFP organisations) perceived technology to be less useful, not as easy to use, and while they did not differ significantly from business students in their attitude to technology, their attitude towards technology had an effect on both their perceptions of the usefulness and ease of use of technology. They also found business students were more likely to act in accordance with their attitudes. This current research examined the same relationships found by Greenfield and Rohde (2009), to determine whether similar differences existed between technological related factors and a person's chosen area of work.

The results from the current study indicate for those persons involved within the marketing role, a significant positive relationship exists between a

Table 2. Effect of group on the overall model and each DV

Source	Df	Mean Squared	F value	Pr > F*	Service Provider Mean@	Marketing Mean@
Combined DV	4		21.849	0.000		
Pu	1	400.48	8.55	0.005	25.513	30.640
Peu	1	288.51	12.11	0.001	17.128	21.480
Att	1	1590.41	80.62	0.000	14.103	24.320
Vc	1	177.17	30.90	0.000	8.590	12.000

@ 1 = Strongly disagree 7 = Strongly agree

person's positive attitude to technology and their perceived usefulness of the technology. Moreover, Figure 2 also indicates a significant negative relationship between a person's negative attitude to technology and their perceived usefulness of the technology. Thus, the worker's attitude to technology directly affects how they perceive technology to be useful. The results also indicate that the worker's attitude (whether positive or negative) is not associated with their perception of the ease of use of technology. While no significant relationships existed between attitude and perceived ease of use, there is a significant positive relationship between perceived ease of use and perceived usefulness. This relationship indicates that those participants who considered technology easier to use found it more useful and vice versa. These results were consistent with previous research of persons entering into a business career.

The results indicate for participants involved in the provision of service delivery a significant positive relationship between a person's positive attitude to technology and their perceived useful-

ness of the technology i.e. persons with a more positive attitude to technology perceive technology to be more useful. Moreover, Figure 3 also indicates a significant positive relationship between a person's positive attitude to technology and their perceived ease of use of the technology. Therefore, participants involved in the provision of service delivery, with a positive attitude to technology perceive technology to be useful and easy to use. The results also indicate that for participants involved in the provision of service delivery with a negative attitude, their attitude is not significantly associated with their perception of the ease of use of technology or perceived usefulness of the technology and may be a deeper overarching negative view to technology as a whole. There is also a significant positive relationship between perceived ease of use and perceived usefulness indicating that those participants who considered technology easier to use found it more useful and vice versa. These results were consistent with previous research of persons entering into a social science career.

Figure 2. The effect of the factors for participants in marketing role

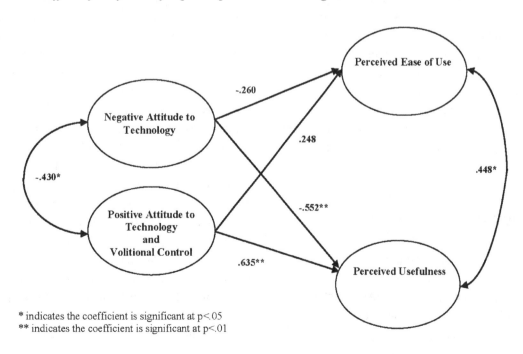

* indicates the coefficient is significant at p<.05
** indicates the coefficient is significant at p<.01

Figure 3. The effect of the factors for participants in the service delivery role

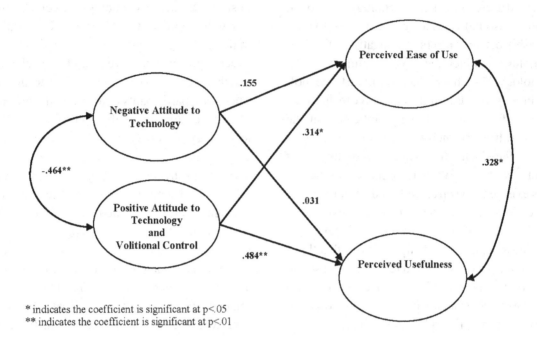

* indicates the coefficient is significant at p<.05
** indicates the coefficient is significant at p<.01

It appears that for persons involved in business based careers; their attitude to technology influences their view of the usefulness of the new technology. If they have a negative attitude then their perception of the new technology's usefulness will be lower and vice versa for the positive attitude. Whereas the view of the person involved in the delivery of services about the usefulness of the technology is not affected by their attitude.

Therefore, from this analysis it appears that participants involved in the provision of service delivery are less likely to act in accordance with their attitudes than the participants involved in the marketing role.

IMPLICATIONS

The findings have potential implications for researchers using models such as TAM. Initially it was suspected that implementing a new technology into a NFP may need to be handled differently from the implementation of the same technology into

a non-NFP. The type of employees within a firm and their attitudes, may also need to be considered when considering technology implementation and acceptance.

One implication of this research is that as organisations are implementing technology they need to recognize that users, even within the same organisation, are not a homogeneous group and that workers performing different business functions within their organisations may have different perceptions that may ultimately affect their technology acceptance of the same artifact. An organisation may need to tailor the implementation of technology to suit the differing attitude of its workers. For example implementing a new artifact in a community centre for their accountants may need to be handled differently to the same technology rollout in a community centre for their social workers.

Recent research in the area of evolutionary psychology (Abraham et al., 2009) and its application to technology acceptance may shed light on these findings. Specifically, the research looked at four drives that provide a stable influence for human behavior. These drives being (1) acquire; (2) bond; (3) learn; (4) defend. Abraham et al., (2009) found that acceptance of new technology mostly occurred when these underlying drives were supported, and failed when they were impeded.

Considering the current research, the drive for the marketing workers is to raise funds for the organisation through telemarketing products. This drive, the drive to acquire (seek, take, control and retain objects and personal experiences), was thus supported by the new system. The drive, however, for the service delivery workers was to support clients through face-to-face interaction and was not supported by the new system. This drive, the drive to bond, is about forming relationships with others and developing mutual caring commitments. A computer impedes their ability to bond with the client. Although the service delivery workers use the system it is in contradiction with their attitude to the technology.

CONCLUSION, LIMITATIONS, AND FUTURE RESEARCH

This research provides additional insight into the relationship between technological related factors and a person's area of work. Prior research has indicated that social sciences students (and thus those more likely to enter careers within NFP organisations) perceived technology to be less useful, not as easy to use and their attitude towards technology had an effect on both their perceptions of the usefulness and ease of use of technology. This research indicates that these relationships are carried over into the workplace.

The usual caveats associated with survey-based research apply. Furthermore, the research only considers a small number of participants within two organisations.

The results suggest some initial understanding of the differences in attitude towards technology in future users and as such a number of opportu-nities exist for future research. Three are briefly mentioned here. First, further research using a larger sample from a larger number of organisations, involving a broader range of sub-disciplines and IT artifacts, is required to verify the findings. Second, with increasing globalization, the combinations of aspects such as culture and career orientations need to be considered jointly when examining the issues of technology acceptance. Thirdly, further research could be undertaken in the application of the tenets of evolutionary psychology to technology acceptance specifically, across different business functions within and across multiple organisations.

ACKNOWLEDGMENT

The authors acknowledge the Editor-in-Chief of the journal, Professor John Wang, and the anonymous reviewers for their indispensable input that improved the paper significantly.

REFERENCES

Abraham, C., Junglas, I., Watson, R. T., & Boudreau, M. (2009). Studying the role of human nature in technology acceptance. In *Proceeding of the Thirtieth International Conference of Information Systems*, Phoenix, AZ.

Australian Bureau of Statistics. (2003). *Australian social trends*. Retrieved from http://www.abs.gov.au/AUSSTATS/abs@.nsf/DetailsPage/4102.02003

Australian Institute of Health and Welfare. (2003). *Australia's welfare*. Retrieved from http://www.aihw.gov.au/

Benamati, J., & Rajkumar, T. M. (2008). An outsourcing acceptance model: An application of TAM to application development outsourcing decisions. *Information Resources Management Journal*, *21*, 80–102. doi:10.4018/irmj.2008040105

Cunningham, B., Nikolai, L., & Bazley, J. (2004). *Accounting information for business decisions* (2nd ed.). Australia: Thomson.

Damanpour, F. (1991). Organizational innovation: a meta-analysis of effects of determinants and moderators. *Academy of Management Journal*, *34*, 555–590. doi:10.2307/256406

Davis, F. D. (1989). Perceived usefulness, perceived ease of use, and user acceptance of information technology. *Management Information Systems Quarterly*, *13*, 319–339. doi:10.2307/249008

Davis, F. D., Bagozzi, R. P., & Warshaw, P. R. (1989). User acceptance of computer technology: A comparison of two theoretical models. *Management Science*, *35*, 36–48. doi:10.1287/mnsc.35.8.982

Davis, F. D., Bagozzi, R. P., & Warshaw, P. R. (1992). Extrinsic and intrinsic motivation to use computers in the workplace. *Journal of Applied Social Psychology*, *22*, 1111–1132. doi:10.1111/j.1559-1816.1992.tb00945.x

Fishbein, M., & Ajzen, I. (1975). *Belief, attitude, intention and behavior: An introduction to theory and research*. Reading, MA: Addison-Wesley.

Greenfield, G., & Rohde, F. (2009). Technology acceptance: Not all organisations or workers may be the same. *International Journal of Accounting Information Systems*, *10*, 263–272. doi:10.1016/j.accinf.2009.10.001

Igbaria, M., Iivari, J., & Maragahh, H. (1995). Why do individuals use computer technology? A Finnish case study. *Information & Management*, *29*, 227–239. doi:10.1016/0378-7206(95)00031-0

Karahanna, E., Ahuja, M., Srite, M., & Galvin, J. (2002). Individual differences and relative advantage: the case of GSS. *Decision Support Systems*, *32*(4), 327–341. doi:10.1016/S0167-9236(01)00124-5

Khalifa, M., & Ning Shen, K. (2008). Explaining the adoption of transactional B2C mobile commerce. *Journal of Enterprise Information Management.*, *21*, 110–124. doi:10.1108/17410390810851372

Kothandapani, V. (1971). Validation of feeling, belief, and intention to act as three components of attitude and their contribution to prediction of contraceptive behaviour. *Journal of Personality and Social Psychology*, *19*, 321–333. doi:10.1037/h0031448

McCoy, S., Galletta, D., & King, W. (2007). Applying TAM across cultures: The need for caution. *European Journal of Information Systems*, *16*, 81–90. doi:10.1057/palgrave.ejis.3000659

Saidel, J., & Cour, S. (2003). Information technology and the voluntary sector workplace. *Nonprofit and Voluntary Sector Quarterly*, *32*, 5–24. doi:10.1177/0899764002250004

Thompson, R. L., Higgins, C. A., & Howell, J. M. (1991). Personal computing: Toward a conceptual model of utilization. *Management Information Systems Quarterly*, *15*, 124–143. doi:10.2307/249443

Venkatesh, V., & Davis, F. (2000). A theoretical extension of the technology acceptance model: Four longitudinal field studies. *Management Science*, *46*, 186–204. doi:10.1287/mnsc.46.2.186.11926

Venkatesh, V., Morris, M., Davis, G., & Davis, F. (2003). User acceptance of information technology: Toward a unified view. *Management Information Systems Quarterly*, *27*, 425–478.

Winter, S., Chudoba, K., & Gutek, B. (1998). Attitudes toward computers: When do they predict computer use? *Information & Management*, *34*, 275–285. doi:10.1016/S0378-7206(98)00065-2

Yang, H., & Yoo, Y. (2004). It's all about attitude: Revisiting the technology acceptance model. *Decision Support Systems*, *38*, 19–31. doi:10.1016/S0167-9236(03)00062-9

Yi, U., Wu, Z., & Tung, L. L. (2005/2006). How individual differences influence technology usage behaviour? Toward an intergrated framework. *Journal of Computer Information Systems*, *46*, 52–63.

Yousafzai, S., Foxall, G., & Pallister, J. (2007). Technology acceptance: A meta-analysis of the TAM: Part 1. *Journal of Modelling in Management*, *2*, 251–280. doi:10.1108/17465660710834453

ENDNOTES

[1]   TAM is widely cited in MIS literature, e.g., Business Source Premier recorded Davis (1989) as cited 601 times in its database.

APPENDIX

Final factors and measurement items

Factor		Measurement items
Perceived usefulness	Pu1	I believe computers will be useful in my future career.
	Pu2	Using computers will increase my productivity in my future career.
	Pu3	Using computers will enhance my effectiveness in my future career.
	Pu4	Using computers will improve my future career performance.
	Pu5	Using computers provides me with information that would lead to better decisions
Perceived ease of use	Peu1	Learning to use computers is easy for me.
	Peu2	I find it easy to get computers to do what I want to do.
	Peu3	It would be easy for me to become skilful at using computers.
	Peu4	I find computers easy to use.
Negative Attitude	Att1	In my future work I would avoid the computer at all possible cost.
	Att4	If it were possible in my future work, I would prefer to delegate computer tasks to someone else.
Positive Attitude and Volitional Control	Att2	If I could choose, in my future work I would prefer to use the computer.
	Att3	If it were possible in my future work, I would computerise most of my tasks
	Vc1	When considering computers in the workplace, I see them as a tool to be used at my convenience.
	Vc2	In my future career I would consider my job is to use a computer.

This work was previously published in the International Journal of Information Systems and Social Change, Volume 2, Issue 2, edited by John Wang, pp. 26-36, copyright 2011 by IGI Publishing (an imprint of IGI Global).

Chapter 8
A Query–Based Approach for Semi–Automatic Annotation of Web Services

Mohammad Mourhaf AL Asswad
Brunel University, UK

Sergio de Cesare
Brunel University, UK

Mark Lycett
Brunel University, UK

ABSTRACT

Semantic Web services (SWS) have attracted increasing attention due to their potential to automate discovery and composition of current syntactic Web services. An issue that prevents a wider adoption of SWS relates to the manual nature of the semantic annotation task. Manual annotation is a difficult, error-prone, and time-consuming process and automating the process is highly desirable. Though some approaches have been proposed to semi-automate the annotation task, they are difficult to use and cannot perform accurate annotation for the following reasons: (1) They require building application ontologies to represent candidate services and (2) they cannot perform accurate name-based matching when labels of candidate service elements and ontological entities contain Compound Nouns (CN). To overcome these two deficiencies, this paper proposes a query-based approach that can facilitate semi-automatic annotation of Web services. The proposed approach is easy to use because it does not require building application ontologies to represent services. Candidate service elements that need to be annotated are extracted from a WSDL file and used to generate query instances by filling a Standard Query Template. The resulting query instances are executed against a repository of ontologies using a novel query execution engine to find appropriate correspondences for candidate service elements. This query execution engine employs name-based and structural matching mechanisms that can perform effective and accurate similarity measurements between labels containing CNs. The proposed semi-automatic annotation approach is evaluated by employing it to annotate existing Web services using published domain ontologies. Precision and recall are used as evaluation metrics. The resulting precision and recall values demonstrate the effectiveness and applicability of the proposed approach.

DOI: 10.4018/978-1-4666-2922-6.ch008

INTRODUCTION

Web services are software components that can enable flexible, low cost and platform-independent application communication and integration (Paolucci, Kawamura, Payne, & Sycara, 2002). The Web service framework is mainly composed of XML-based standards as follows (Curbera et al., 2002):

- SOAP (Simple Object Access Protocol), which is a messaging protocol that facilitates message exchange among services.
- WSDL (Web Service Description Language), which describes the service interface as a set of communication endpoints that enable message exchange.
- UDDI (Universal Description Discovery and Integration), which is a centralized directory of service description.

For Web services to meet the needs of future Web applications, it is essential to enable on-the-fly discovery and composition of services (Agarwal, Handschuh, & Staab, 2003). Unfortunately, the use of existing Web service standards alone does not enable the desired automation and agility of service discovery and composition – primarily because these standards lack the necessary semantic constructs (Sivashanmugam et al., 2003; Sycara, Paolucci, Ankolekar, & Srinivasan, 2003). The utilization of semantics, represented in the form of ontologies, in the area of Web services launched an active research area called "Semantic Web Services" (SWS) (McIlraith, Son, & Zeng, 2001; Sycara et al., 2003).

SWS has attracted increasing attention in computer science and information systems research (Feier, Roman, Polleres, Domingue, & Fensel, 2005; Jacek, Tomas, Carine, & Joel, 2007; Martin et al., 2007). Successful implementation of SWS, however, requires the existence of suitable methods for SWS description (Lara, Roman, Polleres, & Fensel, 2004), catering for service elements such as inputs and outputs annotated using suitable semantic metadata (Verma & Sheth, 2007). In this context, annotation means explicitly referencing the data and functional elements of a service using concepts from shared ontologies. The annotation process is currently performed manually and thus requires comprehensive human involvement. Automating the annotation task is highly desirable as the manual process is tedious, error-prone and difficult (Hepp, 2006; Patil, Oundhakar, Sheth, & Verma, 2004; Rajasekaran, Miller, Verma, & Sheth, 2005).

Few approaches have looked at the problem of semi-automatic annotation. Those approaches that exist can be categorized twofold: First, approaches that automatically build ontologies to represent semantics of given services using learning techniques. Examples of this class of techniques are the approaches of Chifu, Salomie, and Chifu (2007) and Heb and Kushmerick (2003). Second, approaches that require manual development of application ontologies that model implicit semantics of WSDL files. Such application ontologies are then matched against existing domain ontologies using semantic matching techniques in order to find appropriate correspondences that are then used to annotate service data. These approaches are called semantic matching-based approaches. Examples of this category are Patil et al., (2004) and Duo, Juan-Zi, and Bin (2005). Current approaches in both categories have limitations:

- Manual ontology building is difficult and requires extensive technical and domain knowledge. On the other hand, automatic ontology development using learning techniques is still under development and results in ontologies that are of questionable quality.
- Matching-based approaches utilize similarity measurement mechanisms that do not produce precise results when labels of ontological classes and Web service elements are composed of multiple words i.e. Compound Nouns (CNs).

In response to the increasing need for an effective semi-automatic Web service annotation approach; this paper proposes a novel annotation framework that can facilitate easier and more effective WSDL annotation in comparison to other approaches. This approach takes as inputs a WSDL file and a set of ontologies that cover domains of data of the given service. The proposed approach is easy to use because it does not require building application ontologies to represent service data. Rather, Web service data are extracted and used to fill a Standard Query Template - producing query instances that are then executed against shared ontologies using a novel query execution engine. This query execution engine implements name-based and structural matching techniques. The implemented name-based matching mechanism can match labels containing CNs in an effective and precise manner.

In describing this approach the paper is structured as follows: First we provide background information alongside a survey of previous semi-automatic annotation approaches and their limitations. Next, we present a brief analysis of the general structure of WSDL and shows how a WSDL file can be interpreted to extract required service data that should be annotated. We present the proposed annotation framework and then provide evaluation results from the application of the proposed framework. We finally conclude the paper and propose future research avenues.

BACKGROUND AND RELATED WORK

Few approaches and tools have been developed to semi-automatically annotate Web services. What exists can be classified into two categories according to the method used in performing the annotation. These two categories are machine learning-based and semantic matching-based approaches. This section provides a review of different approaches within those two categories and presents their limitations.

Machine Learning-Based Approaches

A framework for learning domain specific taxonomies from textual descriptions of Web pages associated with services is proposed in Chifu et al. (2007). The generated taxonomies represent the domain ontologies necessary for annotating Web services and are used to semantically annotate inputs and outputs of Web service operations. Though useful, the work only explains the process of automatically building an ontology that provides annotations for service parameters but says nothing about the steps required to achieve the annotation.

An approach to automatically creating metadata from training data to semantically describe a Web service is developed by Heb and Kushmerick (2003). The training data comes from HTML pages documenting the service and the WSDL file of the service. Three different interrelated types of metadata are created. The first type is the category taxonomy that categorizes Web services. The second is the domain taxonomy, which describes the functionality of a specific service operation such as 'searching for a book' or 'querying an airline timetable'. The third taxonomy describes the semantic categories of input and output data such as 'book title' or 'destination airport'. Later, Lerman, Plangrasopchok, and Knoblock (2006) proposed an approach that is similar to the approach of Heb and Kushmerick (2003). The approach of Lerman et al. (2006) added a verification stage to guarantee a correct prediction of input data and generation of output data. Lerman's approach, however, has a drawback in that the search for appropriate semantic metadata becomes expensive when the candidate service has more than two inputs (Chifu et al., 2007).

ASSAM is a tool developed by Heb, Johnston, and Kushmerick (2004) to semi-automatically annotate categories, operations and parameters of services. Previously annotated Web services are the training data from which ASSAM learns in order to predict annotations for new services.

Furthermore, ASSAM uses a schema matching technique to aggregate data returned by a number of Web services that are semantically related. Having completed the annotation process, semantically annotated WSDL files can be exported into the OWL-S SWS notation and a concept file that contains complex data types.

In summary, though these learning-based annotation approaches provide clear contributions, they share the following limitations:

1. Ontology learning mechanisms are still under development and thus may produce semantically poor ontologies. The learned ontologies are usually in a form of a taxonomy where important ontological constructs such as properties and axioms are not captured.
2. The resulting application ontologies are representations of sole services instead of being precise representations of shared domain knowledge. Therefore, matching the produced ontologies against shared ontologies is still required either at design-time or at run-time when service related activities such as discovery and composition are to be performed.

Semantic Matching-Based Approaches

The METEOR-S Web Service Annotation Framework (MWSAF), which is part of the METEOR-S Project, is developed by Patil et al., (2004) to add semantics to WSDL documents of Web services. This framework transforms domain ontologies and XSDs of WSDL files into a common representation called SchemaGraph in order to enable structural matching. Once a common representation is achieved, every concept in the WSDL graph is matched against every concept in the domain ontology graph using element level (name-based) and schema level (structural) matching techniques. The element level matching utilizes N-Gram as

a string-based mechanism and synonym–based similarity as a linguistic mechanism. This approach can measure similarities between labels containing multiple terms using basic similarity mechanisms that ignore the linguistic structure of CNs. Other authors have noted that this ignorance may result in imprecise similarity scores however (Kim & Baldwin, 2005).

A framework for generating OWL-S descriptions from WSDL files was developed by Duo et al., (2005). The process of generating OWL-S starts with manually translating XML schema of a WSDL description into an intermediate OWL ontology. The intermediate ontology is then mapped to existing domain ontologies using name-based and structural similarity measures. The name-based matching uses the Levenstein distance to measure similarities between labels containing single terms only. Though effective, this approach cannot measure similarities between labels comprising CNs. Later, a similar approach to Due et al. (2005) was proposed by Zhang, Duan, and Zhao (2008). This later approach utilizes H-MATCH (Castano, Ferrara, & Montanelli, 2006) as a matching tool to find correspondences between an intermediate ontology that represents a service and a set of shared ontologies. H-Match can measure similarities between CNs but it necessitates the addition of CNs, that do not have an entry in WordNet, to a newly constructed thesaurus. Once that is done, a similarity calculation can be carried out between these CNs and other single terms or CNs that already exist in WordNet. This necessity could delay and complicate the semi-automatic annotation because creating new entries requires some degree of human involvement to extend the constructed thesaurus with new entries.

In summary, semantic matching-based annotation approaches therefore suffer from the following limitations:

1. They require manual development of application ontologies to capture implicit se-

mantics of WSDL files of candidate services. Manual ontology building is a tedious and difficult task that requires extensive domain and technical knowledge.

2. Implemented matching approaches cannot provide effective and precise matching results when labels of service data and/or ontological entities contain CNs. The reason for producing inaccurate results is that these matching approaches do not take the linguistic structure into consideration during the similarity calculation process (Kim & Baldwin, 2005) (see Table 1).

WSDL STRUCTURE AND INTERPRETATION

As a precursor to the approach presented here, it is necessary to analyze the WSDL general structure in order to make clear what WSDL elements can be annotated. In overall terms, a WSDL file comprises of an element declaration, type definition, interface, binding and service. The element declaration, type definition and interface provide an abstract definition of a service, while binding and service describe the implementation aspects of a service (Jacek et al., 2007).

Element declaration and type definitions are defined in the schema part of a WSDL file and provide data type definitions for input and output operation messages and their parts. In an XSD, the elements that are direct children of a schema element are called global elements. Other XSD elements are called local elements. Furthermore, sub-elements of a complex type element are called direct child elements of that complex type. Figure 1 presents a WSDL file of a book information service. The binding and service elements of this service are removed due to space limitation.

The data type definition part of this WSDL document defines five global elements: 'Book', 'VendorPrice', 'ArrayOfBookInfo', 'Keyword' and 'Source'. These data types are used to define data of input and out message parts of WSDL operations. The 'Book', 'VendorPrice' and 'ArrayOfBookInfo' are defined as complex types while 'Keyword' and 'Source' are simple types. Every complex type has a set of child elements. For example, the 'Book' complex type element has nine child elements: 'ISBN', 'Title', 'Author', 'PubDate', 'Publisher', 'Format', 'ImageUrl', 'TimeStamp' and 'VendorPrice'. On the other hand, elements that are of a simple type such as 'Keyword' and 'Source' do not have child elements.

Based on the previous brief analysis of WSDL elements, one can conclude that XSD elements including simple types and their attributes as well as complex types and their child elements should be annotated since they describe data of operations' messages. Other WSDL elements such as bindings and service define implementation details and thus do not require semantic annotation.

Table 1. Limitations of previous approaches

Category	Description	Limitations
Machine learning	Based on learning ontologies from textual and/or WSDL descriptions of Services. The resulting ontologies can offer annotations for service category, operation names as well as inputs and outputs of operations.	1. They normally produce semantically poor ontologies. 2. The resulting ontologies are representations of sole services and thus matching is still required either at run time or at design time.
Semantic Matching	A set of transformation rules are utilized to manually transform WSDL files of services into application ontologies. The resulting application ontologies are then matched against existing shared ontologies to find correspondences for service data using different matching techniques.	1. They demand manual building of ontologies which is a difficult task. 2. They use matching approaches that cannot perform effective and automatic CNs matching.

Figure 1. An Example of a WSDL file of book information provider service

```
<?xml version="1.0" encoding="utf-8"?>
<wsdl:definitions>
  <wsdl:types>
    <s:schema>
      <s:complexType name="Book">
        <s:sequence>
          <s:element name="Isbn" />
          <s:element name="Title" />
          <s:element name="Author" />
          <s:element name="Pubdate" />
          <s:element name="Publisher" />
          <s:element name="Format" />
          <s:element name="ImgUrl" />
          <s:element name="TimeStamp" />
          <s:element name="Vendorprice" type="tns:VendorPrice" />
        </s:sequence>
      </s:complexType>
      <s:complexType name="VendorPrice">
        <s:sequence>
          <s:element name="Name" />
          <s:element name="SiteUrl" />
          <s:element name="PricePrefix" />
          <s:element name="Price" />
        </s:sequence>
      </s:complexType>
      <s:complexType name="ArrayOfBookInfo">
        <s:sequence>
          <s:element name="Book" type="tns:Book" />
        </s:sequence>
      </s:complexType>
      <s:element name="Keyword" type="s:string" />
      <s:element name="Source" type="s:string" />
    </s:schema>
  </wsdl:types>
  <wsdl:message name="GetInfoHttpGetIn">
    <wsdl:part name="ISBN" type="s:string" />
  </wsdl:message>
  <wsdl:message name="GetInfoHttpGetOut">
    <wsdl:part name="Body" element="tns:Book" />
  </wsdl:message>
  <wsdl:message name="DoKeywordSearchHttpGetIn">
    <wsdl:part name="keyword" type="s:string" />
  </wsdl:message>
  <wsdl:message name="DoKeywordSearchHttpGetOut">
    <wsdl:part name="Body" element="tns:ArrayOfBookInfo" />
  </wsdl:message>
  <wsdl:message name="GetInfoHttpPostIn">
    <wsdl:part name="ISBN" type="s:string" />
  </wsdl:message>
  <wsdl:message name="GetInfoHttpPostOut">
    <wsdl:part name="Body" element="tns:bookInfo" />
  </wsdl:message>
```

XSD definition embeds implicit semantic information that requires disambiguation however. For example, the relation between a complex type and each of its child elements is similar to an ontological property. Previous research (Due et al., 2005; Zhang et al., 2008) defines sets of rules for interpreting the implicit semantic information embedded in an XSD definition of a WSDL file. We adopt a subset of these rules and implement them for the purpose of disambiguating the semantic information and extracting the set of concepts that will be considered during the annotation process. The rules are presented as follow.

Rule One: Each global complex or simple XSD element is considered as a concept that should be annotated.

Rule Two: Each local complex of simple XSD element is considered as a concept that should be annotated.

Rule Three: The set of child elements of a complex element formulates the set of related elements of a complex type concept.

FRAMEWORK

The proposed framework overcomes two significant limitations of previous semi-automatic annotation approaches: First, it does not require building an ontology to represent semantics of a service. Instead, the concepts and their implicit semantics are extracted from the WSDL file and then used to fill in a Standard Query Template. Second, previous approaches use basic name-based matching methods to compute similarities between labels containing multiple nouns (CNs). These methods are ignorant of the linguistic structure of CNs, which may lead to imprecise matching results. In contrast, the execution engine of the proposed approach utilizes a name-based matching approach that can match labels of concepts including CNs in an effective and precise manner. These two improvements allow the proposed approach to make the automatic annotation of Web services a feasible, easy and effective process. Figure 7 presents the process flow of the proposed annotation framework, which is composed of four phases: (1) Concept Extraction; (2) Query Filling; (3) Query Execution; and (4) Annotation.

Concept Extraction Phase

The XSD concepts and their related concepts are extracted from the given WSDL file based on the

three rules noted earlier. Each extracted complex type concept and its set of related concepts are used to fill in a query instance. When the extracted concept does not have related concepts (i.e. it is of the XSD type simple type), then the set of related concepts is left blank. For example, the complex type 'Book' in the WSDL example of Figure 1 has nine related concepts: 'ISBN', 'Title', 'Author', 'PubDate', 'Publisher', 'Format', 'ImageUrl', 'TimeStamp' and 'VendorPrice'. While, the simple types 'Keyword' and 'Source' have no related concepts.

Query Filling Phase

The candidate service element and its related elements are used to fill in the Query Template (see Figure 2).

The Query Template is designed to provide a standard format for all query instances. This Query Template has place holders for a service element and its related concepts. The template contains two clauses as shown in Figure 2.

When filling a query of a complex type, the label of the complex type is used in Clause (1) and labels of related concepts are used in Clause (2). The resulting query instances for the complex types 'Book' and 'VendorPrice' are given in Figures 3 and 4 respectively. The complex type 'ArrayOfBookInfo' will not have a query instance since it denotes no semantics. This complex type is a syntactic definition of an array of things of type 'Book'. Nevertheless, the 'Book' concept is considered for annotation and therefore it can provide semantics for 'ArrayOfBookInfo' when it is annotated.

Query instances for global simple types contain Clause (1) only because they do not have related concepts (child elements). Figures 5 and 6 show the query instances for the 'Keyword' and 'Source' simple types respectively. The query execution is presented as a part of the framework in Figure 7.

Figure 2. The proposed standard query template

> **Find a concept in an ontology that:**
> **Clause (1): Target concept name is semantically similar to "Given service element name".**
> **Clause (2): Target concept is related by object properties to concepts that are similar to the concepts in the following set**
> **{"Concept one", "Concept two"... "Concept n"}.**

Figure 3. The query instance of Book

> **Find a concept in an ontology that:**
> **Clause (1): Target concept name is semantically similar to 'Book'.**
> **Clause (2): Target concept is related by object properties to concepts that are similar to the concepts in the following set**
> **{'Isbn, 'Title', 'Author', 'Pubdate', 'Publisher', 'Format', 'ImgUrl', 'TimeStamp', 'VendorPrice'}.**

Figure 4. The query instance of VendorPrice

> **Find a concept in an ontology that:**
> **Clause (1): Target concept name is semantically similar to 'VendorPrice'.**
> **Clause (2): Target concept is related by object properties to concepts that are similar to the concepts in the following set**
> **{'Name, 'SiteURL', 'PricePrefix', 'Price'}.**

Query Execution Phase

A query instance is executed during the query execution phase. The inputs of this phase are a query instance and an ontology. The query execution engine is the means that allows the desired similarity calculation.

Each iteration of a similarity calculation takes a query instance and a candidate ontological class as inputs and produces a similarity score in the range [0-1] as an output. This score indicates how similar a query instance concept and an ontological class are. If this score is over a defined threshold, then the corresponding ontological

Figure 5. The query instance of Keyword

> **Find a concept in an ontology that:**
> **Clause (1): Target concept name is semantically**
> **similar to `Keyword'.**

Figure 6. The query instance of Source

> **Find a concept in an ontology that:**
> **Clause (1): Target concept name is semantically**
> **similar to `Source'.**

Figure 7. The process flow of the annotation framework

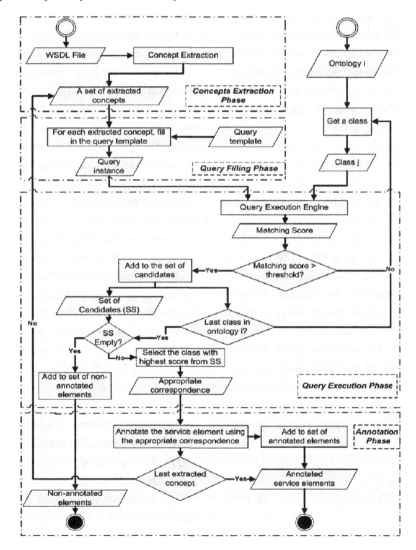

class is added to the set of candidates (SS). Otherwise the matching is ignored. After executing the candidate query instance against each class in ontology *i*, the class with the highest score is selected from the set (SS). This class is taken as the appropriate correspondence of the given query instance concept.

To allow an effective and accurate query execution, a new query execution engine is designed and implemented specifically for the purpose of semi-automatic annotation of Web services. This execution engine implements name-based and structural matching mechanisms to perform the desired similarity calculation. Name-based matching is achieved using CN-Match (AL Asswad, 2011) which is a novel name-based matching tool.

Structural similarity is used to measure similarities between related concepts of a service element and those of the ontological element when executing the query. The following two subsections present CN-Match and the implemented structural matching mechanism.

CN-Match

CN-Match (AL Asswad, 2011) is an automatic name-based matching approach that can calculate similarities between labels containing single terms and compound nouns (CNs). The reason for developing and using CN-Match is that existing name-based matching techniques do not provide accurate matching results when labels of candidates are CNs – primarily because these techniques ignore the linguistic structure of CNs (Kim & Baldwin, 2005). A CN is defined as a noun that is made up of two or more nouns (Girju, Moldovanb, Tatub, & Antoheb, 2005). Each CN is composed of a head and a modifier (Lauer, 1995) - the rightmost constituent is called a head and the leftmost constituent is called a modifier (Girju et al., 2005). For example, the concept 'Name' is the head of the CN 'PlayerName' while 'Player' is the modifier.

The CN-Match similarity calculation mechanism is based on the fact that similarities between any two CNs can be derived successfully from similarities between their constituents (Kim & Baldwin, 2005). Subsequently, similarities between names in CN-Match are calculated as weighted sums of individual similarities between CNs constituents. The weights are assigned based on the CN linguistic structure and a set of experiments to ensure the appropriateness of the selected weights.

String-based and linguistic similarities are the basic similarity techniques that are used to enable similarity calculation performed by CN-Match. The Levensetin distance is used as a string-based similarity measure.

For linguistic similarity, synonym and path length-based similarity measures are utilized. These two linguistic techniques make use of WordNet (Miller, 1995) which is a general purpose English thesaurus. Furthermore, the path length-based measure utilizes the path length-based similarity approach proposed by Wu and Palmer (1994) and makes use of the WordNet hierarchal structure. CN-Match is implemented in the Java programming language version 1.6.0 and tested using published ontological test sets. The results achieved by CN-Match are very promising and prove its effectiveness. Figure 8 presents the evaluation results of CN-Match on the Russia test set (Ehrig & Sure, 2005). Precision, recall and F-measure are used as metrics for the purpose of CN-Match evaluation (AL Asswad, 2011).

Structural Matching

The structural matching is usually performed to enable more accurate similarity measurements between ontological entities (Euzenat & Shvaiko, 2007). Two ontological classes could have the same label but might denote different meanings. Let us assume that the main concept in a query instance is 'Book' which denotes the 'written book' and the candidate ontological concept has label

Figure 8. Precision and recall of Russia test of the CN-match

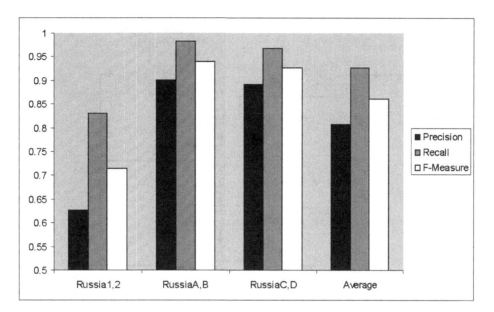

'Book' which means 'reserve'. Matching these two concepts using name-based matching only will provide a full matching score however, they are not similar. Performing structural similarities can figure out that the previous two candidates are not similar since their related concepts are unlikely to be the same. Therefore, it is important to take the structural similarity into consideration.

Generally speaking, measuring structural similarities between two classes belonging to two different ontologies involves matching their super-classes, sub-classes and properties and their domains or ranges (Euzenat & Shvaiko, 2007). In the context of this research, structural similarity between a query instance concept and a candidate ontological class is performed based on calculating similarities between related concepts of the query instance concept (service element) and related concepts of the candidate ontological class. Related concepts of an ontological class are those classes that are linked to this class and its superclasses through object properties. Related concepts of superclasses are taken into account because an ontological class inherits the relations (properties) of its superclasses. Super and subclasses of

a candidate ontological class are ignored in this structural similarity approach because they do not have counterparts in query instances. In other words, the interpretation of an XSD of a WSDL file do not provide neither implicit nor explicit super or sub classing relations between defined data types (Jacek et al., 2007). Related concepts of ontological classes are extracted from candidate ontologies using the OWL API (Euzenat, 2004). The OWL API is a Java API that allows developers to manipulate ontologies represented in the Web Ontology Language (OWL) formalism. Figure 9 presents the process flow of the implemented structural matching approach.

The structural matching process starts by obtaining two concepts - one from the set of related concepts of the query instance concept (Set 1) and one from the set of related concepts of the candidate ontological class (Set 2). Similarity between labels of these two concepts is measured using CN-Match. If the resulting score is higher than the CN-Match threshold then the corresponding concept is added to the set of candidate matches. Otherwise, this matching is ignored. When matching the related concept from Set 1

Figure 9. The implemented structural matching approach

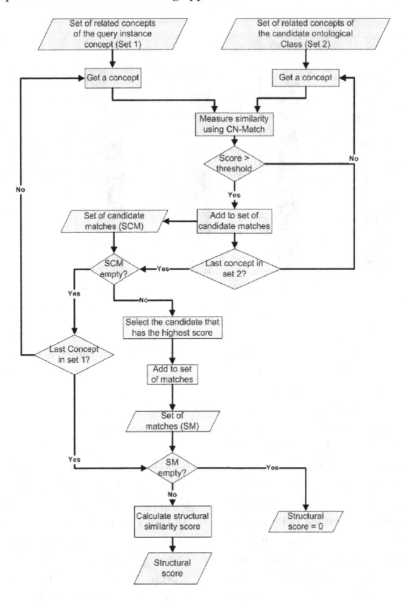

against all concepts of Set 2, the candidate with the highest score is selected as a match of the given Set 1 element. This match is added to the set of matches. If the set of candidate matches is empty, then there is no match for the given related concept. When the matching process is performed for each element of Set 1 and the set of matches is not empty, the final structural similarity score is calculated according to Equation (1).

$$S_{s} = \frac{\sum_{i=1}^{i=n} Si}{n} \tag{1}$$

where Ss is the final structural similarity score, n is the number of elements in set 1, and Si is the highest similarity score between each element of set 1 and all elements of set 2.

The final score of a query execution which represents the similarity score between a query instance concept and a candidate ontological class is calculated as a weighted sum of the name-based and structural similarities scores as given in the following Equation.

$$S = W_n \times S_n + W_s \times S_s \qquad (2)$$

where S is the final score, Wn is the weight of the name-based matching, Sn is the name-based matching score, Ws is the weight of the structural matching, and Ss is the structural similarity score.

Annotation Phase

During the Annotation Phase, the discovered correspondence is used to annotate the candidate service concept. If this concept is the last extracted concept from the WSDL file, then the annotated service can be produced. Otherwise, the next extracted service concept is taken to fill in the Query Template and create a query instance, which is then executed against the target ontology during the Query Execution Phase. The annotated service is produced in the SAWSDL format (Jacek et al., 2007) which is a W3C recommendation for SWS description.

EVALUATION

The evaluation examines the performance of the proposed annotation approach by testing it in practical settings. To avoid any potential bias, existing Web services and ontologies are used for the purpose of this evaluation. The Web services are obtained from Web service repositories such as xmethods.com and soatest.parasoft.com. Testing was limited to services from two domains due to the lack of existing and appropriate domain ontologies that can be used to annotate services from other domains.

Five book information and selling Web services and two camera selling Web services were

chosen. Two ontologies that describe books and cameras were selected. These two ontologies were found using the Google search engine. The search terms used to find these ontologies were 'Book filetype:owl' and 'Camera filetype:owl'. When selecting ontologies from Google list of results, the organization that provides the ontology is taken into account as a criterion for selection. In other words, we selected ontologies provided by well known bodies in the ontology area to ensure the quality of selected ontologies. The selected ontologies are the book ontology provided by daml.org (DAML, 2003) and the camera from the Protégé (Costello, 2006).

To evaluate the quality of automatic annotation results, precision (P) and recall (R) were used as the evaluation metrics. These two metrics are borrowed from Information Retrieval research and have proved to be effective in evaluating the performance of retrieval algorithms (Buckland & Gey, 1994).

In Information Retrieval research, P indicates the purity (cleanness) of retrieval and R denotes completeness of retrieval results (Buckland & Gey, 1994). These two measures are adopted in ontology research to measure the effectiveness of ontology matching approaches (Giunchiglia et al., 2009). In the ontology matching context, P indicates cleanness of similarity results and R means completeness of these results (Giunchiglia et al., 2009). The values of P and R fall in the range [0-1]. These metrics are defined in the following equations.

$$R = \frac{|\{A\} \cap \{M\}|}{|\{M\}|} \qquad (4)$$

where A is the total number of matches automatically found and M is the total number of matches manually found (see Figure 10).

Based on the previous definitions of P and R, one can understand that these two metrics require reference results (M) to compare against in order to find out the correct automatic results (Ehrig & Euzenat, 2005). Figure 11 presents the precision and recall values of the automatic annotation results.

This reference (gold standard) is usually obtained by performing manual matching or annotation. For the purpose of this evaluation exercise, results are obtained by manually annotating the given services using the selected ontologies. The manual annotation process is performed before conducting the automatic annotation tasks.

Using Figure 11, one can conclude that the obtained precision values range from 0.8 for the Store-Wss01 service to 1 for the Books, Camera and CameraPrice Web services. On the other hand, recall values range from 0.72 for the Camera Web service to 1 for the BookInfoPort, Store03 and Store-Wss01 services. The reason for having a low recall value for the Camera Web service test is that some service elements have labels that are not carefully structured. An example is the label "CameraAvailable" which should be named as "AvailableCamera". Nevertheless, the presented annotation results show that predominantly clean and complete results in relation to the manual results can be obtained using the proposed annotation approach.

Though useful, the values of precision and recall do not tell what percentage of the given service elements are annotated using the semi-automatic annotation framework. Therefore, to make this performance test more comprehensive, we calculate the percentage of correctly annotated elements out of the total number of elements of a given service. Figure 12 presents the resulting percentages.

Figure 12 shows that the percentage of annotated elements is moderate and ranges from 22% for the Camera service to 69% for the CameraPrice service. We call this problem the low percentage issue which, importantly, a result of two factors that are irrelevant to the proposed annotation approach:

Figure 10. The intersection between automatic and manual results

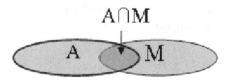

1. The coverage of domain ontologies selected to annotate the given services. In other words, the ontologies used for annotation do not contain correspondences for all elements. This is understandable since the chosen ontologies and services are designed by different parties and differences between them are to be expected. Furthermore, the Semantic Web is still a new research area and the quality of existing ontologies might not be very high. Therefore, we call for a better quality and more comprehensiveness of ontologies. Moreover, effective ontology extension techniques should be developed to aid annotation approaches to extend ontologies when appropriate correspondences cannot be found in these ontologies.

2. The number of service elements that cannot be annotated. Many elements of XSD definitions are syntactic definitions and thus cannot be annotated. Examples are the request and response patterns that define data types for message parts of WSDL operations. These definitions carry no semantic meaning and are usually generated automatically by WSDL generators such as Java2WSDL converters. These elements could be excluded from the annotation process however we include them to highlight their existence.

CONCLUSION

Web service annotation entails referencing Web service data and elements to appropriate

Figure 11. Precision and recall of the proposed annotation approach

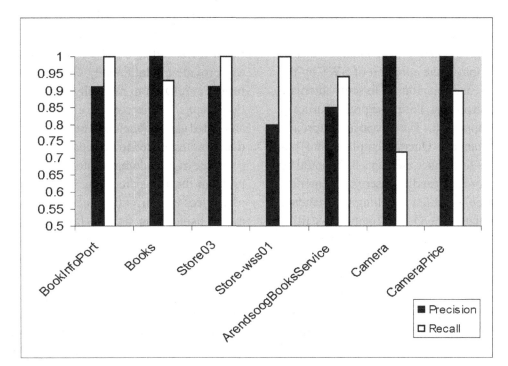

Figure 12. Percentages of annotated services' elements

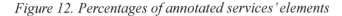
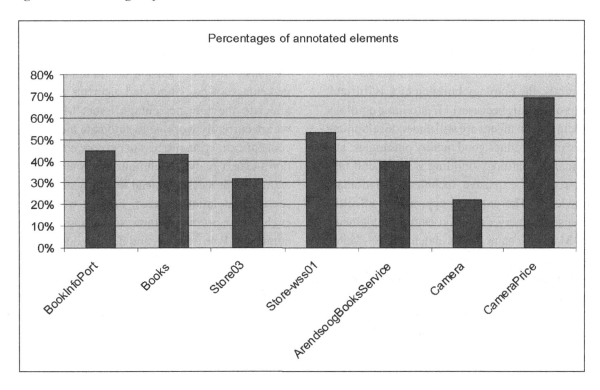

ontological concepts. This annotation process is currently manual and presents numerous issues for Semantic Web service (SWS) developers. Subsequently, automating the annotation task is desired to improve the adoption of SWS by the computer science community. To semi-automate the annotation process, this paper presents a novel annotation approach. The proposed approach utilizes a Standard Query Template that has place holders for service elements that should be annotated. For each candidate service element, a query instance is created by filling the Standard Query Template. The Query Template is filled with labels of candidate service elements and labels of their related concepts. These elements and related concepts are extracted from a service's WSDL file based on defined rules that exploit the WSDL structure. The resulting query instances are then executed using a novel query execution approach to find appropriate correspondences for candidate concepts. The query execution engine implements name-based matching and structural matching techniques in order to perform effective query execution.

The proposed approach is simpler and more effective than previous approaches. It is simpler because it does not demand building application ontologies from WSDL files of candidate services. Currently, the ontology building task is difficult and problematic because: (1) Automatic ontology building mechanisms are still under development and cannot produce good quality ontologies; and (2) manual ontology building is a difficult task that requires extensive technical and domain knowledge. The proposed approach is effective because its query execution engine implements CN-Match for name-based matching. CN-Match can measure similarities between labels containing Compound Nouns (CN) in an effective and precise manner.

The proposed semi-automatic annotation approach was evaluated by employing it to annotate existing Web services using published domain ontologies. Precision and recall are utilized as evaluation metrics. The resulting precision and recall values indicate that the proposed approach can provide predominantly clean and complete results in relation to manually produced annotation results. Some service elements cannot be annotated using the selected ontologies, however, due to issues related to candidate WSDL files and coverage of used domain ontologies. We call this problem the low percentage issue. WSDL files of services are currently generated by automatic means using tools such as Java2WSDL. These tools could assign meaningless names to many WSDL elements which are usually defined using meaningful XSD data types. When these meaningful data types are annotated, they can provide partial semantic descriptions for the non-meaningful elements that are defined using these data types. In addition, many existing ontologies that are used for annotation may have limited domain coverage therefore, these ontologies should be extended when appropriate correspondences for given service elements cannot be found.

Future research will concentrate on developing an ontology extension approach that can extend existing ontologies while maintaining their quality. This ontology extension should minimize the percentage of non-annotated elements and thus alleviate the low percentage issue. Moreover, we are currently working to automate the service concept extraction process using concept extraction and text analysis techniques. Automating the extraction process may lead to a fully automatic service annotation process. Finally, we aim to evaluate the proposed framework using a higher number of Web services and domain ontologies.

REFERENCES

Agarwal, S., Handschuh, S., & Staab, S. (2003). Surfing the service web. In *Proceedings of the 2nd International Semantic Web Conference* (pp. 211-226).

AL Asswad, M. M. (2011) Semantic Information Systems Engineering: A Query-based Approach for Semi-automatic Annotation of Web Services, Brunel University.

Buckland, M., & Gey, F. (1994). The relationship between recall and precision. *Journal of the American Society for Information Science American Society for Information Science*, *45*(1), 12–19. doi:10.1002/(SICI)1097-4571(199401)45:1<12::AID-ASI2>3.0.CO;2-L

Castano, S., Ferrara, A., & Montanelli, S. (2006). Matching ontologies in open networked systems: Techniques and applications. *Journal on Data Semantic*, *5*, 25–63.

Chifu, V. R., Salomie, I., & Chifu, E. S. (2007). Taxonomy learning for semantic annotation of web services. In *Proceedings of the 11th WSEAS International Conference on Computers*, Crete, Greece (pp. 300-305).

Costello, R. L. (2006). *The camera ontology.* Retrieved from http://protege.cim3.net/file/pub/ontologies/camera/

Curbera, F., Duftler, M., Khalaf, R., Nagy, W., Mukhi, N., & Weerawarana, S. (2002). Unraveling the web services web: An introduction to SOAP, WSDL, and UDDI. *IEEE Internet Computing*, *6*(2), 86–93. doi:10.1109/4236.991449

DAML. (2003). *Ontology of book.* Retrieved from http://www.daml.org

Duo, Z., Juan-Zi, L., & Bin, X. (2005). Web service annotation using ontology mapping. In *Proceedings of the IEEE International Workshop on Service-Oriented System Engineering* (pp. 235-242).

Ehrig, M., & Euzenat, J. (2005). Relaxed precision and recall for ontology matching. In *Proceedings of the K-CAP Workshop on Integrating Ontologies* (pp. 25-32).

Ehrig, M., & Sure, Y. (2005). *Test ontologies.* Retrieved from http://www.aifb.uni-karlsruhe.de/WBS/meh/foam/ontologies.htm

Euzenat, J. (2004). An API for ontology alignment. In S. McIlraith, D. Plexousakis, & F. van Harmelen (Eds.), *Proceedings of the Third International Semantic Web Conference* (LNCS 3298, pp. 698-712).

Euzenat, J., & Shvaiko, P. (2007). *Ontology matching.* New York, NY: Springer.

Feier, C., Roman, D., Polleres, A., Domingue, J., & Fensel, D. (2005). Towards intelligent web services: The web service modeling ontology. In *Proceedings of the International Conference on Intelligent Computing,* Hefei, China. (pp. 1-10).

Girju, R., Moldovanb, D., Tatub, M., & Antoheb, D. (2005). On the semantics of noun compounds. *Computer Speech & Language*, *19*(4), 479–496.

Giunchiglia, F., Yataskevich, M., Avesani, P., & Shvaiko, P. (2009). A large dataset for the evaluation of ontology matching. *The Knowledge Engineering Review*, *24*(2), 137–157. doi:10.1017/S026988890900023X

Hepp, M. (2006). Semantic web and semantic web services: Father and son or indivisible twins? *IEEE Internet Computing*, *10*(2), 85–88. doi:10.1109/MIC.2006.42

Heß, A., Johnston, E., & Kushmerick, N. (2004). ASSAM: A tool for semi-automatically annotating semantic web services. In *Proceedings of the Semantic Web Conference* (pp. 320-334).

Heß, A., & Kushmerick, N. (2003). Learning to attach semantic metadata to web services. In *Proceedings of the Semantic Web Conference* (pp. 258-273).

Jacek, K., Tomas, V., Carine, B., & Joel, F. (2007). SAWSDL: Semantic annotations for WSDL and XML schema. *IEEE Internet Computing, 11*(6), 60–67. doi:10.1109/MIC.2007.134

Kim, S. N., & Baldwin, T. (2005). Automatic interpretation of noun compounds using WordNet similarity. In *Proceedings of the 2nd International Joint Conference on Natural Language Processing* (pp. 945-956).

Lara, R., Roman, D., Polleres, A., & Fensel, D. (2004). A conceptual comparison of WSMO and OWL-s. In *Proceedings of the European Conference on Web Services* (pp. 254-269).

Lauer, M. (1995). Designing statistical language learners: Experiments on noun compounds (Doctoral dissertation, Macquarie University). Cornell University Library, 1, 226.

Lerman, K., Plangrasopchok, A., & Knoblock, C. A. (2006). Automatically labeling the inputs and outputs of web services. In *Proceedings of the National Conference on Artificial Intelligence* (pp. 1363-1368).

Martin, D., Burstein, M. H., McDermott, D., McIlraith, S. A., Paolucci, M., & Sycara, K. (2007). Bringing semantics to web services with OWL-S. *World Wide Web (Bussum), 10*(3), 243–277. doi:10.1007/s11280-007-0033-x

McIlraith, S. A., Son, T. C., & Zeng, H. (2001). Semantic web services. *IEEE Intelligent Systems, 16*(2), 46–53. doi:10.1109/5254.920599

Miller, G. A. (1995). WordNet: A lexical database for English. *Communications of the ACM, 38*(11), 39–41. doi:10.1145/219717.219748

Paolucci, M., Kawamura, T. R., Payne, T. R., & Sycara, K. (2002). Importing the semantic web in UDDI. In *Proceedings of the International Workshop on Web Services, E-business and the Semantic Web,* (pp. 815-821).

Patil, A., Oundhakar, S., Sheth, A., & Verma, K. (2004). Meteor-s web service annotation framework. In *Proceedings of the 13th International Conference on World Wide Web* (pp. 553-562).

Rajasekaran, P., Miller, J., Verma, K., & Sheth, A. (2005). Enhancing web services description and discovery to facilitate composition. In *Proceedings of the First International Workshop on Semantic Web Services and Web Process Composition* (pp. 55-68).

Roman, D., Keller, U., Lausen, H., de Bruijn, J., Lara, R., & Stollberg, M. (2005). Web service modeling ontology. *Applied Ontology, 1*(1), 77–106.

Sivashanmugam, K., Sheth, A., Miller, J., Verma, K., Aggarwal, R., & Rajasekaran, P. (2003). *Metadata and semantics for web services and processes* (pp. 1–19). Datenbanken Und Informationssysteme. [Databases and Information Systems]

Staab, S. (2003). Web services: Been there, done that? *IEEE Intelligent Systems, 18*(1), 72–85. doi:10.1109/MIS.2003.1179197

Sycara, K., Paolucci, M., Ankolekar, A., & Srinivasan, N. (2003). Automated discovery, interaction and composition of semantic web services. *Web Semantics: Science . Services and Agents on the World Wide Web, 1*(1), 27–46. doi:10.1016/j.websem.2003.07.002

Verma, K., & Sheth, A. (2007). Semantically annotating a web service. *IEEE Internet Computing, 11*(2), 83–85. doi:10.1109/MIC.2007.48

Wu, Z., & Palmer, M. (1994). Verbs semantics and lexical selection. In *Proceedings of the 32nd Annual Meeting on Association for Computational Linguistics*, Las Cruces, NM (pp. 133-138).

Zhang, M., Duan, Z., & Zhao, C. (2008). Semi-automatically annotating data semantics to web services using ontology mapping. In *Proceedings of the 12th International Conference on Computer Supported Cooperative Work in Design* (pp. 470-475).

Chapter 9

A Broken Supply and Social Chain:
Anatomy of the Downfall of an Industrial Icon

Yanli Zhang
Montclair State University, USA

Ruben Xing
Montclair State University, USA

Zhongxian Wang
Montclair State University, USA

ABSTRACT

There was a time in the history of GM when it was the largest corporation in the US. The history of GM also shows that it was the single largest employer in the world. The announcement of GM's bankruptcy on June 1, 2009 shocked the world and had a tremendous impact on the United States economy. Looking back at the history of GM, there were many indicators which suggested the fate of the company. There were several internal factors that answer the question, what went wrong with GM. These internal factors are management arrogance, not meeting customer demands, the costs and demands of unions, poor forecasting, and internal controls on accounting standards. Similarly, there were several external factors that answer the same question, which include increased competition and loss of market share, rising gas prices and environmental friendliness, and the costs and burdens of meeting government regulations and restrictions. This paper will explore and answer the following questions: What are the fundamental causes of GM's problems? What can be learned from GM's mistakes and experiences? How and why an industrial icon came to ruin?

DOI: 10.4018/978-1-4666-2922-6.ch009

ORGANIZATION BACKGROUND

The history of GM, the world's largest automaker, began at the dawn of the twentieth century. The company was founded by William Durant in the year 1902. The shrewd businessman that he was, Durant realized that the future lay with cars and not carriages. Initially, the company was founded as a holding company for Buick. The latter part of the year was when the company's growth began by acquiring Oldsmobile, followed by the possession of Cadillac, Oakland and Elmore in the very next year.

Many of the motor companies were in dire straits during the difficult years of the early 1900's. The stock market panic in 1907 put a lot of small companies into financial distress. Many of these companies were running on credit from various bankers. This was a golden opportunity for Durant, who proceeded to buy smaller car builders and companies that built car parts as well as car accessories. In 1908, these various companies were folded into a single unit, thus creating the new GM entity. This marked the exciting beginning of the true history of GM.

William Durant was a flamboyant businessman whose curious mix of genius and over-reaching took GM both to its heights as well as plunged it into financial distress. In 1910, bankers were forced to step in to prevent a financial collapse of GM, and Durant was removed from the company he had founded. But by 1911, the company had made enough advances into the international market that the General Motors Export Company was established to handle sales outside the U.S and Canada (Keller, 1989).

Durant managed to use another company he formed, Chevrolet, to come back to power in GM during 1915. The history of GM from 1915-1920 was full of successes as a result. During this time, the Cadillac became wildly popular. In 1918, GM bought the operating assets of Chevrolet Motors. Shortly after, America was hit by a power reces-

sion in 1920 and Durant again found himself out of the company.

During the financial boom in the 1920's, the history of GM virtually glowed with success. Auto sales reached the 4.5 million mark, and the auto industry now had three giants - GM, Ford and Chrysler. GM now had a brilliant engineer turned industrialist at its helm. Alfred Sloan, who was later acclaimed for his marketing genius, had slowly worked his way up among the ranks of GM. It was his marketing genius that breathed a fresh lease of life into GM which was beginning to get overshadowed by Ford.

Ford's philosophy of giving the public the best value for their money offered little variety. But Sloan and GM were interested in providing the public with more than a black box. Stylish colors, features and comfort became the new motto of the company. GM also made a path-breaking offer - the public could now buy a car on credit. The five brands of GM - Pontiac, Cadillac, Buick, Oldsmobile and Chevrolet began changing every year with the focus being directed mainly at looks and style. This strategy paid rich dividends. Ford was pushed to the backseat again by GM.

The great Wall Street crash in 1929 put an abrupt stop to all expansion plans at GM for the time being. Stocks of GM fell rather badly. But, by the early 1930's GM bounced back and bought the Yellow Coach bus company and the Electro-Motive Corporation, the internal combustion engine railcar builder. The next 20 years saw GM powered diesel locomotives running on American railroads. December 31, 1955 is another landmark in the history of GM, when it became the first company to make more than a billion dollars in a single year.

SETTING THE STAGE

In the early twentieth century, GM grew at an incredible rate as well as created a powerful union and steadily had a growth of employment.

Union workers were paid handsome wages, often viewed as overpaid for a line worker. Post World War II, GM has had an uphill battle to stay in the automobile industry as a company that produces quality products and steadily claimed its place as the leader in the industry. Conversely, GM has run aground and has been financially bailed out by the Obama administration.

GM filed for Chapter 11 bankruptcy in 2009. They are the largest industrial company to ever file for bankruptcy as well as the fourth largest company to ever file for bankruptcy in the United States. The federal government has a majority stake in GM since the bailout has been completed. In addition, GM has close to 80 billion dollars in assets and double that amount of debt.

Some claim that GM could have seen some of these issues coming fifty years ago. For example, GM often took the approach that their products were simply good enough for the consumers. Therefore, the company was not furthering efforts to compete with other leaders in the industry or listen to the demands of customers. At the same time, Japan was solely focused on producing quality at an efficient rate. Stephenson (2008) states, "By the 70's and 80's, build quality was GM's weakest point. Fit-and-finish became the standard used for measurement against the Japanese competition" (p. 2). Consumers now place a high value on Japanese car makers, which polls confirm by showing steep and relentless declines in American opinions of GM and Chrysler, their workers, their unions, their Cadillac wages and benefits, their bailouts and their favored status with the Obama administration. Similarly, a recent report shows that GM had a favorable opinion from only 32%, down from 69% two years ago (Buss, 2009). This rapid decline in public opinion may be a result of GM's faulty marketing and erroneous assumptions of consumer values over the past several decades. GM's neglect to understand customer demand goes hand in hand with increased competition as factors in their significant decline in favorability.

CASE DESCRIPTION

The stage for the demise of GM was forty years in the making. For far too long, GM had been ignoring factors that are often directly responsible for the success or failure of any corporation. The automobile industry within the United States has grown tremendously since the 60's with several new competitors entering the market. The biggest portion of competition arose from the Japanese automakers. Toyota, Nissan, Honda, Mitsubishi, Mazda, and Subaru are some of the major players that GM has had to contend with for the past five decades. GM needed to become more dependent on the design and production of automobiles that the public wanted to buy. In order to do this, GM would have to stay in close touch with its current consumer base, as well as try to see what the consumer around the corner would be interested in and understanding why. The United States auto company would have to understand its customer's wants, dislikes, likes, lifestyles, pain points, etc. GM wound up being a weak market competitor in the realm of taking interest in what its customer base was looking for.

Specifically, one of the great transformations in the 1970s, 1980s, and 1990s was the switch from a company-driven economy to a customer-driven economy. The post World War II era in America was one of much prosperity being driven by companies developing products for consumers craving such products. This push economy made many companies complacent, of which the automotive industry is a poster child example for. All these companies believed they had to do was put products on the shelves and the consumer would buy them. While seeking to differentiate drivers based on styling, the car companies never looked closely at the needs of their customers-until the Japanese car companies came along and started thinking not about what cars they could produce, but about what cars consumers wanted to buy.

INTERNAL FACTORS

Arrogant Leadership

The executives at General Motors hold most accountability for the internal factors that led to its bankruptcy. With better forecasting and more humility among other internal factors, management could have perhaps taken better stances on the external issues affecting the company. GM is known for thinking whatever is good for GM is good for the United States and was even articulated by a former CEO (Ou, Massrour, & Noormohamed, 2009). It is arguable to state that success breeds insularity, arrogance, and ultimately hubris, which is precisely what happened with what was once America's largest employer, GM. GM was plagued with extreme bureaucracy where even the minor decisions were determined by several committees. GM also inherited a self-defeating management style developed during its overwhelmingly successful times. They assumed that superior management would always anticipate and control change. By contrast, many top managers in younger companies accepted that disruptive surprises are always part of the business environment and they could, unless successfully countered, handle and overcome them. The difference has consistently left GM in the lurch. Its latest cut backs are the third since the beginning of the 1980s. With each downsizing, GM has had to fight to catch up with changes that it badly misjudged -- the demand for smaller cars in the late 1970s; the superior quality and production techniques of Japanese manufacturers in the 1980s; and now the demand for sharper looking cars as well as ones with better fuel efficiency. The conceit that GM could manage change often served as an excuse to stand firm, until change was unavoidable.

Furthermore, GM's management focuses on the company's bottom line, business performance, and profitability. George "Rick" Wagoner, Jr. of GM exemplifies the organizational only perspec-

tive seeking rewards in his statement that, "we [GM] believe sustainable development can be a competitive advantage for us, if we move fast and take a leadership role in applying the principles and lessons of sustainability in all facets of our business around the world" (Weber, 2010, p. 17). The arrogance of GM's management is why they consistently focus on the company and how they will benefit or avoid harm. In contrast, Toyota's focus is on how the consequences of its actions affect personnel and society. It is imperative that GM adapts to thinking similar to Toyota if they want to remain at the top of the automobile industry.

The more prolific example of GM's bad management comes from Rick Wagoner, CEO, Chairman, and president from June 2000. Under Wagoner's leadership, GM suffered terrible losses due to his lack of decision making skills under pressure. He was also criticized for his lavish living during the hearings for government loans. In total, GM incurred more than 85 billion in losses while Wagoner was in charge. He was finally ousted by the US government in late March of 2009 and replaced by Chief Operating Officer, Fritz Henderson (Taylor, 2010).

GM rewarded people who followed the old way of doing things and those who challenged that thinking found themselves on the outs -- causing them to lose opportunities for promotion. So the smart thing for those seeking promotion within GM was to praise the CEO's wisdom and carry out his orders.

Unhappy Customers

GM needs to better understand and manage brand loyalty to best battle competition. Long replacement cycles for buyers create challengers for automobile manufacturers like GM to ensure consumers will repeat their purchase within the same company franchise when replacement time occurs. These repeat purchases within the same company exemplify brand loyalty. If GM systematically tracks brand loyalties of its brands over time, they may find early warning indicators about downward trends in some brand strengths and determine which competitors are primarily stealing sales away (Che & Seetharaman, 2009). Once again, understanding its customers plays an important factor in maintaining loyalty and possibly what will sway a person from one brand to another.

Kiley and Welch (2009) clearly articulate the importance of meeting customer demands by stating, "In the coming years, automakers will compete for the next generation of American drivers, 73 million 21- to 33-year-olds who have shown little inclination to buy Detroit" (p. 28). This is a huge battle for GM because their products are not popular amongst generation Y. If today's younger generation cannot see their friends or peers in GM manufactured cars, they certainly will not be able to see themselves driving them. The arrogance of GM's management coincides and even drives the lack of focus on the customer's needs and wants. Just because GM launches a vehicle, customers should not be expected to like, let alone purchase it. The customers' voices must be heard and demands met for GM to prosper and improve sales. In order to reach equilibrium and achieve optimization, GM needs to supply enough cars to meet the demand of customers and the customer need to demand GM cars. Since the demand for GM cars is low, it costs GM more than it should to produce each vehicle, particularly the gas guzzling SUVs in a present environmentally friendly society plagued with rising gas prices.

"Generous Motors"

Prior to competition creating more efficiency in the marketplace, GM had the reputation of mistreating its workers in the 1930s. This abuse would eventually lead to the development of the United Auto Works union (UAW). The UAW would organize strikes and orchestrate production slowdowns in protest. They would negotiate contracts, win wage increases, and gain several additional benefits. Benefits included fully paid healthcare, legal advice, child care, vacation, education benefits, pensions, and job security. Such negotiations allowed for low income families to move into a higher level of lifestyle. The UAW saw no problem in utilizing their numbers to realize much higher benefits from its employer. Soon, however, it would add to the demise of GM as the extensive benefits being paid out, were not economical as it related to remaining competitive. Prior to the 1960's, GM had such a large market share that they were able to pass the extensive cost of their auto works on to the consumer. The company, which in the 1980s employed more than 800,000 people worldwide, brought labor peace by giving its union workers lavish pay and benefits packages. From this point on, GM became known as "Generous Motors" as a result of their generosity towards its union workers. Thus, it was these labor contracts that aided the crippling of the company's ability to cut costs.

However, when Toyota, Honda, Mitsubishi, and Mazda began offering their products to the American consumer and taking substantial market share, the UAW became a disadvantage. Most Japanese transplants are located in anti-union states and use younger, less expensive workers (Holstein, 2009). GM also carries a heavy amount of legacy costs for past employees. Full benefits are still being paid and continue to offer a competitive disadvantage for the American auto company (Tong & Tong, 2009). Lastly, clear examples of the negative cost impacts unions have on GM is a comparison to foreign automobile industry employees. The average hourly wage and benefits for GM workers is $72.21 compared to Toyota and Honda's employees average $48 in hourly wage and benefits. In addition, union contracts and pensions boost the cost of each American-made vehicle by approximately $2,300 per vehicle (Edmund, 2009).

Poor Forecasting

Lack of proper forecasting also contributed to the demise of GM. Forecasting includes several different factors including sales, demand, and products. GM failed to plan either. With such a changing economy, it is baffling why such a strong company would not take the proper measures to assure continued success. Hartung (2009) states, "GM never undertook scenario planning to help it anticipate market shifts" (p. 8).

Around 1970, new car prices were doubling every six years and GM seemed to completely ignore the demand for economy cars. This made it fairly easy for Japanese automakers to enter the US market seizing many first time buyers. Also, the rise in oil prices did not come as a huge surprise. Energy costs in general were well on the rise with oil consumption exceeding domestic production leading to increases in home fuel, oil, and electricity. After 2004, gas prices had nearly reached four dollars a gallon! Had they foreseen this steady rise in gas prices and market shift towards smaller vehicles perhaps GM's story would have been different.

But forecasting does not end there. After spotting the possible trend, it is the company's responsibility to enact changes that align with the new trend. With the rising gas prices, GM should have taken focus off of their sport utility vehicles and pickups. Hartung (2009) comments that GM was blatantly reluctant to move from big profitable vehicles to building the smaller cars. It would have been less profitable to invest in smaller, energy efficient cars but the results would not have been as devastating.

Forecasting proves to be very crucial for successful businesses. It helps manage inventory, cash, and expenses. Failure to evaluate the economic environment of the auto industry cost GM a lot. Their unwillingness to change production in alignment with current demands caused them to fall behind, and lose over 50% of their market share.

Casual Controls

Alfred P. Sloan was to follow with a widely criticized accounting system for inventories. His system valued inventories the same as cash, therefore resulting in a large build up of excess inventory and the impression of more cash (Waddell & Bodek, 2005). He served as president, chief executive officer, and chairman until 1956.

Due to numerous accounting scandals that stunned corporate America at the turn of the 21st century, the U.S. Government passed the Sarbanes-Oxley Act of 2002 (SOX). This act established standards for all public company boards, management, and public accounting firms in the United States. These controls in turn give publicly traded companies a much greater understanding of internal controls and the need for such controls. GM identified and detailed their material weaknesses including actions to address previously reported material weaknesses that no longer exist at December 31, 2008. In the 2008 annual report filed with the SEC, Deloitte & Touché LLP revealed doubt about GM's ability to generate sufficient cash flow to meet obligations and sustain its operations and thus continue as a growing concern (Pineno & Tyree, 2009). Furthermore, MLC (Motors Liquidation Company, the new name of the "Old GM") determined that its internal controls over financial reporting were not effective. It is this lack of effective internal controls that could materially adversely affect their financial condition and ability to carry out their business plan (Blumer, 2009). GM's failure to have a handle on things when the government started bailing them out will certainly affect their ability to be competitive in the long-run relative to efficiency, reliability, credit, and decision-making. The company made a statement in its annual 10-K report of results for 2006, filed with the Securities and Exchange Commission. GM said in the filing that management has recognized the problems and is taking steps to correct them. It acknowledged that the company had not maintained a sufficient complement of personnel with an appropriate level of technical accounting expertise and said it would make greater use of outside experts while it added staff with that necessary level of knowledge, but it was not to be. A company as large as GM must have financial responsibility if it wants to stay at the top of its industry. Currently, it is arguable whether or not GM in fact has the financial conditions necessary to be the top automobile manufacturer.

EXTERNAL FACTORS

Severe Competition

The influx of foreign cars into the US market has had huge ramifications for the Big Three (Ingrassia, 2010). In the early to mid-1980s, consumers perceived that Japanese manufacturers were building superior cars, which on average had 1.5 defects per car while U.S. automakers had three defects per car (Edmund, 2009). The quality of both its trucks and cars did not rival that of its main Japanese competitors, Toyota and Honda. Consumer Reports magazine has a rating system that they use to compare competitors' products, which published that over the past twenty five years overall consumers consistently felt the Japanese automakers offered better quality for the money. They felt that GM had lower quality safety features, their cars and trucks would break down more easily, require repair services more often, and would take more maintenance time out of their schedules. In contrast, consumers felt that Japanese auto makers had better features geared toward safety, did not break down as easily or often, and offered overall better quality and less headaches. American manufacturers, like GM, are inconsistent in the main focus categories of quality, reliability and styling-and as a result, repeatedly leave their customers heartbroken from the lack of understanding and respecting them. In turn, millions of customers, loyal to GM

for generations, finally got tired of tinny doors, keys that didn't fit both the door and the trunk, and instrument panels that simply looked cheap. Despite the improvements, vows, and promises that GM has made, the list of flaws in its cars continues to exceed that of its rivals.

Foreign companies like Toyota and Nissan are poised to dominate the mainstream, small- to medium-sized car market with their flexibility, cheaper prices, and (in Toyota's case), early investments in fuel efficiency (Dewar, 2009). In an attempt to gain market share, companies such as Honda, Toyota, Mitsubishi, Mazda and Subaru continued to think and act on new mechanisms to minimize production time and effort in order to maximize profits. GM did not. GM's leaders believed every automobile manufacturer wanted to be like them and therefore were not a threat. They ignored the small car manufacturers because of the misconception that they could not match GM's resources or capabilities. In such a dynamic, competitive environment, any corporation that desires to remain a force to be reckoned with must continuously think about whether or not it can better its production methodology. The foreign manufacturers developed a lean manufacturing system that was completely different from the mass-assembly-line techniques GM was still using. GM's fractured structure meant that each division had its own manufacturing processes, its own parts, its own engineering, and its own stamping plants. This meant significant procedure fragmentation, cost, and efficiency compared to its ever growing competition. Even large organizations must adhere to changes in their marketplace. Complacency will cripple even an enormous corporate giant such as GM.

Back in 2005, Toyota cars topped the initial quality surveys year after year. And people were willing to pay for that quality. For example, Toyota charged 14 percent more for their average vehicles ($24,500) than GM ($21,000) according to various consumer reports (Dewar, 2009). Toyota also built cars faster and at lower cost. Toyota could build a car 7 percent faster (in 21.63 hours) than GM (23.09 hours). Moreover, Toyota enjoyed a $300 to $500 per vehicle cost advantage over GM. Part of this advantage was in health care costs. While GM complained about its $1,500 per vehicle health care charge, Toyota made most of its cars in countries where governments picked up much of the health care bill.

Even though it charged higher prices, demand for its cars was so high in 2005 that Toyota was operating at full capacity and it appeared poised to take over the North American market share lead in 2006. Meanwhile, Toyota earned an average of $1,488 per vehicle in profit, while GM lost $2,300.

On average, about fifty percent of all Americans are loyal consumers and buy the same brand as last time. The comparatively small variation in retention performance between brands of vastly different market share is largely due to some brands being easier to buy repeatedly and some not. The larger the market share, the more dealers, better geographic coverage, broader product rangers, and larger advertising budgets which all lead to higher consumer retention. GM's lack of smaller model cars has hindered its retention and eroded its market share (Sharp, 2009).

Relentless Environment

Another cause of GM's failure that was more or less out of their control was the intense cost of gasoline. Gasoline prices hit a national average of four dollars in the second quarter of 2008. Also, the global financial crisis of 2008 caused GM's vehicle sales to drop extensively. There was speculation that GM was caught completely off guard by the significant spike in oil prices and therefore, did not possess a backup plan when truck sales cratered and ultimately destroyed their business model for North America. The macroeconomic forces at work were no fault of GM's, but, already heavily leveraged, the company had left itself no room to maneuver. Former CEO Rick Wagoner suggested that it was the fuel prices along with

other factors led to the bankruptcy of GM; however the development of their demise started over half a century ago. Their business model was a catastrophe waiting to happen; the credit and fuel crisis was simply the straw that broke the camel's back (Tong & Tong, 2009). GM's management arrogance compared to the younger management of foreign companies who understood disruptive surprises are a part of business was the controllable factor in adapting to the unexpected. Always be prepared was not a motto GM understood well.

Asian automobile manufacturers insightfully reacted to the rising gas prices by creating green cars. As a result of the escalating interest in the environment and continuing oil price spikes, the number of consumers purchasing green cars has been on the rise and the Asian auto manufacturers were able to reap the rewards by meeting the growing consumer demand. Specifically, Toyota's Prius is already on its third generation and GM is just launching its first green car, the Chevrolet Volt. Since the United States plans to intensify current regulations by introducing the 35.5 miles per gallon rule, which is a 42 percent increase, by 2016, American auto manufacturers have to produce vehicles to meet this ever growing demand if they want to remain competitive (Lim, 2010). Competition continues to release green cars such as hybrid electric vehicles (HEV), plug-in hybrid electric vehicles (PHEV) and electric vehicles (EV). GM needs to follow suit, which will create more relevant technologies that competition often brings.

Intense Regulations

The United States government imposes several regulations and requirements on companies, many of which can often negatively impact a company. One example is the federal income tax, which GM paid $37.16 billion for in 2007 versus Toyota's payment of $7.61 billion (Edmund, 2009). Another example is all of the safety standards required by the National Highway Traffic Safety Administration (NHTSA). As early as 1916, U.S. courts recognized that motor vehicle manufacturers have an obligation to avoid defective designs or materials that may cause a crash. Fast forwarding to the 1960s, magistrates extended this rule of avoiding defective designs or materials to failure to include safety devices that might minimize the risk or severity of injury when a crash occurs. GM was one of the first automobile manufacturers to feel the pain of this law and responsibility. In Larsen vs. GM, a federal appellate court applied this doctrine of crashworthiness to injuries caused when the steering column of a 1963 Chevrolet Corvair moved rearward during a crash and struck the driver's head (Vernick, Rutkow, & Salmon, 2007). The decision ruled in favor of the plaintiff based on the assumption that an automobile manufacturer is under no duty to design an accident-proof or fool-proof vehicle, however such a manufacturer is required use reasonable care in the design of its vehicle to circumvent the driver to an unreasonable risk of injury in the event of a collision. It was around this time that Congress enacted the National Traffic and Motor Vehicle Safety Act of 1966, which publicized safety standards for automobiles. Today's minimum safety standards include requiring seat belts, air bags, and conspicuous brake lights. The more standards that the government mandates increases costs for automobile manufacturers like GM. Labor, production, parts, research and design, and time to build each car are affected with each and every regulation.

Moreover, in a proactive effort to minimize the threat of lawsuits, GM routinely provides greater safety features than the standards require. For example, before air bags were mandatory, numerous lawsuits were filed by people injured in crashes of cars without air bags claiming that their injuries would have been less severe had the car been equipped with air bags. Thus, GM is investing significantly in reinventing its vehicles with emphasis to further improve safety.

CURRENT CHALLENGES/ PROBLEMS FACING GM

As of the week of April 19th, 2010 GM announced it has repaid the $6.7 billion in loans from the United States and Canadian governments in full with interest years ahead of schedule. Establishing a pristine balance sheet was the ultimate goal of the giant automaker, which led to financial and political decision making. The government bailout was almost $60 billion, of which GM is only obligated to pay back $6.7. Nearly 90% of the bailout does not need to be paid back and the company's credibility is slightly tarnished because Ed Whitacre, chairman and CEO of GM, neglected to mention this in his statement that GM has paid back its debt almost four years early (Ingrassia, 2010). Getting the remaining $52 billion back from GM requires an initial public offering, which means convincing the investing public to buy the 73% of the company's stock owned by the U.S. and Canadian governments. The other 27% is owned by the United Auto Workers union and by bondholders in the old company. In an effort to raise the remaining $52 billion, GM is curtailing its habit of producing too many cars and selling them with discounts that exhausted all the profits. However, the company lacks marketing clout and a strong dealer network, so its U.S. market share has been perpetually stuck between 2% and 3% (Ingrassia, 2010).

GM has lost the trust of the general public. Although GM has paid back the $8.1 billion received in 2007 from the government as part of the TARP program, taxpayers will continue to perceive the company as the enigma of corporate irresponsibility until it proves that it can produce cars that people want to buy, take control of its labor forces, earn its way out of government intervention, and adequately adapt to competitive demands (Talton, 2010).

CONCLUSION

GM has been leading the automotive industry for a very long time. Arising out of the ashes of the great depression it became one of the largest companies in the world, creating jobs, and boosting up the American economy. Over the years, GM faced a number of financial hurdles, and these past issues had an impact on the status of GM today. GM's troubles can be attributed to a myriad of internal shortcomings and challenging external forces. The decline of the U.S. automobile market, rising fuel prices, high pension costs, and a failure to innovate and develop products to meet consumer needs, were all factors that led to the failure of GM.

Many argue that if GM responded differently to these issues, especially when it came to handling employee benefits and compensation, their present financial troubles could have been avoided. Others state that powerful forces such as fuel costs and economic recessions were too vast for even General Motor to avoid. Regardless of these varied opinions, it is critical to look at General Motor's past in order to fully understand their present situation.

GM's failure after 101 years is an indictment of American management in general. It highlights the damage to our economy that results when finance becomes the tail that wags the economic dog. And it shows what happens to any company that rests on its laurels and fails to adapt to change.

REFERENCES

Blumer, T. (2009). *Story on new GM, with AP help, buries news about financial non-disclosure, unique risks.* Retrieved from http://newsbusters.org/blogs/tom-blumer/2009/08/10/ap-story-new-gm-ap-help-buries-news-about-financial-non-disclosure-uniqu

Buss, D. (2009). Museums: Rise and fall of a car town. *Wall Street Journal (Eastern Edition)*, p. D7.

Che, H., & Seetharaman, P. (2009). Speed of Replacement: Modeling brand loyalty using last-move data. *JMR, Journal of Marketing Research*, *46*(4), 494–505. doi:10.1509/jmkr.46.4.494

Dewar, R. J. (2009). *A savage factory: An eyewitness account of the auto industry's self-destruction* (pp. 74–98). Bloomington, IN: AuthorHouse.

Edmund, M. (2009). The big three: Will a bailout be enough? *Quality Progress*, *42*(1), 14–15.

Hartung, A. (2009). *The fall of GM: What went wrong and how to avoid its mistakes* (pp. 1-15). Upper Saddle River, NJ: Financial Times Press.

Holstein, W. (2009). *Why GM matters: Inside the race to transform an American icon*. New York, NY: Walker Books.

Ingrassia, P. (2010, April 23). Two cheers for General Motors; Yes, it's paid back $6 billion and is more efficient. It still owes taxpayers about $52 billion. *Wall Street Journal (Online Edition)*, p. A17.

Keller, M. (1989). *Rude awakening: The rise, fall, and struggle for recovery of General Motors*. New York, NY: William Morrow.

Kiley, D., & Welch, D. (2009). The hard road ahead for government motors. *Business Week, 28*.

Lim, T. (2010). Rapid emergence of rechargeable car battery market. *SERI Quarterly*, *3*(2), 23–29.

Ou, Y., Massrour, B., & Noormohamed, N. (2009). Putting the pedal to the metal: Forces driving the decision-making process toward American-made vehicles by consumers in Taiwan, China, and Thailand. *Competition Forum*, *7*(2), 343–353.

Pineno, C., & Tyree, M. (2009). The changing public reports by management and the auditors of publicly held corporations: An updated comparative study of General Motors Corporation and Ford Motor Company. *Competition Forum, 7*(2), 465–472.

Sharp, B. (2009). Detroit's real problem: It's customer acquisition, not loyalty. *Marketing Research, 21*(1), 26–27.

Stephenson, S. (2008). What went wrong with GM took half a century. *Motor Age, 124*(8), 2.

Talton, J. (2010, April 21). *GM's challenge isn't repaying debt but making sexy cars; worry about the second half of '10.* Retrieved from http://seattletimes.nwsource.com/html/soundeconomy-withjontalton/2011662521_gms_challenge_isnt_replaying_d.html

Taylor, A. III. (2010). *Sixty to zero: An inside look at the collapse of General Motors--and the Detroit auto industry* (pp. 31–34). New Haven, CT: Yale University Press.

Tong, C., & Tong, L. (2009). GM: Problems, solutions and lessons. *Competition Forum, 7*(1), 136–141.

Vernick, J., Rutkow, L., & Salmon, D. (2007). Availability of litigation as a public health tool for firearm injury prevention: Comparison of guns, vaccines, and motor vehicles. *American Journal of Public Health, 97*(11), 1991–1997. doi:10.2105/AJPH.2006.092544

Waddell, W., & Bodek, N. (2005). *Rebirth of American industry - a study of lean management.* Vancouver, WA: PCS Press.

Weber, J. (2010). Assessing the "Tone at the Top": The moral reasoning of CEOs in the automobile industry. *Journal of Business Ethics, 92*(2), 167–182. doi:10.1007/s10551-009-0157-2

This work was previously published in the International Journal of Information Systems and Social Change, Volume 2, Issue 2, edited by John Wang, pp. 55-65, copyright 2011 by IGI Publishing (an imprint of IGI Global).

Section 3
Informatics and Semiotics in Organisations

Chapter 10
Actability Criteria for Design and Evaluation:
Pragmatic Qualities of Information Systems

Göran Goldkuhl
Linköping University & Stockholm University, Sweden

ABSTRACT

Information systems actability theory builds on a communicative action perspective on IS. Information systems are seen as instruments for technology mediated work communication. Human actors are communicating (i.e. sending and/or receiving messages) through an information system. Information systems actability emphasises pragmatic dimensions of information systems. The paper presents 19 actability criteria divided into three groups: 1) criteria concerning user-system interaction, 2) criteria concerning user-through-system-to-user communication, and 3) criteria concerning information system's contribution to workpractice processes. These actability criteria should be possible to use in design and evaluation of information systems.

INTRODUCTION

Information systems (IS) cannot be seen just as repositories of facts of the world. An IS is a communicative instrument in organisations. Actors can perform communicative actions by support of an IS. An IS is thus a mediator of communication and action between organisational actors. An IS is also an agent with capabilities to perform predefined communicative actions. This gives IS

a dual role of an instrument for users and an agent interacting with users. These roles raise demands on pragmatic qualities of information systems.

Information systems actability theory (ISAT) is a conceptualisation of information systems emphasising their pragmatic dimensions. It can be seen as a practical theory according to this notion by Cronen (2001). Practical theories have a function of *directing* actors' *attention* towards certain types of phenomena. Cronen describes practical theories in the following way: "Its use

DOI: 10.4018/978-1-4666-2922-6.ch010

should, to offer a few examples, make one a more sensitive observer of details of action, better at asking useful questions, more capable of seeing the ways actions are patterned, and more adept at forming systemic hypotheses and entertaining alternatives" (Cronen, 2001, p. 30). Goldkuhl (2008a) has elaborated the notion of practical theory and divided it into several constituents: Conceptualisation, patterns, normative criteria, design principles and models.

ISAT has evolved over several years with contributions from many scholars (e.g. Ågerfalk, 2003; Ågerfalk & Eriksson, 2004; Broberg, 2008; Cronholm & Goldkuhl, 2002; Goldkuhl 2008b, 2009; Goldkuhl & Ågerfalk, 2002; Sjöström, 2008). Information systems actability has also been operationalised into 1) methods for specification and design of IS (e.g. Ågerfalk, 2003; Ågerfalk & Eriksson, 2004; Cronholm & Goldkuhl, 2002; Sjöström & Goldkuhl, 2004) methods for evaluation of IS (e.g. Ågerfalk, 2004; Ågerfalk et al., 2002; Cronholm & Goldkuhl, 2003; Sjöström, 2008). ISAT and its methodological operationalisations have ambitions to cover several aspects of information systems and their designs; as relations to business process, human-computer interaction, conceptual modeling and database design. An information system is considered as a technology-based system for communication and information processing including storage and transfer. This means that within ISAT, we use the concept of IS in a restricted sense corresponding to IT-system or IT artefact.

ISAT gets its current theoretical backing from theories and knowledge traditions like pragmatic philosophy, speech act theory, classical semiotics, social action theories, affordance theory, semiotic HCI engineering, conversation analysis, discourse theory and activity theory. Confer Goldkuhl (2009) for an overview of how these different background theories have influenced ISAT. As a practical theory, ISAT comprises a conceptualisation of IS and several models (Goldkuhl, 2009). ISAT also comprises normative criteria of pragmatic charac-

ter (quality ideals) which can be used for design and evaluation. There have earlier been several contributions of actability criteria (e.g. Ågerfalk, 2003, 2004; Ågerfalk et al., 2002; Cronholm & Goldkuhl, 2002; Röstlinger & Cronholm, 2009; Sjöström, 2008).

The *purpose* of this paper is to make a further contribution to actability criteria. Nineteen actability criteria will be presented. These criteria will be structured in three different groups. The different criteria groups depend on what pragmatic scope is applied (confer Figure 5). There are 1) criteria associated with the user interacting with the system (interaction quality). There are 2) criteria for a broader scope, the user-via-system-to-user communication (communication quality). There are 3) criteria for an even broader pragmatic scope; the use of IS as part of a business process (process quality). The nineteen different actability criteria that have been identified will be articulated and clearly related to ISAT conceptualisations and models of information systems.

The *research approach* is a combined theory-informed (deductive) and empirically based (inductive) endeavour (see Figure 1), following the principles of Goldkuhl (2004). The conceptualisation of IS in actability theory is of course a main source of inspiration. I have also looked into earlier actability criteria; some references men-

Figure 1. Combined research approach

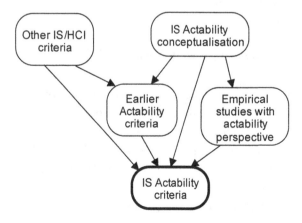

tioned above. However, my partial discontent with some of these criteria has been a driving force for conducting this research, i.e. developing a new set of actability criteria. I will not go through these earlier criteria and make any critical review of them; this is beyond purpose and scope of this paper. I acknowledge also an influence from other IS theories, especially from human-computer interaction. There are many other criteria concerning IS usage and human-computer interaction (HCI); e.g. well-known usability criteria from Shneiderman (1998) and Nielsen (1993). There has been an influence from such criteria on earlier ISAT criteria.

I have also used learnings from several empirical studies. I have (together with colleagues and students) studied several existing information systems based on the actability perspective. When using an actability lens we have identified some feature of a system that seemed to make the system less actable. Such an empirical observation was a basis for generation of an actability criterion. The IS deficiency was turned the other way around and we stated how the system would look like if it should be actable. This gave rise to formulation of new IS actability criteria as quality ideals.

In the next section of the paper I will go through some fundamentals of information systems actability theory. Some models will be presented as a basis for the formulation of criteria, which is the main part of the paper. The paper is ended with conclusions.

ACTABILITY FUNDAMENTALS

Information systems actability theory builds on a communicative action perspective on IS. *Information systems are seen as instruments for technology mediated work communication.* Human actors are communicating (i.e. sending and/or receiving messages) through an IS (see Figure 4). Sending a message through an IS means per-

forming a communicative action. The IS affords a communicative action repertoire to its users. This repertoire enables and constrains the users in their communicating.

Such a communicative action perspective does, however, not dismiss the perspective of a human utilising and interacting with an IT artefact. ISAT comprises aspects of human-computer interaction. One important notion within ISAT is the Elementary InterAction Loop (EIAL) (Ågerfalk, 2003; Cronholm & Goldkuhl, 2002; Goldkuhl, 2009; Goldkuhl et al., 2004). This loop describes the smallest kind of interaction between a user and the IT-system (see Figure 2). EIAL consists of four phases. First the user interprets what can be done in this particular interaction situation. This is a phase of investigating the action repertoire and finding out what do; often called the informing phase or pre-assessment. This is followed by the intervention phase when user does something with the system; entering some kind of input to the system. The system reacts, which implies some internal processing and some exposure of messages (feedback) to the user. The fourth and concluding phase is a "consummation" phase where the user interprets the feedback from the system. The user tries to find out (in a post-assessment phase) if the system did what was expected. EIAL builds on a fundamental action-cyclic model of pre-assessment, intervention and post-assessment (Goldkuhl, 2008b; Mead, 1938). An EIAL can be a reading loop or a formulation loop or a navigation loop (Goldkuhl, 2009; Goldkuhl et al., 2004).

The elementary interaction loop will be framed by what is afforded by the current screen document. Different user actions will be possible to conduct given the repertoire afforded through the current screen document. There may be reading or formulation options. There may also be possibilities to move to other interaction situations (screen documents); i.e. navigation options. This means that information systems usually hold a repertoire of interaction situations (screen documents). One can talk about a document space and that the user will move around in this document space (see Figure 3).

Figures 2 and 3 have been concerned with the interaction level of information systems. I will now turn back to the essential communication level. Information systems are used for communication between different users. This is described in Figure 4; which is modification of earlier ISAT figures presented in Goldkuhl and Ågerfalk (2002) and Goldkuhl et al. (2004). A focused user may read messages exposed by the system. These messages may have origin from other users (F). Messages exposed by the system to the focused user may have a direct origin from other users' formulated messages or these messages can be aggregations, combinations or other modifications (performed by the system in accordance with its programmed rules) of such original messages (Sjöström, 2008). The database(s) of the system (called workpractice memory in the ISAT terminology) will keep input messages or modifications of such messages over time. This will enable the communication to be asynchronous. The user can formulate and enter messages into the system. This makes the user a sender of messages. Such messages may be kept by the system (in its workpractice memory) as such or processed and transformed in some way according to the processing and storage repertoire of the system. Messages in the workpractice memory may be forwarded to other users, i.e. reading users (R). This means that the focused user can act as a communicator to other users by the support of the systems. The system functions thus as a medium for communication. Usage of an information system means *a user-via-system-to-user communication*. The receiving users (R) may use messages from the system as an informative basis for subsequent actions (Goldkuhl, 2008b, 2009). This means an external use of the system (messages from the system).

What a user can do with a system (reading, formulation, navigation) is dependent on what the system affords to the user. The user must interpret and understand the afforded action repertoire. This afforded action repertoire should be seen as communication from those responsible for the design of the system; i.e. a designer-to-user communication (Andersen, 2001; De Souza, 2005). IS usage can thus be differentiated in several types of situations (Goldkuhl, 2008b):

- Reading situation.
- Formulation situation.
- Automatic processing situation.
- Navigation situation.
- Subsequent action situation.

Information systems are parts of workpractices. They are communicative and information-processing instruments used for the benefits of the workpractice. The use of an IS should improve user actions within the workpractice (see Figure 5).

Figure 2. A user interacting with an information system: an interaction situation (an elementary interaction loop - EIAL)

Figure 3. Navigation between different interactive situations in a system (moving in the document space)

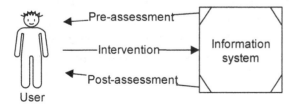

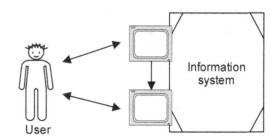

Figure 4. An information system as communication between different actors

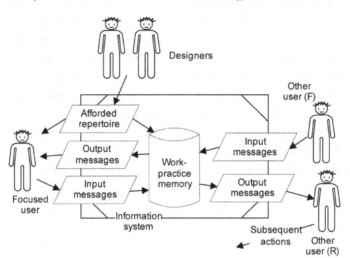

Figure 5. Three layers of actability criteria (action quality layers) (based on Goldkuhl, 2009)

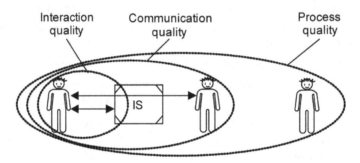

The description above has focused IS interaction and IS communication. Sjöström and Goldkuhl (2004) introduced the concept pragmatic duality to express and emphasise that an IS user within the same action performed a dual action of 1) interacting with an IT artefact and 2) communicating with other users. This concept of pragmatic duality has been taken one step further into the concept of pragmatic multi-functionality (Goldkuhl, 2009). Besides system-interaction and communication, pragmatic multi-functionality adds work-process contribution. This implies that one particular user action at the same time means 1) that the user interacts with the system and 2) that the user communicates with other actors and

3) that the user contributes to business process performance (Goldkuhl, 2009).

The notion of pragmatic multi-functionality implies a differentiation into three layers of pragmatic qualities (see Figure 5). Process quality is the broadest layer. This layer comprises the layer of communication quality. To say that the process layer comprises the communication layer means that communication is instrumental in relation the process quality. What is communicated (through the IS) should be beneficial to the workpractice. The communication layer comprises in turn the layer of interaction. This means that the quality of interaction should contribute to the quality of user-to-user communication and also to process quality.

ACTABILITY CRITERIA: A TYPOLOGY

Criteria for information systems are normative. They express values and quality ideals associated with IS. These criteria can be used both for design of IS and for evaluation of IS. The criteria described below are based on the four sources described in Figure 1; i.e. IS actability theory, earlier ISAT criteria, other IS and HCI criteria and empirical studies on IS based on an actability perspective. As said above there is some influence from HCI criteria. However, the actability conceptualisation of IS are different from main-stream IS and HCI views which entails formulation of new criteria and re-formulation of old ones. The actability criteria presented here are divided into three groups following the action quality layers in Figure 5:

- Interaction quality criteria.
- Communication quality criteria.
- Process quality criteria (i.e. context quality).

Interaction criteria are concerned with what the user does *with* the system. Communication criteria are concerned with what the user does *through* the system. Process criteria are concerned with what the user does *outside* the system.

There are specific criteria, which can be said to relate to more than one layer. However, each criterion has been associated with only one layer (group). Nearly all criteria are associated with the two first layers. Only one criterion is associated with process quality and this will commented below. The group interaction quality has been sub-divided into:

- Fundamental interaction criteria.
- Navigation criteria.

The group communication quality has been sub-divided into:

- Reading criteria.
- Formulation criteria.

The pragmatic qualities of an information system are dependent on 1) the afforded action repertoire and 2) what messages have been provided by its users. It is important to recognize that the stated criteria express desirable properties of the system. These properties are seen as affordances of the system, i.e. they offer possibilities for user conduct of diverse actions (Gibson, 1979; Norman, 1988; Goldkuhl, 2008b). These kinds of properties are what make the system actable to its users. The content and quality of a user action is not only dependant on system features. Action quality (in IS usage) is also dependant on user characteristics (skill and other properties) and different characteristics of the workpractice.

INTERACTION CRITERIA

These criteria are concerned with the user's interaction with the IT artefact (see Figure 2).

Fundamental Interaction Criteria

There are five fundamental interaction criteria which are based on Figure 2 and the elementary interaction loop. Confer also the interaction quality dimension in Figure 5.

I1:Clear Action Repertoire

This criterion is concerned with the first step in the EIAL; the informing (pre-assessment) phase where the user tries to understand what the system can do or in other terms what the user can do with the system (see Figure 2). The system affords a repertoire of possible actions to the user. In order to conduct any actions, the user needs to identify and understand what the action potential is. The action repertoire can be divided into 1) reading repertoire (what the user can get to know through the system), 2) formulation repertoire (what the user can say to others through the system), and 3) navigation repertoire (what kind of interaction situations/document types the system affords and how to reach these interaction situations). All these three types of action repertoires need to clear to the user.

This criterion states that the system should *expose its action repertoire* in a clear way to the user. A clear action repertoire enables the user to be well-informed what do with the system and that he can perform subsequent actions with confidence. An actable system should have a clear action repertoire.

I2: Intelligible Vocabulary

This criterion is in one sense more general than the earlier one. It is concerned with exposed action repertoire, but not only this. It is concerned with the language used in the whole user interface of the system. For the user to understand what to do and also to accomplish different tasks, it is necessary to understand the language used. The vocabulary of the system needs to be harmonious with the workpractice language of its users. As much as possible the same terminology and concepts should be used in the system as in the surrounding workpractice. The system should also provide explanations for the different used concepts, so the user can check the meanings.

The criterion states that the used vocabulary should be intelligible and *correspond with the users' workpractice language*. This enables the user to act with confidence in many interactive situations and to interpret information from the system accurately. An actable system should have an intelligible vocabulary.

I3: Action Transparency

This criterion is concerned with the afforded actions of the system. The system affords actions (e.g. formulation actions) to the user. For the user it is important to understand what the consequences

of a conducted action will be. What will happen after the user has pressed a certain button on the user interface? What does the system execute after the user has entered a certain message? In order to act with confidence (i.e. to be sure what will happen subsequently in and through the system) the user needs to have a clear picture of subsequent actions after the interaction has been performed.

The criterion states that the user should *understand in advance* what will happen when he performs different IT-mediated tasks. This enables the user to act in confidence when interacting with the system. The user becomes well-informed about the action repertoire and can anticipate the consequences of different actions. An actable system should be action transparent.

I4: Clear Feedback

This criterion is concerned with the post-assessment phase of the EIAL (see Figure 2). The user obtains a clear feedback of what has been performed through an interaction. The user will not be in doubt of what has happened after his intervention. Depending on what type of interaction (reading, formulation or navigation loop) there will be different kinds of feedback (Goldkuhl et al., 2004). A formulation action needs explicit feedback, so the user can be ensured that his formulated message has been registered. A reading interaction will give messages to read as an apparent feedback. A navigation demand should bring the user to the requested screen document. This criterion on feedback coincides with what has been stated in many HCI publications, e.g. Shneiderman (1998). It has here been clearly related to the phase conceptualisation in ISAT (the fourth phase in the EIAL loop).

The criterion states that the user should receive a *clear response* (a feedback) to his intervention to the system. This feedback enables the user to post-asses his intervention properly and thus to be sure about the results of his earlier action. An actable system should produce clear feedback to the user.

I5: Amendability

This criterion is concerned with possibilities to correct earlier actions made to the system. Errors are made during IS usage. This is the case even if there is an information system with high actability. How does the system permit the user to undo what has been done earlier? There will of course sometimes be actions which are not possible to reverse due to workpractice reasons. However, a good feature of an IS is that it capable of being corrected. The user should be able to make amendments to the system. In order to do this the user should have access to what earlier actions have been performed (confer criterion C1 below) and how to undo this action. This means that amendability is a special kind of action repertoire of the system; that it comprises both the doing and undoing of a certain type of action. This criterion is also well-acknowledged in the HCI literature (e.g. Nielsen, 1993; Shneiderman, 1998).

The criterion states that the user should be able to *correct an earlier erroneous action*. This enables the system to be more accurate and permit the user to make certain errors without severe consequences (a "forgiving" system). An actable system should have amendability; i.e. to be possible to recover from identified erroneous actions.

Navigation Criteria

Besides these fundamental interaction criteria there four criteria especially concerned with navigation. This means movement from one interaction situation to another (see Figure 3).

I6: Easy Navigation

This criterion is concerned with the ease of moving from one interaction situation to another. This criterion is concerned with the navigation repertoire of the system (cf. criterion I1). It should be clear for the user what interaction situations there are in the system and how to reach such a situation; i.e. a screen document where some desired action can be conducted. It should be easy to move around in the document space of the system. There are different navigation principles in information systems as hierarchical, sequential and direct navigation. Adequate, intelligible and accessible navigation principles should be used. In order not to get lost in the document space of a system, it should offer some navigation transparency and traceability to the user.

The criterion states that the user can *move around* in the system in a controlled manner. This enables the user to get to the desired spot in a system easily. An actable system should be easy to navigate.

I7: Action Stage Overview

This criterion is concerned with the movement from one situation to another in a defined sequence of action stages in a work process. To facilitate such a movement, the system should present an action stage overview of the process and it should also inform where the user is in the process at the moment. This contextual information makes a work process more transparent. This may contribute to process quality. Since this system feature is concerned with movement from one interaction situation to another, it has been grouped as an interaction and navigation criterion.

The criterion states that the user should get an *overview of* the IT-mediated *work process* and where he is in this process at the moment. This makes the user aware of the subsequent stages of a work process and thus what is expected from him. An actable system should (when relevant) give action stage overviews.

I8: Conceptual Consistency

This criterion is concerned with relations between different interaction situations. The different situations should offer similar structure and use of concepts. The meaning of terms should be similar through the system in order to avoid confusing the user. Different symbols in the system should be used in a consistent way. This criterion can be found (with other formulations) in the HCI literature (e.g. Nielsen, 1993; Shneiderman, 1998).

The criterion states that the user should meet a *consistent terminology* and other symbology in the system. This prevents to user to be confused when using the system. The user can act with more ease and comfort. An actable system should be conceptually consistent.

I9: Action Accessibility

This criterion is concerned with the allocation of action possibilities in different interaction situations. It is a challenging IS design task to make meaningful groupings of different tasks into one interaction situation (one screen document). Each interaction situation should contain appropriate action affordances. It is desirable that certain action alternatives are accessible within one interaction situation. If it appears a certain need for action in a particular situation, then it should be easy for the user to access this action affordance. It is not actable if the user needs to move around in the system in order to search for and find the requested affordance and then has problems to return back to the original place. This is not only the case of easy navigation (I6), but to put action affordances in proper places (interaction situations).

The criterion states that the user should have *easy access* to action affordances when needed. This enables the user to perform his tasks fluently and avoid unnecessary movements within a system. An actable system should have proper action accessibility.

COMMUNICATION CRITERIA

These criteria are concerned with user-to-user communication mediated through the IT-system. Confer Figure 4 and the communication quality dimension in Figure 5. Communication criteria are divided into reading criteria (i.e. the user interpreting messages that have origin from other users) and formulation criteria (i.e. the user is communicating messages through the systems intended for other users). There may of course be different kinds of processing and aggregation of messages before they reach the reading user (Sjöström, 2008). The user will may not see the messages with original content, but rather something refined (i.e. processed by the IT system). Communication in both directions is mainly considered as a formalised communication according to a predefined message types (in a formalised workpractice language).

Reading Criteria

This group of criteria is concerned with the user getting informed through the IT-system. What the user can get to know through reading messages that are presented by the system to him. This should not be seen as the user being passive receiving whatever messages are exposed to him. In most cases, reading is the result of an active retrieval made by a user with information needs.

C1: Clear and Accessible Workpractice Memory

This criterion is concerned with the possibility for users to reach information stored in the system. In ISAT such storage in databases etc is called a workpractice memory. Messages are stored over time because there are requirements that different actions and observations need to be remembered. A workpractice memory is seen as an external and collective memory for actors within a workpractice; confer the notion organisational memory (e.g. Ackerman & Halverson, 2004; Randall et al., 2001). A workpractice memory will therefore be a source for users to obtain information which might be relevant for conducting actions in the workpractice. To use workpractice memories of an IT-system is a way they get informed by others. It is also a way to be reminded by the system about earlier actions performed by the user himself.

A clear and accessible memory is thus a primary means for users to get informed by other users through the system. The reading user needs to get a good view of the reading repertoire of the system/workpractice memory (cf. criterion I1) and obtain possibilities to access the desired information. What is presented to the user needs to be intelligible to him (cf. criterion I2).

The criterion states that the user should have *easy access to the workpractice memory* of the system (i.e. messages from other users and also earlier messages from the actual user himself). This enables the user to get informed about different relevant issues in the workpractice. An actable system should have a clear and accessible workpractice memory.

C2: Information Accuracy

This criterion is concerned with the information content of the exposed/retrieved messages. What is said and conveyed to the user must be accurate. The information must not be false or inadequate. The user should not be miss-guided by the information. Messages need to be correct. There should be sufficient detail in the information and if needed there should be relevant abstractions or aggregations. The temporal dimension of the information should be stated explicitly. Is the information valid at this moment? Is it up-to-date? If there are any limitations in the conveyed information, then these limitations should be stated explicitly. Information quality has been addressed in many IS studies; a well-known framework can be found in DeLone and McLean (1992).

The criterion states that *the information should be accurate* and that it should be possible for the user to judge this accuracy. This enables the user to trust the messages and to act in accordance what has been said. An actable system should convey accurate information to its users.

C3: Actor Clarity

This criterion is concerned with the possibilities to "see" the sender behind messages. What is said is always said by someone. IT-systems have often, unfortunately, a tendency to "objectify" messages and strip away the senders from the messages. What is kept in databases (workpractice memories) is often only the message content and not information about message origin (i.e. the sender). To know about the senders (i.e. the originators) of different messages is a way for reading users to be better informed about what is conveyed. There may arise needs for more information which leads to a need for the user to get into contact with the sender; either through the system or besides the system. To have visible senders means more transparent systems. This criterion is a call against anonymity in information systems. It is a call for actor clarity. Actor clarity can either mean clarity about organisational actors, roles or distinct persons. Actor clarity is about identification of senders.

The criterion states that the user should as far as possible be aware of *who has said what* through the system. This may enable the user to reach contact with originators of messages. An actable system should have actor clarity (i.e. visible senders of messages).

C4: Intention Clarity

This criterion is concerned with the precision and clarity of messages. It states that messages should not only contain a propositional content, but also a clear intention; or stated in terms of speech act theory (Searle, 1969) a clear illocutionary force. IT-systems have often, unfortunately, a tendency to "objectify" messages and strip away the intention from the messages. What is kept in databases (workpractice memories) is often only the propositional content and not information about message intention. To be sure about the meaning of a message there is a definitive need to understand the message intent; i.e. the kinds of relations that are established between senders and receivers.

This criterion is classified as a reading criterion; the reader must understand the intention of messages presented to him. There is of course also a need for senders to be aware of the presumed intention in the communication repertoire of the system.

The criterion states that the user should be aware of *intentions* of the conveyed messages. This enables the user to properly understand the full meanings of read messages. An actable system should have intention clarity.

Formulation Criteria

This group of criteria is concerned with the user communicating through the IT-system. What the user can say to others through the system. This is concerned with role of the user as a formulator and sender of messages through the system.

C5: Satisficing Communication Needs

This criterion is concerned with the possibility to serve the communication needs of the user. Are there possibilities (within the communication repertoire of the system) for the user to formulate desired messages? The criterion is concerned with the semantic expression capability of the system. Can the user say what he desires through the system or must he use other communication channels? Is it possible, not only to express routinized messages, but also to express the exceptional and unexpected?

The criterion states that the user should be able *to realize communication needs* through formulating messages into the system. This facilitates communication in the workpractice and puts it into organised and recognised patterns. The users will know where and how to communicate which contributes to both communication efficiency and communication comfort. An actable system should satisfy the users' communication needs.

C6: Relevant Communication Demands

This criterion is concerned with the demands on the user when entering messages into the system for communication purposes. Communication through the system is not only concerned with user needs but also with demands on him. When communicating something through the system, there should not be put too much demands on entering information to the system that are already there. Necessary identification of information objects needs to be done, but this identification process needs to be done with as much ease (guidance) to the user as possible. The user should not be demanded to re-register (repeat) information that is already kept in the system.

The criterion states that the user should *not be demanded to re-register information* into the system that is already kept by the system itself. This will not put unnecessary burden on the user. It will instead facilitate the task of entering information into the system. An actable system puts relevant communication demands on the user.

C7: Workpractice Memory Addition

This criterion is concerned with adding messages into the workpractice memory of the system (see C1). The formulation of messages (as described in C5) should lead to a deliberate use of these messages in the system (processing, transfer, storage). This criterion is concerned with aspects that are directly subsequent to the formulation of messages (C5). The entered messages may need to be processed in proper ways and stored in the workpractice memory for later use. The addition of messages to the workpractice memory establishes proper communication conditions of the IT-system. These communication conditions need to be managed dynamically by the system through its continual addition of messages into the system. The workpractice memory needs to be accurate and up-to-date. Workpractice memory addition enables the system to be used for retrieval and reading (C1) and directing messages to targeted addressees (see C8 and C9).

The criterion states that the system should process and store input messages in proper ways in order to *establish good communication conditions* through the workpractice memory of the system. This enables later retrieval and reading of messages and intended distribution of messages to targeted addressees. An actable system should have an updated workpractice memory, through addition (processing and storing) of input messages which functions as communication conditions of the system. An actable system should have a well functioning workpractice memory addition.

C8: Addressee Relevant Communication

This criterion is concerned with distribution of messages to users, i.e. to actors that are seen as intended addressees of the messages. Formulation of messages into the system (C5) shall in some defined situations be followed by the exposure of these messages or some related messages (that has been processed) to appointed receivers. The system should be able to distribute messages to such intended addressees and that these messages are considered relevant for these addressees. We are here concerned with intended communication effects of formulating messages; that messages will reach the addressees as presumed by the communicator.

This criterion is also related to I3 above about action transparency. The distribution of messages

to addressees must be transparent to the sender. When formulating messages he must understand and foresee the distribution of messages to targeted receivers.

The criterion states that *relevant messages* should *be presented to intended users* (addressees). This creates a full communication process from sender to receiver. The intentions of the sender to communicate something to targeted receivers will be fulfilled. Receivers will obtain relevant information. An actable system should have addressee relevant communication.

C9: Addressee Adapted Communication

This criterion is also concerned with distribution of messages to users, i.e. to actors that are seen as intended addressees of the messages. This criterion focuses not what is said (as in C8) but how messages are distributed to receivers. It may not be sufficient that a certain message is distributed to an addressee. The way an addressee receives the information may be crucial for an efficient communication process. It can concern when (i.e. appropriate time) or where (i.e. appropriate distribution place) something is presented to the receiver. It can also concern the distribution media or the form of the message. The distribution way should be adapted to the addressee and his possibilities to receive and make use of the intended message.

The criterion states that messages should be *presented to intended users* (addressees) *in appropriate ways*. This enables the addressee to be reached by the message and use it accordingly. An actable system should have addressee adapted communication.

PROCESS CRITERIA

This criteria group is concerned with the effects of the IS usage to the workpractice processes. Confer the process quality dimension in Figure

5. In this group there is only one general criterion stated. Process quality is concerned with particular characteristics of each specific workpractice process. Goals and values concerning the specific IS and its workpractice context need to be stated in order to create workable process criteria for design and evaluation. This means that this general actability criterion (P1) needs to be supplemented by domain-specific criteria (Cronholm & Goldkuhl, 2003).

P1: Subsequent Action Support

This criterion is concerned with the use of information mediated through the IT-system. What is presented to the user may be used to enable or improve actions in the workpractice. The user should be supported by the system in the course of his actions. These actions can be performed outside the system or they can be performed subsequently through the system. The key issue is that the user obtains an action support relevant for tasks in the workpractice. The action support can consist of explicitly recommended actions, i.e. the information contains a very clear and action-directing support. The information can also be more indirectly action-supportive. There can be information forming the basis for decisions or giving suggestions for aspects to observe or to be attentive about. Historical information can be action-supportive in the sense that it informs the actor/user what to take into account when acting. There may also be regulative information guiding the user of what rules to follow when conducting some action; e.g. stating what kinds of action that should not be conducted.

The criterion states that information from the system should be *useful to its users*. Presented information should give users action support in order to improve the workpractice. Information from the system should thus contribute to process quality of the workpractice through enabling or improving actions that are seemed pivotal. An actable system should give appropriate action support to its users.

CONCLUSION

The presented actability criteria are based on:

- The actability conceptualisation of information systems.
- Actability models of IS (Figures 2 through 5).
- Other IS/HCI criteria.
- Earlier actability criteria.
- Empirical studies on IS based on the actability perspective.

Nineteen criteria have been presented and described in similar ways. These criteria have been divided in three main criteria groups: 1) interaction criteria, 2) communication criteria and 3) process criteria. These criteria can be seen as a further development of the practical theory of information systems actability. A practical theory and its normative criteria should guide people to perceive and consider the values of studied phenomena (Cronen, 2001; Goldkuhl, 2008a).

These criteria should be possible to use in the design of information systems. The functions of the criteria are here to express possible quality ideals to strive for. It is of course the responsibility of designers and other stakeholders participating in the IS development process to judge and evaluate which of these criteria should be taken into account. The criteria can be used as inspiration when designing the system together with domain-specific goals. They can also be used in formative evaluation of design proposals during the IS development.

These criteria should also be possible to use in post-evaluations of information systems. There may be many reasons for evaluations of IS in use. There are also different strategies and approaches for IS evaluation (Cronholm & Goldkuhl, 2003; Lagsten & Karlsson, 2006). In any evaluation there are needs for explicit criteria to be used as a yardstick when investigating the IS. Criteria can be captured and generated from workpractice itself. External criteria of general character can also be used (Cronholm & Goldkuhl, 2003). These presented actability criteria can be used as such general criteria in post-evaluations of IS. As said above, it is important to supplement the actability criteria with domain-specific process criteria.

The use of the actability criteria in design and evaluation may be a source for further research. It is important to inquire the practical uses of these criteria. This may lead to validation or re-formulation of criteria. Theoretical development of ISAT and empirical studies on IS and their development and use may also contribute with new or re-formulated criteria.

This paper is written as a contribution to the practical theory of information systems actability. What has been said should however not be interpreted as something restricted to this specific theory. The paper gives ultimately a contribution to our understanding of pragmatic qualities of information systems.

ACKNOWLEDGMENT

The author is grateful to the research colleagues, Pär Ågerfalk, Stefan Cronholm, Annie Röstlinger and Jonas Sjöström, who also have contributed to IS actability criteria in different collaborations.

REFERENCES

Ackerman, M., & Halverson, C. (2004). Organizational memory as objects, processes, and trajectories: An examination of organizational memory in use. *Computer Supported Cooperative Work, 13*(2), 155–189. doi:10.1023/B:COSU.0000045805.77534.2a

Ågerfalk, P. J. (2003). *Information systems actability: Understanding information technology as a tool for business action and communication.* Unpublished doctoral dissertation, Linköping University, Linköping, Sweden.

Ågerfalk, P. J. (2004). Investigating actability dimensions: A language/action perspective on criteria for information systems evaluation. *Interacting with Computers, 16*(5), 957–988. doi:10.1016/j.intcom.2004.05.002

Ågerfalk, P. J., & Eriksson, O. (2004). Action-oriented conceptual modelling. *European Journal of Information Systems, 13*(1), 80–92. doi:10.1057/palgrave.ejis.3000486

Ågerfalk, P. J., Sjöström, J., Eliason, E., Cronholm, S., & Goldkuhl, G. (2002). Setting the scene for actability evaluation – Understanding information systems in context. In *Proceedings of 9th European Conference on Information Technology Evaluation*, Paris, France.

Andersen, P. B. (2001). What semiotics can and cannot do for HCI. *Knowledge-Based Systems, 14*(8), 419–424. doi:10.1016/S0950-7051(01)00134-4

Broberg, H. (2008). Understanding IT-systems in practice: Investigating the potential of activity and actability theory. In *Proceedings of the 5th International Conference on Action in Language, Organisations and Information Systems*, Venice, Italy.

Cronen, V. (2001). Practical theory, practical art, and the pragmatic-systemic account of inquiry. *Communication Theory, 11*(1), 14–35. doi:10.1111/j.1468-2885.2001.tb00231.x

Cronholm, S., & Goldkuhl, G. (2002). Actable information systems - Quality ideals put into practice. In *Proceedings of the 11th International Conference on Information Systems Development*, Riga, Latvia.

Cronholm, S., & Goldkuhl, G. (2003). Strategies for information systems evaluation – Six generic types. *Electronic Journal of Information Systems Evaluation, 6*(2).

De Souza, C. S. (2005). *The semiotic engineering of human-computer interaction*. Cambridge, MA: MIT Press.

DeLone, W., & McLean, W. (1992). Information systems success: The quest for the dependant variable. *Information Systems Research, 3*(1), 60–95. doi:10.1287/isre.3.1.60

Gibson, J. J. (1979). *The ecological approach to visual perception*. Boston, MA: Houghton Mifflin.

Goldkuhl, G. (2004). Design theories in information systems – a need for multi-grounding. *Journal of Information Technology Theory and Application, 6*(2), 59–72.

Goldkuhl, G. (2008a). Practical inquiry as action research and beyond. In *Proceedings of the 16th European Conference on Information Systems*, Galway, Ireland.

Goldkuhl, G. (2008b). Actability theory meets affordance theory: Clarifying HCI in IT usage situations. In *Proceedings of the 16th European Conference on Information Systems*, Galway, Ireland.

Goldkuhl, G. (2009). Information systems actability - tracing the theoretical roots. *Semiotica*, (175): 379–401. doi:10.1515/semi.2009.054

Goldkuhl, G., & Ågerfalk, P. J. (2002). Actability: A way to understand information systems pragmatics. In Liu, K., Clarke, R. J., Andersen, P. B., & Stamper, R. K. (Eds.), *Coordination and communication using signs: Studies in organisational semiotics – 2*. Boston, MA: Kluwer Academic.

Goldkuhl, G., Cronholm, S., & Sjöström, J. (2004). User interfaces as organisational action media. In *Proceedings of the 7th International Workshop on Organisational Semiotics*, Setúbal, Portugal.

Lagsten, J., & Karlsson, F. (2006). Multiparadigm analysis – Clarity of information systems evaluation. In *Proceedings of 13th European Conference on Information Technology Evaluation*, Genoa, Italy.

Mead, G. H. (1938). *Philosophy of the act*. Chicago, IL: University of Chicago Press.

Nielsen, J. (1993). *Usability engineering*. San Diego, CA: Academic Press.

Norman, D. A. (1988). *The psychology of everyday things*. New York, NY: Basic Books.

Randall, D., Hughes, J., O'Brien, J., Rouncefield, M., & Tolmie, P. (2001). 'Memories are made of this': Explicating organisational knowledge and memory. *European Journal of Information Systems, 10*, 113–121. doi:10.1057/palgrave.ejis.3000396

Röstlinger, A., & Cronholm, S. (2009). Design criteria for public e-services. In *Proceedings of the 17th European Conference on Information Systems*, Verona, Italy.

Searle, J. R. (1969). *Speech acts: An essay in the philosophy of language*. Cambridge, UK: Cambridge University Press.

Shneiderman, B. (1998). *Designing the user interface: Strategies for effective human-computer interaction* (3rd ed.). Reading, MA: Addison-Wesley.

Sjöström, J. (2008). *Making sense of the IT artefact – a socio-pragmatic inquiry into IS use qualities*. Unpublished licentiate thesis, Linköping University, Linköping, Sweden.

Sjöström, J., & Goldkuhl, G. (2004). The semiotics of user interfaces – A socio-pragmatic perspective. In Liu, K. (Ed.), *Virtual, distributed and flexible organisations: Studies in organisational semiotics*. Dordrecht, The Netherlands: Kluwer Academic.

This work was previously published in the International Journal of Information Systems and Social Change, Volume 2, Issue 3, edited by John Wang, pp. 1-15, copyright 2011 by IGI Publishing (an imprint of IGI Global).

Chapter 11
Collective Construction of Meaning and System for an Inclusive Social Network

Vânia Paula de Almeida Neris
University of Campinas (UNICAMP), Brazil

Leonardo Cunha de Miranda
University of Campinas (UNICAMP), Brazil

Leonelo Dell Anhol Almeida
University of Campinas (UNICAMP), Brazil

Elaine Cristina Saito Hayashi
University of Campinas (UNICAMP), Brazil

M. Cecília C. Baranauskas
University of Campinas (UNICAMP), Brazil

ABSTRACT

Information and Communication Technology has the potential of benefiting citizens, allowing access to knowledge, communication and collaboration, and thus promoting the process of constitution of a fairer society. The design of systems that make sense to the users' community and that respect their diversity demands socio-technical views and an in-depth analysis of the involved parties. The authors have adopted Organizational Semiotics and Participatory Design as theoretical and methodological frames of reference to face this challenge in the design of an Inclusive Social Network System for the Brazilian context. This paper presents the use of some artifacts adapted from Problem Articulation Method to clarify concepts and prospect solutions. Results of this clarification fed the Semantic Analysis Method from which this paper presents and discusses an Ontology Chart for the domain and the first signs of the inclusive social network system.

INTRODUCTION

In the process of exercising citizenship, communication is a fundamental component. Organized groups and communities have a proper way to spread news among participants, share information and exchange knowledge. In this context, tools for communication and expression should provide all users - in their vast variety of gender, age, abilities and disabilities - with an opportunity to establish a digital culture making the system part of their social life.

DOI: 10.4018/978-1-4666-2922-6.ch011

In Brazil we face a situation characterized by vast differences with regard to socio-economics, culture, and geographical region differences as well as access to technology and knowledge. Social indicators presented by the Committee of Entities Combating Hunger and for Life (COEP in its Portuguese acronym) show 30.1% (approximately 52.5 million) of the population live below the poverty line (minimum annual level of income deemed necessary to achieve an adequate standard of living). Furthermore, according to the 2008 National Survey by Household Sample, 11.2% of the population are considered illiterate (approximately 14.2 million). Moreover, the census of 2000 found that 24.6 million Brazilians or 14.5% of the population have impairments.

In this scenario, Information and Communication Technology (ICT) represents a hope for benefiting citizens, allowing access to knowledge, communication and collaboration, and thus promoting the process of constitution of a fairer society. Within this context, it is necessary to investigate how to design systems that consider the diversity of users taking also the digitally excluded into account. Although there are many computational systems developed to support people in activities of communication and cooperation - known as social networks, they were built for the digitally literate. They do not address the demands of people with different interaction abilities, including inexpert users, elderly, illiterates, people with disabilities and others. A review about 14 current systems that support social networks shows that the resources that social networks make available today are not enough to allow access to the variety of users cited above (Santana *et al.*, 2009).

Hendler *et al.* (2008) have pointed out that social applications cannot be specified, designed and built based on the software engineering traditional practices. They argue in favor of a combined view that allows the study of the social phenomenon and the engineering process and emphasize the importance of knowing and considering the protocols involved in the interaction among users in society.

The design of systems that make sense to the users' community and that respect their diversity demands socio-technical views and an in-depth analysis of the involved parties. As far as we know, there is a lack in methodological and technical solutions to support this design process. The research challenge of formalizing new techniques and methods that allow a holistic view of the problem, considering the individuals and their relations in society and with technology are being addressed in the *e-Cidadania* Project (e-Cidadania, 2008)..

In this sense, as a frame of reference for problem understanding, modeling of the organizational context, as well as gathering of user and system requirements we based our approach in Organizational Semiotics – OS (Stamper *et al.*, 1988; Liu, 2000) and some of its methods and artifacts, adapted to participatory practices. Participatory Design – PD (Schuler & Namioka, 1993) inspires our practice to investigate the users' interaction needs through their voices during the design process.

This paper aims at presenting an approach to designing systems that consider the social relations and the meanings that the involved parties share about the domain. The object of study was the problem of designing an Inclusive Social Network (ISN) system. Our results are based on the Stakeholder Analysis (Kolkman, 1993) and the Evaluation Framing (Baranauskas *et al.*, 2005) artifacts. Also a Semantic Analysis (Liu, 2000) was conducted from which we present and discuss an Ontology Chart for the domain.

This paper is organized as follows: the Second section presents some related works; the Third section summarizes the methodological references; the Fourth section describes how the participatory activities were conducted; the Fifth section presents the main results; Sixth section presents how the results were materialized and the Seventh section discusses them; last Section concludes.

RELATED WORKS

Other similar initiatives for developing social networks have been proposed. Based on Friedman's Value Sensitive Design (Friedman, 2006), Cotler and Rizzo (2010) respond to the challenge of designing social networks by proposing the use of a Value Sensitive Framework. This approach should provide a deeper understanding of all stakeholders and the discovery of potential implications and benefits to them. In the process of analyzing the conceptual, technical and empirical aspects of a problem (the social network), they distinguished three important human values associated with social networks: the values of privacy, visibility and what they called informed consent – about participants being aware of the open disclosure of their information. The authors suggest that the design of social networks demands a balance between the provision of privacy in an informed way, maintaining, at the same time, the visibility that allows the social network to grow. The authors point out the importance of having social networks that are sensitive to the values of stakeholders and that more methods are needed to determine how to provide these required values.

Venkatesh (2009) outlines a model for the design of social systems. Concerned with the importance of the actors and actions in a social organization, he proposes the combination of two design activities: the constitutive design (with its normative base, designing the institutions as value-imbued cultural abstractions, and attentive to the actions or the "facts in the world") and the instrumental design. Still, he concludes stating that there is still a need for "new models for interpreting and representing the socio-technical space of problems and solutions".

The research problem that motivated Jung and Lee (2000) was related to the creation of a socio-technical environment that would promote social interactions in electronic market places. They started their investigation by performing some informal studies of Web sites in order to collect the elements that were effective or that were lacking in communities. This resulted in a set of design needs and a set of design constrains. The solution that they propose involves taking the design needs and iteratively refining them until they are turned into design ideas. This process should be guided by the previously identified design constrains. The needs identified in their context were: visibility of people, activities and social interaction; and mechanisms for social interaction. The design constrains that guided the formulation of solutions to these needs were: sociability, scalability, spatiality and imageability. Their iterative process consisted of loops, each composed of 4 design considerations: design idea, design criteria, design options and design rationale. This approach starts with little or no contact with the community of end users of the system. In their process, the contact with users is only on phase 6.

Treiber, Truong, and Dusdar (2009) based their work in FOAF (Friend of a Friend): an ontology that describes persons and their activities, allowing groups to describe their social networks. The authors then propose SOAF (Service of a Friend): they extend FOAF by adding to its network structures the service related information and linking services and humans in the same network. There are three basic relations between the entities in a SOAF network: knows users and provides. Their diagram allows the relations between entities to be dynamic. For example, when a service is no longer used, the relation between a person and a service is moved to the knows state. This creation of a dynamic ecosystem of services, from a bottom up approach, is pointed by the authors as one of the major benefits of the SOAF network. They focus on a different perspective from ours, giving emphasis to the technical aspects, and to services rather than to people.

Regarding methodological issues, a related approach is the Soft System Methodology - SSM (Checkland & Scholes, 1991). Firstly proposed as a seven-step inquiring process, it was later on condensed to four main activities related to finding out the problem situation, formulating some purposeful activity models, comparing the model to the real situation and taking actions to a possible solution. Although the activities proposed are steps to clarify a problem and improve the chances of finding a solution, SSM relies on abstract models which do not really focus on the information systems development. The theoretical-methodological reference adopted here is based on Semiotics with focus on users' requirements, as described in the following sections.

METHODOLOGICAL REFERENCES

OS is a discipline that has roots in Semiotics applied to organizational contexts. OS studies the nature, characteristics, function and effect of information and communication within organizational contexts. Organization is considered a social system in which people behave in an organized manner by conforming to a certain system of norms. These norms are regularities of perception, behavior, belief and value that are exhibited as customs, habits, patterns of behavior and other cultural artifacts (Stamper *et al.*, 1988; Liu, 2000). The Human-Computer Interaction (HCI) understood through Semiotic's lens reveals the complex processes of making sense of computational systems through their user interfaces. Such processes, analyzed only according to the perspective of engineering, have been interpreted as purely syntactic phenomena. The analysis using Semiotics reveals the primary function of computer systems as vehicles of signs and supplies an adequate vocabulary that makes possible the understanding of computer systems in function of other types of sign systems (Nadin, 1988).

In the late 80's, Stamper defined a set of methods named MEASUR (Methods for Eliciting, Analyzing and Specifying Users' Requirements) (Stamper *et al.*, 1988; Liu, 2000) to support the use of OS concepts in information systems organiza-

tions. In this work we used part of two of such methods: the Problem Articulation Method (PAM) and the Semantic Analysis Method (SAM). PAM helps in the identification of issues and articulation of problems raised by interested parties (Stamper & Kolkman, 1991). SAM contributes to clarify the meanings of the language used to define the problem and its solution in operational terms. From PAM, we used the Stakeholder Analysis Chart (Kolkman, 1993; Liu, 2001) which supports the designer in clarifying the stakeholders and their potential specific interests – information forces – in the prospective system.

Another artifact used was the Evaluation Framing Chart proposed by Baranauskas *et al.* (2005), which allows the elicitation and discussion of problems and issues the stakeholders would face, as well as ideas and solutions for these problems.

These artifacts and the concept of ontology, as used in OS, are presented in the following sections.

Stakeholder Analysis

The Stakeholder Analysis Chart has evidenced the importance of understanding the context into which the problem is immersed. When the stakeholders are clearly identified it becomes easier to visualize the extension of the issue by knowing what the particular interests of the involved parties are and how they affect or are affected by the project.

Following Kolkman (1993) and Liu (2001), the layers distribute the stakeholders into four categories, which represent different information fields: the community, market, source and contribution layers.

The first category is the "Contribution" layer, where the "Actors and Responsible" are the stakeholders who directly contribute to the system and who are mostly affected by it. The Responsible are the main characters, the ones who conceived the main idea or issue in place. They can either coordinate other stakeholders to take actions and fulfill their ideas or they can play the role of the

Actors themselves. The Actors are those who actually perform the desired idea, materializing or producing the consequences.

"Clients and Suppliers" make the "Source" layer. They are those who want to make use of the system. They will feed the system with the data/information/services needed. The clients are usually those who benefit from the consequences of the actions taken or those who receive the outcomes. They are the main concern of the Responsible, since they are the reason why the system exists in the first place. Suppliers are responsible for providing the necessary conditions for the system to function.

The category "Partners and Competitors", in the "Market" layer, includes all members of the market related to the prospective system. Partners collaborate by dividing resources and joining forces in order to solve issues together. On the other hand, Competitors often represent a challenge as they are in conflict with the system's interests – but not necessarily in a unfriendly manner.

Finally, the last layer, "Spectator and Legislator", or the "Community", not only comprehends those responsible for establishing the rules (be these rules official or not), but also the whole community that will receive the gains or losses as consequence of the implementation of this system, in the broader scope of the social context. Both spectator's and legislator's actions are usually detached from the system as they are not directly related to it.

Evaluation Framing

In addition to the list of stakeholders and their relation with the problem domain, Baranauskas *et al.* (2005) identified the need of also elicit the issues and problems that these stakeholders may face to reach the problem solution. In advance, they proposed a chart that incentives the discussion about the problem and also the identification of possible solution.

The chart is formed by lines for each stakeholder layer – community, market, source and contribution – and two columns: one for the problems and another one for the solutions. As the chart is generally applied at the beginning of a project, it is possible not to have solutions for all the problems pointed out. However, a set of initial ideas start to be elicited.

With this chart, it is possible to extract the main issues that should be considered while developing the system and the problems and solutions suggest important characteristics of the problem situation that have impact in requirements, architecture, design, selection of technology to name a few.

Although the Evaluation Framing is not part of the original PAM, we have used it to support the clarification of the problem situation. This chart have been used in several projects related to digital inclusion (Bonacin *et al.*, 2006; Neris & Baranauskas, 2009a), education (Miranda *et al.*, 2007), interactive digital television (Miranda *et al.*, 2008), and geographical information systems (Baranauskas *et al.*, 2005).

Ontology Charting

An ontology chart is the result of applying SAM. However it is important to emphasize that the OS approach rescues the original sense of ontology as part of the philosophy that studies the nature of reality. It adopts a radical subjectivism stance and an agent-in-action perspective for ontology; this means that each word or expression used is a name for invariant patterns of behavior in the set of actions and events which the agents experience. SAM focuses on agents' actions and this analysis is performed by observing these actions and detecting invariants, *i.e.*, general behavior patterns (Liu, 2000).

SAM is divided into four phases: Problem definition, Candidate affordance generation, Candidate grouping and Ontology charting (Liu, 2000). Considering a statement that defines the (design) problem, the main affordances in the domain are elicited. After identifying the affordances and agents and grouping them, an ontology chart is drawn. The Ontology Chart presents, among others, the following concepts:

- **Affordances:** Initially proposed by Gibson (1979) as a word for the behavior of an organism made available that "implies the complementarities of the organism and its environment". As Gibson defined it, "the affordances of the environment are what it offers the animal, what it provides or furnishes, either for good or ill". For Stamper (1996, 1988), the word affordance in Gibson's theory is related to the invariants we perceive that are significant for physical and biological reasons. Stamper (1988) generalized the term to include the invariants that we perceive in our social word. According to Norman (2008, 1988), Gibson invented the word affordance "to refer to a relationship: the actions possible by a specific agent on a specific environment". In the ontology chart, affordances are presented as rectangles. Specifics are used to stress some particular affordances related to a general one. In the chart, specifics are presented as rectangles in the lower region of an affordance;
- **Agents:** A special type of affordance which refers to those who are capable of assuming responsibilities. In this context, computational agents are not included. In the chart, agents are presented as ellipses;
- **Ontological Dependency:** Express existential dependency between two affordances. In the ontology chart, ontological dependencies are represented by links between two affordances where the existence of the element in the right is ontologically dependent on the existence of the element in the left, its ontological antecedent (*i.e.*, the element in the right can only exist during the existence of the element in the left).

Participation

In the e-Cidadania project we are applying these artifacts following the PD approach which proposes effective users involvement throughout the software lifecycle promoting benefits such as mutual learning between designers and users; leverage of comprehension of users characteristics, preferences and needs; and the democratic design of new technologies (Schuler & Namioka, 1993).

The PD approach began in the early 70s with manifestations in Scandinavia and in England. In Scandinavia, it emerged for the development of strategies and techniques to support participation of workers and syndicates in the decision making process related to development of new technologies for workplaces. In England, it started with the Tavistock Institute with a proposal for a democratic socio-technical approach to work organization. Later on, Enid Mumford (1924-2006), inspired by these ideas, started to develop information systems in a participatory way (Mumford, 1964; Mumford & Henshall, 1979).

Recently, Muller (1997) proposed a taxonomy of participatory practices to guide designers while choosing participatory practices to be employed along the software lifecycle and considering the players involved in those activities.

While these authors usually address work environments, the e-Cidadania project faces the challenges of involving ordinary people got from their life environments, into a system design situation. To construct a technical information system by considering the interaction requirements present in a diverse population demands a proper way of using the participative approach. We have been working with heterogeneous groups, which demand artifacts adaptation (Bonacin *et al.*, 2009), warming environment, and mutual respect among other qualities for the participatory setting (Neris *et al.*, 2009b).

The next section presents the way we designed, planned and conducted participatory workshops, joining end user representatives, community leaders and other parties, together with the researchers towards the social sharing for ISNs.

PARTICIPATORY WORKSHOPS

Seeking to understand the meanings the prospective users build for ISNs, we conducted the first workshop in the project using the Stakeholder Analysis Chart (see Figure 1b) and the Evaluation Frame (see Figure 1c), both in a participatory approach, as proposed by Baranauskas *et al.* (2002). The combination of the practices from PD and artifacts from OS enabled collaborative work and effective interaction with end users. Moreover, this approach offered a practical way of sharing vocabulary, beliefs, aesthetics etc. among the involved parties which are fundamental aspects of design, especially in our context of a system to support ISNs.

Thirty individuals took part in the workshop, from which 18 were from the research team and 12 represented the local community. We classified the participants from the community in three groups: 1) representatives of Non-governmental Organizations - NGO (*e.g.*, the coordinator of a communitarian course that prepares students to be accepted in public universities, the head of an environmental NGO concerning the environmental causes); 2) citizens in the society (*e.g.*, members of an art cooperative, housewives, laborers from the informal economy); and 3) people from government (*e.g.* member of the city hall). The entire set gave us a broad range of participants and thus a diverse and rich source of opinions and ideas during the workshop. Among the participants, there were those who had had little or no experience with technology, as well as those with different schooling levels and work experience. These diverse points of view were combined in

Figure 1. Three moments of the workshop: (a) brief introduction of the OS artifacts; (b) Stakeholder Analysis Chart and (c) Evaluation Framing filled out in the workshop

a democratic and respectful way, where no opinion was more important than the other, but together they contributed to the construction of the artifacts.

The activities lasted three hours and they took place at the *CRJ – Centro de Referência da Juventude* (Youth Reference Center), a public space supported by the local government that is composed of a Telecenter (from where the community can have free access to computers, including the internet), a public library and rooms for communitarian courses. This place is strategically located, as it is close to a main bus station, public schools and associations. Its location makes CRJ a center that aggregates different community groups. For the activities, chairs were arranged in a semi-circle in front of the artifacts (Stakeholder Analysis Chart and Evaluation Frame), which were hung on the wall. Post-its were given to the participants to write their ideas and then paste them on the artifacts.

At the beginning, each participant introduced himself/herself, sharing with the group some information on their backgrounds. Following this introduction, objectives of the workshop were exposed and a brief explanation of the artifacts was made. This warm up phase was important to make the group more engaged and familiar with the activity. Next, the group started to elicit the stakeholders, discussing each stakeholder's roles and responsibilities. Sometimes, the same stakeholder can play different roles in different levels of the chart; or sometimes an improbable stakeholder comes up, raising polemic discussions. Though these are very rich and insightful discussions, it is important that the discussions are controlled so as to keep the time of the activity as well as to have the focus maintained. At a second moment, the participants raised some problems that might occur concerning the prospective social network system. For each problem, possible solutions were proposed from the different points of view. Each problem and solution was, as in the other artifacts, written on the post-its, which filled out

the poster - in this case, the Evaluation Frame. In the last moment of the workshop, participants were invited to write, individually, a short description on what they understood Inclusive Social Network meant. Figure 1 illustrates some moments of this participatory workshop. Details of the artifacts are described in the next section.

ANALYZING THE RESULTS

The OS artifacts used in the workshop contributed to build an understanding of the main questions concerning Inclusive Social Networks, by sharing experiences and ideas about potential problems surrounding social networks. This activity enabled the group to collaboratively envision possible solutions for the identified problems and discuss potential consequences. From the Stakeholders Analysis the group was conducted to build the Evaluation Frame, which is the starting material for the researchers to elicit system requirements and to understand aspects that will directly impact design decisions. Also, the research team analyzed the participants' definitions for "Inclusive Social Network" as an input for the semantic analysis (SAM).

Stakeholder Analysis Chart

In the original method as proposed by Kolkman (1993), the Stakeholder Analysis Chart was intended to be used by the developers of the system considering the action courses. According to the author, an action course can be understood as a unit system, i.e., a "canonical structure (...) which (...) can be a single activity by one person or a complex set of tasks by a group" (Kolkman, 1993). This unit system definition technique considers the action courses as the cores of the Stakeholder Analysis Charts. In the version used in the activities of the workshop, we have considered the whole system – the Inclusive Social Network - as being our single action course. As shown in Table 1 the

Table 1. Classification of the stakeholders

	Stakeholders
Core	Brazilian society, citizens, final user/community
Contribution	Telecenter users, digital community, network administrator, moderators, weavers, craftswomen, dressmakers, families, pre-college students, housewives, elderly, retired people, people with disabilities, pregnant women, children and adolescents, communitarian representatives, community managers, people from cooperatives, habitation cooperative, telecenter monitor, social educators, health agents, technicians, psychologists, teachers, physicians, people from Jovem.com
Source	Actors from the contribution layer, Herbert de Souza school, cooperative of technology, culture and music cooperative, Google company (by Open social), software companies (produce evaluation tools), lavadeiras[1], maidservant, boleiras[2], hairdressers, traders, service providers, producer/consumer, sponsors
Market	Organizations and software companies (by Orkut, Convers, Sonico, GIGOH, Tocadigital, Hi5, Facebook, WAYN, Yahoo!, eMule, Google, Wikipedia), health centers, researchers, financing institutions for social programs, financing institutions for projects and programs, Jovem.com, popular courses, instructors/educators, "Escola Viveiro"
Community	Leaders of community groups, neighborhood association of Vila União, directors, ONG's coordinator, ONG's, community center Santa Lúcia (ONG), craftswomen cooperative, Cidarte, Rede Mucambos, Educafro, Grupo de Capoeira, Cine Clube Resistência, Grupo de Oração Cristo é Paz, Seicho-no-iê, non-organized classes, City hall, telecenters, secretaries, public administration, responsible ones for accessibility legislation (federal, state and municipal)

Stakeholders elicited during the activities include the Brazilian society, its citizens and final user/community.

The stakeholders experience the action course having, in each layer, a different point of view of the action course, depending on the role he/she has, in the context of its norms. The identification of the stakeholders along the layers of this chart allowed the group to discuss on the different interdependencies that they share regarding Inclusive Social Networks. In this way, the Stakeholder Analysis Chart helped us to put in evidence the social relationships among those directly or indirectly involved.

The discussions that rose during the activity also showed us the different sorts of concerns that the participants have, as the discussion brought up stakeholders from different fronts, including agents from religious and spiritual communities. However, other stakeholders, as important and evident as they might seem, were not mentioned. One example is the broadcast media, which represents a very powerful stakeholder when it comes to making the system known and commented. One may speculate on this and suppose that their context would not include such players

- who would come through other means, as from the higher ranks of the city hall, for example. Or simply, that the discussions led to different fronts that made the group neglect the broadcast media inadvertently.

During the post-activities phase, researchers gathered in meetings in order to review and analyze the products from the workshop, discussing the findings and extracting information relevant for the project. All contributions that came from the workshop were studied and developed into new artifacts.

Starting with the Stakeholder Analysis, the researchers' team tried a new organization of the post-its in the chart, arranging the similar ones closer together, but maintaining their original layer. This work resulted in a classification of the involved parts into groups (see Table 1). As result of the discussions, for example, two new names for stakeholders arouse and they were added to the chart ("software companies (produce evaluation tools)" and "WAYN" - a social network system). In order to identify the contributions from the Post-workshop activities, this new addition was made in post-its of a different color from the color used during the workshop.

179

In the "contribution" layer it is possible highlight three focus points: community representatives (*e.g.*, community managers, communitarian representatives, telecenter monitors, social educators), local professions (*e.g.*, weavers, craftswomen, dressmakers), and people under differentiated conditions (*e.g.*, elderly, pregnant women, people with disabilities). On the external layer "community" participants indicated some local organizations (*e.g.*, Grupo de Capoeira, Cine Clube Resistência, Seicho-no-iê), NGOs that act in the community, and city hall units. The focus points were took into account in the participants' selection for the next participatory practices (*e.g.*, Hayashi & Baranauskas, 2009; Neris & Barnauskas, 2009b).

Participants suggested that every stakeholder cited in the "contribution" layer should be considered as a source of information too. This corroborates a principle of Participatory Design regarding the participation of people along the whole software lifecycle. Recently, in an interview conducted with some of the community participants of the workshops, some of the interviewed people mentioned their feeling of being co-authors of the social network and how they perceived they are exposing their products, services, events and ideas to the world.

The Evaluation Frame

The elements on each category of the Stakeholders chart have their own interest, responsibilities and concerns. The questions and problems related to these groups were explicitly identified with this artifact, which helped participants to view and discuss ideas and possible solutions that will result in system requirements.

From the problems/questions raised, 17 were in the Contribution layer; 8 in the Source layer; 9 in the Market layer and 5 in the community layer.

The main concerns in the Contribution layer were related to the use of the system (*e.g.* how the use of technology can affect children's behavior,

ethical and non-ethical uses, appropriateness of use, etc) and abilities and specificities of the users of the system (*e.g.* how to cope with the difficulties presented by the illiterate and the digitally illiterate users, resistance to the use of technology, etc). The Source layer brought mainly the sustainability of the system: ways to provide support to the maintenance and to the use of the system. The result for the Market layer was interesting, as participants looked at the competitors with a positive view: how to make them partners for social projects, instead of rivals. Another issue raised when talking about other online social networks (competitors), was that there are many different tools for communication and it makes it hard for the new users to learn them all.

As the stakeholders of the Community layer are usually not directly affected by the system, it was reasonable that it projected less problems/ questions. Among those, there were questions relating to the communitarian use of the system (*e.g.* community radios, communal awareness of other communities, etc.).

Some examples of the problems and ideas resulting from the workshop are illustrated in Table 2. Full results are presented in Neris *et al.* (2009a).

Semantic Analysis

An important result of this work is a shared understanding for the domain of ISNs. This section presents how we constructed an ontology for the domain using the Ontology Chart proposed in OS. To support the construction of the Ontology Chart (see Figure 2) we conducted a Semantic Analysis in order to provide means to represent agents and their behavioral repertories (Liu, 2000).

For the Problem Definition phase we considered the definitions generated by the participants of the workshop. Some examples of these definitions include:

- ISNs is the interconnection between community and systems where they

Table 2. Part of the evaluation framing

	Problems/Questions	**Ideas/Solutions**
Contribution	How to keep communication among people with different functional restrictions? How to avoid non-ethical use of the network? How to maintain appropriate use of the network? How to define what an "appropriate use" is?	Audio-visual, popular language, solutions that help break the low literacy problem; provide methods to identify what is not ethical; system alerts for not addicting people: "get out, you've got more to do"
Source	How to make the system sustainable? How to form community representatives that will maintain the system? How to articulate the dissemination strategy? Are there qualified persons to support the realization of actions?	Mechanisms to disseminate events and services, models to support communities management (sustainability); allow input/output in different media;, there should be ways in which the community could take part in the system's evolution/development of new functionalities
Market	Too much information. How to select the most important? What is the distinguishing feature of this network that makes it different from the others? Why didn't the other network implement the recommendations for accessibility? Is the development of systems focused on "average users"?	Sponsors, economic partners, independent economical collaborative initiative, Open Social
Community	Are there conflicts of interests among ONGs? How to guarantee freedom for each community? Are there lacks of appropriate information?	Articulate interests among ONGs, promote interaction among people, find new partners to solve environment related problems

¹ Lavadeiras in Brazil are those who manually wash clothes for a living.

² Boleiras: usually housewives who prepare eatables (e.g. cakes, pies etc.) for selling.

mixture composing one single mass of communication;

- A group of people that interact sharing different elements without discriminating participants, i.e., when we mention "inclusive" that means that everyone is part of that network and that the network has a common objective;

- An ISN is a way to "connect" common interests of people/communities offering no barriers to their participation (as they are part of the network or they can become part of it), without forgetting that there are rules for participating/acting.

From those definitions we extracted the major phrases and concepts resulting in the set of candidate affordances corresponding to the phase 2 of the method. In the Candidate Grouping phase we identified some synonym concepts and classified each item in one of the following components: affordance, agent, role, specific and determiner.

Figure 2 represents the ontology chart resulting from first three phases of semantic analysis. The root element "society" affords "person", "group" and "thing". "Person" and "group" afford "membership". This relation is important to represent the digital inclusion process. In this scenario, "group" represents any set of people, including that group that has access to information and communication through computers. This represents that any technical system that intends to support inclusive social networks should make "membership" possible which implies in important design issues regarding accessibility and universal design.

"Person" and "group" also afford "interacts". Furthermore "interacts" and "thing" afford "shares". These relations represent that interacting, in such modes as communicating, cooperating or collaborating; the groups are able to share things that can be products, services or even information. Therefore, an ISN can be synthesized as the social network where every person can integrate a group and interact which allow them to share different sorts of things.

Although simple, this chart includes the main ideas of all the participants who joined the first workshop, let's consider two examples: a) "*a group*

Figure 2. Ontology chart for the domain of inclusive social networks

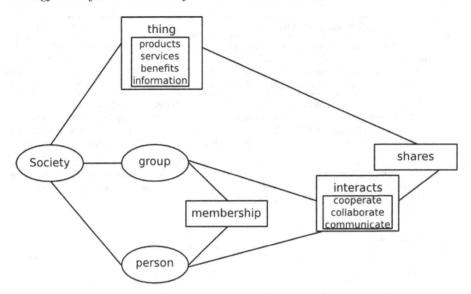

of people that interact sharing different elements without discriminating participants[...]"- as the affordances person and group have no specifics, which means we are considering all people, the chart says that everyone should be able to share, what implies in no discrimination; or b) *"Group of people that have common interests that interact in a way to promote the knowledge and reach personal and group goals"* – further than allowing interaction, an ISN should promote knowledge, which relies on sharing information.

MATERIALIZING THE RESULTS

Based on the initial results obtained with the use of OS artifacts adapted to a participatory approach - as described in the previous Sections - other activities involving OS and PD were held during all the system's development lifecycle, making it possible the conception, design and implementation of an inclusive social network system. Some relevant activities were the development of an agile process model (Bonacin *et al.*, 2009) that fits the project characteristics and, in addition, addresses, research questions related to HCI and

inclusion (Neris *et al.*, 2009b). We conducted a participatory workshop aiming at eliciting norms present in the target information system (Neris & Baranauskas, 2008b), and involving the user in the design of functionalities and layout for the social network system (Almeida *et al.*, 2009; Neris & Baranauskas, 2009b).

Concomitant to participatory requirements elicitation we conducted an analysis of the current most popular social network systems, and extracted requirements for social network systems development considering the Brazilian context (Santana *et al.*, 2009). Based on this work we developed a first prototype and released it in the Web. Recently other participatory activities identified potential new features and improvements. One example is the online conversation tool that is currently being developed to be embedded in the social network system (Hayashi & Baranauskas, 2009).

Although the details of these activities are not in the scope of this article, these activities led to *Vila na Rede* (2009) - the inclusive social network system founded by the approach presented. It is worth mentioning that not all issues raised in these workshops have already been resolved, since there may be items with lower priority, as well as challenges that are still being investigated in our research (still in progress). However, many of the challenges raised have already been worked and can now be checked directly in the current version of the production system. Figure 3 shows the current version of the system (in Portuguese). From this image it is possible to see how some of the issues presented in the OS artifacts contributed to the development of the system.

From the Evaluation Framing we selected some questions to show how they were treated within the current version of the system.

How to deal with participants without access to computers? In the social context in which our research is situated, a considerable range of potential users of the product resulting from our study may not have access, in fact, to computers. Thus, a solution was to provide a way to access the system via other media. Currently, we have implemented a solution - Cel-Cidadania (Silva & Baranauskas, 2009) - which allows users to read, edit and comment advertisements posted on this system directly via mobile phone, thus with no need of using computers. The choice of media, besides other factors, was driven by statistical data relevant to our research context, since these data indicate some important information about the context of use of Information and Communication Technologies (ICTs) in Brazil. For example: 1) In Brazil, 75% of households do not own computers (CETIC.BR, 2009b) and 82% of households have no Internet access (CETIC.BR, 2009a); and 2) 67% of Brazilians use cell phone (CETIC.BR, 2009c).

How to keep communication among people with different functional restrictions? This is one of the big differences of the resulting research product - the inclusive social network: *Vila na Rede* - since the implementation of such a system has been conducted according to the principles of Design for All. Thus, to provide access to our target audience, composed by people with limited interaction with Web applications or even digitally illiterate – we have adopted the strategy of using multiple media for input/output content in system, *i.e.*, textual, audio, video, and LIBRAS (Portuguese acronym for the Brazilian Sign Language) (Figure 3d).

How to cope with the low literacy problem? An approach to not exclude potential users was to use a "virtual presenter". This feature allows textual content posted to the system be automatically read by the system. A virtual presenter is a solution that provides, in addition to text-speech conversion, processing of phonemes to produce a facial animation based on real images (Costa *et al.*, 2009).

How to make the system sustainable and How to prepare community representatives to maintain the system? These two issues are being addressed along our research as we maintain active participation of the different parties of the target audience at all stages of system development. In the Participatory Workshops Section we described how members of the target community are being engaged directly with the development of the system through the use of artifacts from the OS workshops. We believe that this approach contributes to the future autonomy of community members to maintain the system by the end of the project.

How to cope with the difficulties of using the system (to offer products/services)? Understanding that this difficulty is faced by our audience when using the Web in general, we have implemented some functionalities to support the digitally illiterate users to use the system at the same time they are learning the "digital language". An example of this was to offer orientation arrows as an alternative to using the scroll bar (Figure 3e).

How to provide methods for identification of ethical problems? What users consider to be ethical is decided by them; in some cases system managers could provide some content moderation to solve possible conflicts.

How to articulate the dissemination strategy? Throughout our research we are carrying out different actions to motivate members of our audience

Figure 3. Some of the functionalities in the social network system Vila na Rede. (a) announcement types (b) buttons for the features Collaborate and Comment (c) announcement privacy control (d) media options for input of contents (e) optional arrows to easy the scrolling activity

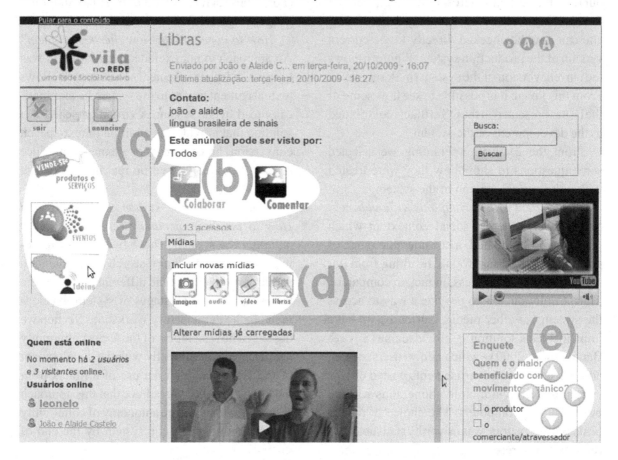

to spread out the use of the system in their region, and also encourage users to use different ICTs as a "new form" of reaching the digital culture. This approach has promoted a more effective use of ICT by our target users.

How to select the most relevant information? The system should provide mechanisms to organize the information. To address these issues in the system we offer different ways of ordering content. In the current version users can apply different filters on content, for example, allowing announcement listing ordered by the more recent (default option), the most accessible, or even by the most discussed.

What is the distinguishing feature of this network that makes it different from the others? As the various social network systems that are currently available on the Web do not offer access to the audience we are considering, we believe that one of the main differences of the product resulting of our research, is a system design anchored by a set of guidelines that indicate some key points for developing such social systems. Note that these guidelines were based on the results of the artifacts previously presented and an analysis of nine different social network systems available on the Web, as described in Santana *et al.* (2009).

DISCUSSION

From social aspects resulted in the analysis, in the Brazilian society, citizens and the final user/community were cited as main agents, formalizing the perception that an Inclusive Social Network system must benefit everyone. Another important aspect is that advanced users were cited, as they are part of the entire information system, but also many of those that are not experienced in computers were cited as well, *e.g.*, weavers, maidservants etc. The diversity of users was also present with the elderly, retired people, people with disabilities, teenagers, among others mentioned.

The Stakeholder Analysis activity also initiated a discussion about the costs to maintain the system. Although Funding Institutions (for social programs or research projects) were cited as partners, the self sustainability was discussed as an important aspect in any social project. As we intend to develop a technical system that will be maintained by the community, we should consider ways towards sustainability.

Regarding technical aspects, other systems to support social networks were mentioned as competitors. These other systems have been technically analyzed regarding the social context of the Project (Santana *et al.*, 2009). Also, other web systems (email clients and searching sites) were cited as well, pointing out that it is necessary to investigate their relation in the community. In addition, tools for evaluating accessibility and initiatives that intend to provide communication between social network systems were mentioned. These tools should be considered in the design of an inclusive system for social networks.

With the Evaluation Frame, it was possible to raise other important concerns regarding an Inclusive Social Network system. Directly related to responsible actors, the participants reinforced the low literacy level of the target population. For this problem, the use of audio-visual solutions besides text was suggested. Other problem is related to the resistance to the unknown. Participants reported that the acceptation of new technology solutions is not easy. As an idea to deal with this problem, the design should be attractive and participants should be motivated to participate in the network. Also, Telecenters have a special role in this question. It is necessary to provide a warm atmosphere where people do not feel intimidated and where they can take their time to learn how to use the computer and the system. Previous research has shown that learning to use the mouse, in particular, is a challenge for many. Nevertheless, training and a design made to minimize mistakes related to motor control skills may help. Another requirement, elicited with the Evaluation Frame, is to allow the use of the system to prospect partners for social projects. Participants would like to obtain support to write projects (for submitting to government or financial institutions) and be aware about dates and opportunities. An idea to tackle this problem also considers the community involvement – the system should support the collaboration of volunteers (people that could correct texts or even translate them to other media, among other activities) to help the less skillful. Considering partners and competitors, participants complained that actual systems present many kinds of communication tools but they do not work together, asking for extra interactive efforts. This turned to be a special challenge – communication should go naturally and in an integrated way.

The Ontology Chart built from participants' definitions for Inclusive Social Networks made clear what people intend with this kind of system. They expect a solution that allows them to interact, communicate, cooperate, collaborate, participate and learn. Besides the peculiarities in each of these terms, sharing (products, services or even ideas and information) seems to be the central idea. Another interesting aspect is that people interact under certain protocol, so the technical system will only be successful if it reflects and supports the rules of the group. Therefore, it is necessary to elicitate which norms are these and how they interfere in the interaction among people.

The combined use of PD techniques and artifacts from OS allowed a socially shared construction of the system, including even those less familiar with the use of technological artifacts. Our experience shows that, although the OS artifacts were initially proposed to be used by analysts, the approach using frames, post-its and a warming environment allows eliciting information from those that are directly influenced by the technical solution.

This participation of the diversity of parties during the workshops reflects in the results, which show a more holistic understanding of the problem, taking into account the norms that direct the users' actions and the relationships between users and technology. This leads to the development of systems that make sense to its community, promoting the use by those not yet within the digital culture.

The Ontology Chart resulted from the Semantic Analysis provided us with an initial view for social networks as them occur in the real world. During the activity people expressed their interest in sharing and exposing their work and ideas reflecting the participants' characteristics. Consequently, this special interest were materialized by focusing the design of the social network system to allow people expose "things" and engage in communication acts dynamically, according to their interest in such "things", instead of establishing static contact lists.

The affordances and relationships present in the Ontology Chart guided our efforts in the design of a computer-mediated system. Some elements from the Ontology Chart can be identified in the online social network system.

Things: This affordance was materialized in three announcement categories, *i.e.*, products/services, events, and ideas (see Figure 3a). Products and services were directly mapped; information and benefits were mapped into "ideas". So people are able to put in the system their subject interests and navigate through the others too.

Interaction: Each specific of this affordance, *i.e.*, Collaborate (Coll), Cooperate (Coop), and Communicate (Comm) was mapped into system functionalities:

- **Conversas Online (Comm):** An instant messaging tool embedded in the social network that allows users to talk using different media simultaneously;
- **Comments (Comm, Coop):** Allow users to register multimedia comments about an announcement (see Figure 3b);
- **External Contacts (Comm):** This is an optional resource that allows users to register their phone number in their profiles. The phone number is automatically published in the users' announcements, and;
- **Collaboration (Coll):** One important feature of the social network is to provide features to promote the sustainability of the system and collaboration among users. For this purpose the feature "Collaboration" (see Figure 3b) allows users to add contribution to other announcements as, for example, someone could add a caption to an image or video, other could add a video using sign language explaining the content of the announcement.

Shares, and of: As we have mentioned in the affordance "things" people are able to register their announcements and, additionally they can choose how they want to share them *i.e.* make them visible to all visitants or only to registered users (see Figure 3c). Similarly, users can choose how they allow collaborations to their content *i.e.* every visitant can collaborate, only registered users, no one.

People: One important premise of our work is the inclusion of every person in the system. This means not providing "especial" areas for different people but promoting the coexistence of differences. Aiming that, the system provides users with different input and output media, besides accessibility conformance with available guidelines (W3C, 2009).

Finally, the aspects of simplicity and generality of the ontology chart can be discussed. Although few affordances and agents are present in the chart, the ontological dependencies between them demand adequate HCI solutions for any technical system that would support an ISN. For instance, if a social network system requires that a user has an email account to become a member, people that do not have an email account cannot join the group and the affordance "membership" does not exist. If only text can be used to communicate, then those that do not write cannot interact. Several other (counter) examples could be mentioned.

The focus with the semantic analysis was to conceptualize an ISN considering opinions arising from a heterogeneous group, without focusing on any specific social network domain. If applied to a specific domain, the chart could be extended and specific affordances would be specified. If considering the domain of professional recommendation, affordances as organization or determiners as how many years of professional experience a person has would appear. Therefore, the actual stage of the chart represents the generality coming from the participants' concept for ISN.

CONCLUSION

The complexity of developing a system to support Inclusive Social Networks that make sense to the users requires a socio-technical vision of the problem and a participatory and inclusive approach for the proposal of solutions. We adopted OS and PD as theoretical and methodological frames of reference that permitted a treatment of the problem that equally considers the informal, formal and technical levels of the social group and the system in question.

This paper presented the activities conducted within the target community in a participatory

workshop, where the Stakeholder Analysis Chart and the Evaluation Framing were filled out by representatives from the community, designers and developers in a participatory approach. Also, participants described an Inclusive Social Network from their point of view. From the participants' speeches and the variety of stakeholders that were elicited, it was possible to notice how the community demands a system to be inclusive. The descriptions written by the participants were analyzed and from them, the researchers built an Ontology Chart to represent the affordances related to Inclusive Social Networks.

Therefore, the OS and PD approach contributed to the clarification of the problem, identification of stakeholders, and delimitation of the solution scope with regard to technology and social/digital inclusion for the design of Inclusive Social Networks. The results regarding ethics, literacy, volunteering, and sustainability are examples of important points that could have been neglected with other approaches. This shared understanding of the problem helped us to search for solutions that make sense to the community and that were materialized in the software application.

ACKNOWLEDGMENT

This work is funded by Microsoft Research – FAPESP Institute for IT Research (#2007/54564-1), and partially by the State of São Paulo Research Foundation – FAPESP (#2006/54747-6 and #2007/02161-0), the National Council for Scientific and Technological Development – CNPq (#141489/2008-1), and the Brazilian Federal Agency for Support and Evaluation of Graduate Education – CAPES (#01-8503/2008). The authors also thank colleagues from IC/UNICAMP, NIED/UNICAMP, InterHAD, Casa Brasil, CenPRA and IRC/University of Reading for insightful discussion.

REFERENCES

Almeida, L. D. A., Neris, V. P. A., Miranda, L. C., Hayashi, E. C. S., & Baranauskas, M. C. C. (2009). Designing inclusive social networks: A participatory approach. In *Proceedings of the 3rd International Conference on Online Communities and Social Computing* (pp. 653-662).

Baranauskas, M. C. C., Bonacin, R., & Liu, K. (2002). Participation and signification: Towards cooperative system design. In *Proceedings of the Anais do V Simpósio sobre Fatores Humanos em Sistemas Computacionais* (pp. 3-14).

Baranauskas, M. C. C., Schimiguel, J., Simoni, C. A. C., & Medeiros, C. M. B. (2005). Guiding the process of requirements elicitation with a semiotic approach – A case study. In *Proceedings of the 11th International Conference on Human Computer Interaction* (pp. 100-110).

Bonacin, R., Melo, A. M., Simoni, C. A. C., & Baranauskas, M. C. C. (2009). Accessibility and interoperability in e-government systems: outlining an inclusive development process. [REMOVED HYPERLINK FIELD]. *Universal Access in the Information Society, 9*(1), 17–33. doi:10.1007/s10209-009-0157-0

Bonacin, R., Rodrigues, M. A., & Baranauskas, M. C. C. (2009). An agile process model for inclusive software development. In *Proceedings of the 11th International Conference on Enterprise Information Systems* (pp. 807-818).

Bonacin, R., Simoni, C. A. C., Melo, A. M., & Baranauskas, M. C. C. (2006). Organisational semiotics: Guiding a service-oriented architecture for e-government. In *Proceedings of the 9th International Conference on Organisational Semiotics* (pp. 47-58).

CETIC.BR. (2009a). *Proportion of Brazilian households with Internet access.* Retrieved from http://www.cetic.br/usuarios/tic/2008-total-brasil/rel-geral-04.htm

CETIC.BR. (2009b). *Proportion of Brazilian households with computers.* Retrieved from http://www.cetic.br/usuarios/tic/2008-total-brasil/rel-geral-01.htm

CETIC.BR. (2009c). *Proportion of Brazilian individuals who used cell phone in the last three months.* Retrieved from http://www.cetic.br/usuarios/tic/2008-total-brasil/rel-semfio-01.htm

Checkland, P., & Scholes, J. (1991). *Soft systems methodology in action.* New York, NY: John Wiley & Sons.

Costa, P. D. P., De Martino, J. M., & Nagle, E. J. (2009). Speech synchronized image-based facial animation. In *Proceedings of the International Workshop on Telecommunications*, São Paulo, Brazil (pp. 235-241).

Cotler, J., & Rizzo, J. (2010). Designing value sensitive social networks for the future. *Journal of Computing Sciences in Colleges, 25*(6), 40–46.

e-Cidadania. (2008). *E-cidadania project: System and methods for the constitution of a culture mediated by information and communication technology.* Retrieved from http://www.nied.unicamp.br/ecidadania

Friedman, B. K. (2006). Value sensitive design and information systems. *Human-Computer Interaction and Management Systems, 6*, 348–372.

Gibson, J. J. (1979). *The ecological approach to visual perception.* Boston, MA: Houghton Mifflin.

Hayashi, E. C. S., & Baranauskas, M. C. C. (2009). Communication and expression in social networks: Getting the "making common" from people. In *Proceedings of the IEEE Latin American Web Congress, Joint LA-WEB/CLIHC Conference* (pp. 131-137).

Hendler, J., Shadbolt, N., Hall, W., Berners-Lee, T., & Weitzner, D. (2008). Web science: An interdisciplinary approach to understanding the web. *Communications of the ACM, 51*(7), 60–69. doi:10.1145/1364782.1364798

Jung, Y., & Lee, A. (2000). Design of a social interaction environment for electronic marketplaces. In *Proceedings of the 3rd Conference on Designing Interactive Systems: Processes, Practices, Methods, and Techniques* (pp. 129-136).

Kolkman, M. (1993). *Problem articulation methodology*. Unpublished doctoral dissertation, University of Twente, Enschede, The Netherlands.

Liu, K. (2000). *Semiotics in information systems engineering*. Cambridge, UK: Cambridge University Press. doi:10.1017/CBO9780511543364

Liu, X. (2001). *Employing measure methods for business process reengineering in China*. Unpublished doctoral dissertation, University of Twente, Enschede, The Netherlands.

Miranda, L. C., Hornung, H. H., Solarte, D. S. M., Romani, R., Weinfurter, M. R., Neris, V. P. A., & Baranauskas, M. C. C. (2007). Laptops educacionais de baixo custo: Prospectos e desafios. In *Proceedings of the Anais do 18th Simpósio Brasileiro de Informática na Educação* (pp. 358-367).

Miranda, L. C., Piccolo, L. S. G., & Baranauskas, M. C. C. (2008). Artefatos físicos de interação com a TVDI: Desafios e diretrizes para o cenário brasileiro. In *Proceedings of the Anais do 8th Simpósio Brasileiro de Fatores Humanos em Sistemas Computacionais* (pp. 60-69).

Muller, M. (1997). Participatory practices in the software lifecycle. In Helander, M., Landauer, T. K., & Prabhu, P. V. (Eds.), *Handbook of human-computer interaction* (pp. 255–297). Amsterdam, The Netherlands: Elsevier Science.

Mumford, E. (1964). *Living with a computer*. London, UK: Institute of Personnel Management.

Mumford, E., & Henshall, D. (1979). *A participative approach to computer systems design: A case study of the introduction of a new computer system*. London, UK: Associated Business Programmes.

Nadin, M. (1988). Interface design: A semiotic paradigm. *Semiotica, 69*(3-4), 269–302. doi:10.1515/semi.1988.69.3-4.269

Neris, V. P. A., Almeida, L. D. A., de Miranda, L. C., Hayashi, E. C. S., & Baranauskas, M. C. C. (2009a). Towards a socially-constructed meaning for inclusive social network systems. In *Proceedings of the 11th International Conference on Informatics and Semiotics in Organisations* (pp. 247-254).

Neris, V. P. A., & Baranauskas, M. C. C. (2009a). Designing e-government systems for all – a case study in the Brazilian scenario. In *Proceedings of the IADIS International Conference on WWW/Internet* (pp. 1-8).

Neris, V. P. A., & Baranauskas, M. C. C. (2009b). Interfaces for all - a tailoring-based approach. In *Proceedings of the 11th International Conference on Enterprise Information Systems*, Milan, Italy (pp. 928-939).

Neris, V. P. A., Hornung, H. H., Miranda, L. C., Almeida, L. D. A., & Baranauskas, M. C. C. (2009b). Building social applications with an agile semio-participatory approach. In *Proceedings of the IADIS International Conference on WWW/Internet* (pp. 1-8).

Norman, D. A. (1988). *The psychology of everyday things*. New York, NY: Basic Books.

Norman, D. A. (2008). Signifiers, not affordances. *Interaction, 15*(6), 18–19. doi:10.1145/1409040.1409044

Santana, V. F., Solarte, D. S. M., Neris, V. P. A., Miranda, L. C., & Baranauskas, M. C. C. (2009). Redes sociais online: desafios e possibilidades para o contexto brasileiro. In *Proceedings of the Anais do XXIX Congresso da Sociedade Brasileira de Computação* (pp. 339-353).

Schuler, D., & Namioka, A. (1993). *Participatory design: Principles and practices*. Mahwah, NJ: Lawrence Erlbaum.

Silva, F. B., & Baranauskas, M. C. C. (2009). Cel-cidadania: Uma aplicação para celulares integrada a redes sociais inclusiva. In *Proceedings of the 17ᵗʰ Congresso Interno de Iniciação Científica da UNICAMP*.

Stamper, R. (1988). Analysing the cultural impact of a system. *International Journal of Information Management, 8*, 107–122. doi:10.1016/0268-4012(88)90020-5

Stamper, R. (1996). Signs, norms, and information systems. In Holmqvist, B., Andersen, P. B., Klein, H., & Posner, R. (Eds.), *Signs at work* (pp. 349–397). Berlin, Germany: Walter de Gruyter.

Stamper, R., Althaus, K., & Backhouse, J. (1988). MEASUR: Method for eliciting, analyzing and specifying user requirements. In Olle, T., Verrijn-Stuart, A., & Bhabuts, L. (Eds.), *Computerized assistance during the information systems life cycle* (pp. 143–163). Amsterdam, The Netherlands: Elsevier Science.

Stamper, R., & Kolkman, M. (1991). Problem articulation: A sharp-edged soft systems approach. *Journal of Applied System Analysis, 18*, 69–76.

Treiber, M., Truong, H. L., & Dustdar, S. (2009). SOAF - Design and implementation of a service-enriched social network. In *Proceedings of the 9th International Conference on Web Engineering* (pp. 379-393).

Venkatesh, M. (2009). The constitutive and the instrumental in social design. In *Proceedings of the 4th International Conference on Design Science Research in Information Systems and Technology*.

Vila na Rede. (2009). *Vila na rede: An inclusive social network*. Retrieved from http://www.vilanarede.org.br

W3C. (2009). *Web content accessibility guidelines 2.0*. Retrieved from http://www.w3.org/TR/WCAG/

This work was previously published in the International Journal of Information Systems and Social Change, Volume 2, Issue 3, edited by John Wang, pp. 16-35, copyright 2011 by IGI Publishing (an imprint of IGI Global).

Chapter 12
A Semiotic Analysis of a Model for Understanding User Behaviours in Ubiquitously Monitored Environments

Keiichi Nakata
University of Reading, UK

Stuart Moran
University of Reading, UK

ABSTRACT

Improvements in electronics and computing have increased the potential of monitoring and surveillance technologies. Although now widely used, these technologies have been known to cause unintended effects, such as increases in stress in those being observed. Further advancements in technology lead people towards the 'pervasive era' of computing, where a new means of monitoring ubiquitously becomes possible. This monitoring differs from existing methods in its distinct lack of physical boundaries. To address the effects of this kind of monitoring, this paper proposes a model consisting of a series of factors identified in the monitoring and pervasive literature believed to influence behaviour. The model aims to understand and predict behaviour, thereby preventing any potential undesirable effects, but also to provide a means to analyse the problem. Various socio-technical frameworks have been proposed to guide research within ubiquitous computing; this paper uses the semiotic framework to analyse the model in order to better understand and explain the behavioural impact of ubiquitous monitoring.

INTRODUCTION

Being watched or monitored has always been a part of human life to assist and enhance security, safety, health and even productivity assessment. Over time, advancements in electronics and computing have seen improvements to monitoring technologies, resulting in an increase in their adoption and application. *Even though these technologies are now widely used and easily accessible, they have been often known to cause undesirable effects* (Vorvoreanu & Botan, 2000), *such as increases in stress and distrust in those being observed.*

DOI: 10.4018/978-1-4666-2922-6.ch012

Advancements in mobile and wireless technologies have led towards what can be referred to as the 'pervasive era' of computing. This brings intelligent pervasive spaces (IPSs) and their manifestations, such as 'intelligent buildings', closer to reality, as they will be some of the first adopters of the technology. The principle of pervasive or ubiquitous computing is to embed computers into the environment, removing them from the focus of our daily lives. In order to function as intended, the embedded devices need to continuously collect significant amounts of data regarding user behaviours (Albrechtslund, 2007).

While existing monitoring technologies can be used to achieve this level of data collection, in order to accomplish true pervasiveness, the monitoring devices themselves must act ubiquitously. This new Ubiquitous Monitoring (UM) differs from existing methods mainly due to the absence of physical restrictions such as walls and other physical attributes of the environment.

Insufficient research has been applied to the study of the effect of UM on human behaviour (Jonsson, 2006). In light of this, a model has been created, consisting of a series of factors drawn from the pervasive/ubiquitous computing and monitoring literature which are believed to influence behaviour. The model is augmented by the Theory of Planned Behaviour (TPB) (Ajzen, 1991), and could allow user behaviours to be understood and possibly predicted in an IPS.

The purpose of this paper is to analyse how these behaviour-influencing factors can be understood in terms of how they relate to the data collected, how data is processed and assigned meanings, and what social and behavioural effects are expected as a result. To this end, a semiotic analysis is applied to analyse the factors identified and the proposed model, which enables us to further examine how these factors contribute to the understanding of problems related to UM.

This paper is structured as follows. First, we review the background and related work in this topic, including discussion of the behaviour influencing factors related to monitoring. In the following section, we propose a model that relates the identified factors to one another and to the TPB. This is followed by an analysis of the factors from three levels—physical, technical, and social. The model is then analysed from a semiotic perspective of the problem using the semiotic ladder, followed by a discussion on these analyses in relation to the information field paradigm, and a conclusion.

BACKGROUND MONITORING

The current application of information technology for monitoring has been known to result in unintended effects on users (Vorvoreanu & Botan, 2000). These undesirable effects are often found to outweigh the benefits of such systems, creating an overall negative impact (Botan & Vorvoreanu, 2005). Simply having awareness of the monitoring changes behaviour, and a well known, but often contested example of this, is the Hawthorne effect (Roethlisberger & Dickson, 1939). Employees in the Hawthorne works were monitored as part of an experiment to establish a relationship between their productivity and lighting levels. The employee's productivity was unexpectedly found to increase regardless of the light intensity; this was attributed to the fact that they were aware that they were being observed. Awareness of monitoring has even been shown to cause changes in a person's writing style and internet browsing habits (Botan & Vorvoreanu, 2005). While previous studies provide an insight into the likely effects of UM (Hayes et al., 2007; Zweig, 2005), its pervasiveness remains an area that has had little attention. This may result in an overall increase in the negative effects found in existing monitoring systems, or even generate previously unseen effects.

Ubiquitous Monitoring

In this paper we use a working definition of an IPS as an adaptable and dynamic environment that optimises user services and management processes using intelligent systems and ubiquitous technologies. IPSs are often controlled by software known as intelligent agents, who monitor the users and alter the environment according to 'perceived' or stated preferences, such as those related to temperature, lighting and humidity. Once sufficient data has been collected, the system takes into account the requirements of all the users. By collecting this data, many tasks and actions can be automated and forced into the background, thus contributing toward Weiser's idea of invisibility (Weiser, 1991).

Ubiquitous Monitoring is the use of pervasive devices for collecting data in an IPS or other environment. These devices are generally unrestricted by physical boundaries, and as such their capabilities are significantly greater than existing monitoring technologies (Langheinrich et al., 2005).

Given that UM is a monitoring technology, it is anticipated that many of the effects caused by existing monitoring technologies will also occur in ubiquitously monitored environments. In addition to this, previously unseen effects may occur due to the increased coverage of the monitoring (Vorvoreanu & Botan, 2000).

Among existing work related to IPS and UM, Tiburcio and Finch (2005) have looked at the positive impact an intelligent classroom has on pupil behaviour, and Clements-Croome et al. (2006) conducted a study which found that occupants like their environments to be both controllable and adaptable. Live-in laboratories such as the Place Lab (Intille et al., 2005) have been constructed in an attempt to create a natural environment in which to study the behaviour of individuals in intelligent homes. Some of the data collected through laboratory studies have shown that ubiquitous technology does cause a change in human behaviour (Beaudin, Intille, & Tapia, 2004). There are, however, limitations to this method of study as the environment places constraints on behaviour variability (Intille et al., 2005) and is unlikely to generate many behaviours that would occur in real life scenarios (Konomi & Roussos, 2007). Even with this research, little is still known about the effects of UM (Jonsson, 2006).

BEHAVIOUR INFLUENCING FACTORS

Based on the examination of the monitoring and ubiquitous/pervasive computing literature, a series of seven factors related to UM was identified that are believed to affect human behaviour. These are: the *context* in which the monitoring takes place, whether or not it has been *justified*, the levels of *awareness*, *intrusion* and *control* felt by the user implied by the technology used, the *boundaries* of the monitoring and whether those being monitored *trust* who is collecting the data and how it is used.

In order to understand the true consequences of UM, studies must be conducted in multiple contexts. IPSs can be designed to function in almost any context and the technology used is likely to be dependent on the context. The value of a technology to a user is also likely to vary according to the context in which it is used (Sheng, Nah, & Siau, 2006). Justification depends strongly on the context, for example monitoring in a prison is justified as it ensures that people are prevented from committing further crimes. Without such justification, levels of trust are likely to decrease (Stanton, 2000). The more intrusive an UM device, the more aware a user will be of the monitoring. Consider the intrusiveness of wearable sensors: a user could not wear such a device without having a heightened awareness of the monitoring taking place; there is also a risk of intruding on a person's personal space. If UM technologies are to be accepted and used for lengthy periods, they must not be perceived as intrusive (Intille et al., 2003).

Entirely autonomous environments are likely to find undesirable responses from users (Intille et al., 2003), which implies that IPSs should provide a means of manual control. With this manual control will come an increased sense of awareness of both the monitoring and the environment. Having control over monitoring has been shown to be beneficial to users (Stanton, 2000); but with increased ubiquity and levels of data collection, there is the question of how much control a user should be given.

Unless a user trusts who is carrying out the monitoring, how the data is collected and for what purpose, the monitoring is unlikely to be accepted (Zweig, 2005). Current monitoring technologies are restricted to defined boundaries, such as the range of a CCTV camera or the actions carried out on a desktop computer. With the introduction of UM these boundaries are extended. This factor may heavily influence the other factors, and could be the main contributor toward any previously unseen undesirable effects. Each of the above factors may affect the behaviour of those being monitored to some degree.

Some additional factors have been identified which could also cause behavioural changes in an IPS. *Privacy, ethical, economic, cultural* and *legal* issues are important topics which have been studied in various contexts. Unfortunately these are not easily separated from the other factors, and as such, are difficult to integrate into the proposed model. However, their influence can be represented within one or more of the identified factors. When considering different cultures, interpretations of the purpose and intention of the monitoring are likely to vary. The impact of this can be explained via the social factors of trust and justification. The physical attributes of a technology, such as the level of control it provides or its physical boundaries, are purely objective and can be considered independent of culture.

EXPLAINING BEHAVIOUR

We are interested in how UM affects the behaviour of those being monitored. Arguably, the effects on behaviour can be analysed by studying whether ubiquitous technology is accepted by users. As such, we examined the approaches in technology acceptance studies as the method of modelling user behaviours. We assume that, if systems are not accepted, they are unlikely to be used as intended, if at all. The Technology Acceptance Model (TAM) (Davis, 1989) is an influential theory, based on Theory of Reasoned Action (TRA) (Ajzen & Fishbein, 1980), which models how users come to accept and use a technology. While the model initially seemed an appropriate choice to theoretically link the factors to behaviour, in the context of this research there are more behaviours, in addition to acceptance, that could be displayed in a ubiquitously monitored environment. The Theory of Planned Behaviour (TPB) (Ajzen, 1991), also based on TRA, can be used to predict and explain human behaviour by examining its relation to intentions, attitudes and beliefs towards behaviour. A person's intention is influenced by their attitude, subjective norms and perceived control over behaviour. These are in turn determined by behavioural, normative and control beliefs, which are beliefs about the likely consequences of the behaviour, the opinions of valued people about the behaviour, and factors that may assist or hinder performance of the behaviour.

MODEL

The model presented in Figure 1 depicts the relationship between the behaviour influencing factors identified earlier and the TPB. Using this model we are able to explain the occurrence of behaviours by investigating how the factors, which act as external variables, influence salient beliefs. The effects of this influence may then be propagated through the model. Including the

TPB not only creates a more relevant behavioural analysis tool but may also provide more insights into why certain behaviours have occurred in the first place.

The *context* in which monitoring resides and the *justification* behind it will affect which technology is chosen for monitoring users (A, B). *Justification* is only likely to be meaningful if it is *trusted (C)*. Levels of *awareness*, *control* and *intrusion* felt by the user, as well as the *boundaries* of the monitoring, are all defined by the technology itself (D, E, F, G). These factors in turn are believed to influence user behaviour through their salient beliefs.

Depending on the *context*, the consequences of certain behaviours may change and through this, a person's behavioural beliefs can be explained (H). Whether or not the monitoring can be *justified* may impact a person's behavioural beliefs (I). If someone knows why they are being watched, they are more likely to have an increased awareness of the consequences of particular behaviours. *Awareness* of being watched could influence a person's normative beliefs, with the identity of the person/machine carrying out the monitoring altering their adherence to social norms (J). Without an awareness of being monitored, the user may be less aware of the consequences of certain

behaviours (K). Having *control* over aspects of an environment, or even the monitoring itself, could be seen as having control over behaviours and even control over their consequences. For example, if a person has control over heating and lighting it may influence their energy saving behaviour, explaining both behavioural and control beliefs (L, M). With different technologies come different levels of *intrusion*, perhaps changing a user's perception of how much control they have over the monitoring, and therefore their behaviour (N). UM is potentially *unbound* and with this comes a lack of personal space, which could lead to a change in a user's perception of how much control they have over a behaviour (O). There may be an inherent relationship between levels of intrusion and awareness, where intrusive methods of monitoring result in a heightened sense of awareness over unobtrusive methods (P). Having control over the monitoring or the environment could also have a similar effect on awareness (Q). The behaviours displayed by users of an IPS are either intentional or unintentional from a designer's perspective (R, S); with unintentional behaviour being either desirable or undesirable (T, U). It is also possible that there is no effect on behaviour, which is also explained by the model (V).

Figure 1. Model for understanding and predicting the effects of ubiquitous monitoring on human behavior

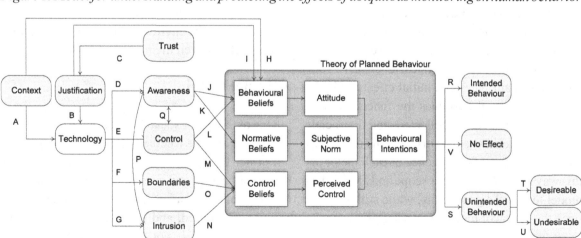

Our model acts as a first step to identifying the likely impact of particular factors in the implementation of UM on human behaviour, establishes a means of understanding this behaviour and provides a basis for investigating how these factors can be used to predict behaviours. The current model is limited in the sense that it only identifies the potential relationships between the factors and the TPB, without empirical evidence to support them. However, the model does provide a means for guiding the analysis of the problem of behavioural change. In order to enhance the model, a series of experiments and surveys will be conducted to discover to what extent the factors discussed affect behaviour and the strength of the relationships between them. With the information gained from the completed model, designers would be able to improve existing and future IPS systems, preventing any of the potential undesirable effects likely to be associated with UM (Davis, 2002; Langheinrich, Coroamă, Bohn, & Mattern, 2005; Zweig & Webster, 2002)

INFLUENCES OF FACTORS

Ubiquitous computing consists of three interrelated environments: social, technical and physical, where a change in one will affect the other (Nguyen & Mynatt, 2002). Analysing the problem using this framework creates the opportunity to view it from different levels. Each factor described above influences behaviour in different ways and can be categorised into the three different levels based on these environments. The physical level focuses on the physical attributes and initial effects of UM, the technical level examines the functions of a device and what data is collected, and the social level focuses on the social attributes and use of the data. Categorising the influence of each factor in this way gives them a wider scope and allows the model to be used generically, while still leaving open the option of focusing on a specific context or system.

Physical Level

Pervasive sensors and other devices will be used for UM, and their physical presence will act as a visual cue to the monitoring and data collection taking place. As devices become smaller, unobtrusive monitoring becomes more realistic; however, the physical actuations as a result of the monitoring will still act as a sign of the monitoring taking place. The physical boundaries of ubiquitous devices could also cause decreases in perceptions of private space. Whether in a building, outside or even in a vehicle, the physical context of the monitoring could influence behaviour.

Technical Level

The devices used for UM vary in their purpose and functions. Different types of data are required for automating particular aspects of a person's life, and require different methods of collection. Such data could include users' preferences on heating, their natural behaviours or even their bank account details; no information in an UM environment need be considered useless. Ideally this data should be collected passively but in some cases the devices will be placed near or on a user, and this presence could be perceived as an intrusion of personal space. Even when data is collected passively, the data itself could equally be considered intrusive of personal privacy. In some scenarios, UM devices will provide or restrict user control over the environment and other variables, and perhaps even the monitoring itself.

Social Level

One of the major social issues related to any form of monitoring is whether or not it can be justified. Further to this, the types of data, and levels of data collection and access, all need justification. This is also dependent on the nature of the people being monitored, and the shared understanding of what is and is not *normally* accepted in the society and the social context to which they belong. In relation to the data, people should be aware of, have control over and trust how much data is collected, who has access to this data, the types of data and how it is used. Some of the data collected could be perceived as intrusive of personal privacy. This data is likely to be made widely available in a person's life, shared across home and work environments, and cause issues with data access. Hospitals, schools, workplaces and the home are all social contexts in which UM could be implemented, and will have a significant impact on users' behaviour and acceptance of the monitoring.

Through the analysis of UM in these three levels, we can observe the parallels with the semiotic framework.

AN ORGANISATIONAL SEMIOTICS PERSPECTIVE

Organisational semiotics (Liu, 2000) applies the principles of semiotics—a study of signs—to analyse and understand technology in organisational and social settings. It provides a range of methods to analyse the requirements of information systems, but also offers a way to understand what social effects technology is intended to produce. The semiotic framework is used to explain signs, and the ability to analyse an information system through the semiotic ladder (Stamper, Althaus, & Backhouse, 1988) allows us to examine sign-based systems at different levels within a socio-technical system. It is for this reason that organisational semiotics is a useful tool for studying UM and its effects. Elements of our model, including some of the relationships between factors, can be found in the semiotic framework. In order to use the framework effectively, we must first take the semiotic viewpoint that a device can act as a sign.

One of the central concepts in Peircian semiotics (Hartshorne & Weiss, 1960) is the notion of *semiosis*. Semiosis can be understood as a sense-making process of completing the linkage between three elements: a sign that stands for something, an object that is signified by the sign, and an *interpretant* which is a process of assigning a meaning to the sign to signify the object. As such the same sign can stand for different objects depending on the interpretant. Moreover, an 'object' above is better identified as a *referent*, since it can also be a concept or even an action that is represented by a sign.

In the case of a monitoring device, it can be considered to be a sign that signifies the act of monitoring. Depending on certain factors within an IPS, someone who sees this sign will interpret the visual presence of the device in different ways. For example, in a workplace an occupant may consider a CCTV camera as a surveillance device, a manager may consider it a device for increasing productivity and a researcher may only consider it as a method of data collection. Due to the different interpretations of the device and its purpose, varying behaviours are likely to be seen. The physical actuations caused by a device can act as a sign of the monitoring, and is also likely to result in behavioural changes. The physical environment itself, be it a building, a vehicle or simply outdoors can provide a context in which the intention of the existence of the device can be interpreted. Figure 2 illustrates an example of semiosis of a monitoring device as a semiotic triangle.

The contextuality of meanings to signs is captured in semiotics by three aspects of meaning assignment to signs. The syntactic aspect of a sign examines whether the sign or combinations of signs is well-formed to be understood as signs. The semantic aspect assigns meanings to the syntactically well-formed signs. However, as observed earlier, the same sign can have different meanings depending on the context - this is the pragmatic aspect of the sign. In communicative acts, signs are used to convey meanings with an intention of the initiator of the sign to create an effect on the recipient of the sign in a specific

Figure 2. An example of semiotic triangle for the understanding of a monitoring device

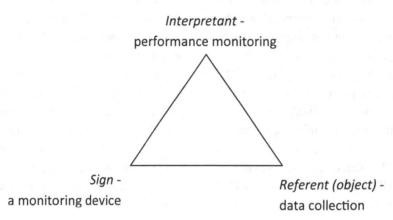

context. As such the pragmatic aspect is also associated with the *intensional* aspect of the sign.

To these three aspects (or levels), Stamper added two more levels leading up to the syntactic level, namely the physical level and the empiric level, and another level beyond the pragmatic level, namely the social world (described in Liu, 2000). The physical level (or the physical world) is concerned with the physical properties of the sign. The transmission of the signs and their assemblage are the concern of the empiric level. At the other end, while the pragmatic level is concerned with the intensionality of signs, the social world is concerned with the actual social effect of the sign. This results in the six levels in the semiotic framework (Figure 3). Since signs can be anything that stands for something, this framework can be used to analyse not only a simple exemplification

of a sign but also more complex human activities such as business processes (Liu, 2000).

Hence, it is possible to examine UM within the framework as a whole, thus consider the following: The device itself (*Physical World*), coupled with other sensors and technologies are introduced into an environment (*Empirics*). These devices may or may not be appropriate in terms of their placement within their surroundings (*Syntactics*) and in some cases could be considered intrusive. Depending on the stakeholder, the devices are interpreted as a monitoring system (*Semantics*) and will have been implemented for a specific purpose or intention (*Pragmatics*). The intrusiveness of a device could be interpreted as a sign of intention, e.g., choosing a particular device over another, for instance, a smart watch against a camera, could easily change a person's perception of the purpose and intentions of the

Figure 3. The semiotic framework and its relation to the seven behaviour influencing factors (based on Stamper et al., 1998)

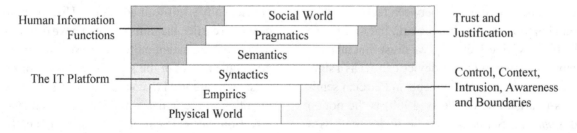

device, leading to greater acceptance. Users and other stakeholders then form ideas such as about how comfortable they are with the monitoring and whether or not they think it is acceptable according to their social norms (*Social World*). What is clear from this traversal through the semiotic framework is that we are able to view and understand the problem from different levels, at each of which a behavioural change is possible. This phenomenon is caused by the nature of monitoring devices as signs, which lend themselves to interpretation.

BEHAVIOUR INFLUENCING FACTORS AND THE SEMIOTIC FRAMEWORK

After examining how semiotics can be applied to UM and understanding the behavioural factors at different levels, it is evident that parallels exist. Figure 3 describes how the factors that form our model relate to the semiotic framework.

Intrusion is associated with the physical nature of the device, the effect of which is propagated up to the social level creating a *sense* of intrusion for the user. The same can be observed of awareness, although this is intensified by the empiric nature of the device, again propagated to create a perceptive effect of being aware of the device in the social level. Control is enabled by the physical device, and boundaries are also strongly associated with the capability of the device. These again have an impact on user's perceptions regarding the capability of devices at the social level. Context, in particular the physical context, would have influence across all levels of the semiotic framework, and all types of environment in ubiquitous computing. Boundaries present an illustrative example of influences across the levels. Boundaries can be purely 'physical', constrained by the extent to which the devices cover. However, this is not the same as what one may perceive, i.e., the

physical boundaries may be perceived as greater (or smaller) than what is physically possible. This is carried up the levels to the concept of what is and is not expected as boundaries according to the social norm. These five factors have an effect at both the physical and social levels, e.g., there is a possible relationship between the physical/syntactical and social context, in which placing a device in an environment may influence or define the social context formed around it.

The influence of trust and justification, however, are only effective within the social aspects of Ubiquitous Computing, and are therefore restricted to the upper levels of the semiotic framework. In particular, trust is associated with the social norm of the environment in which the users encounter the monitoring device. Similarly, justifications, or whether justifications make sense to the users, depend on the social norm. In certain situations or cultures it could be more justifiable to place surveillance on people than in others, which depends to a great extent on the shared social values.

By linking the factors believed to influence behaviour in UM with the semiotic framework, a new perspective of the problem, and therefore understanding, has been achieved. What this reveals is that the factors themselves are subject to variations in interpretations at different semiotic levels, adding another dimension to the model. Even with the same monitoring device, it might carry different meanings to those being monitored and thereby affect the level of trust depending on how it is placed in the surroundings. The framework has identified issues regarding stakeholder viewpoint in the model and the semiotic way of thinking may be able to help rectify this. With further analysis of the link between organisational semiotics and UM, it is possible that previously unconsidered issues may be found and investigated. In this sense, the work presented in this paper is a first step towards a more comprehensive analysis based on semiotic approaches.

DISCUSSION

Although the proposed model requires empirical validation for it to be used as a predictive model, it provides a research model to analyse and understand the problem. The two approaches to the analyses of behaviour influencing factors in UM described above support each other. When seen from each layer of physical, technical, and social it is evident that the ubiquitous nature of monitoring devices encompasses all three layers. In the semiotic analysis, factors of trust and justification were associated with the human information function levels of the semiotic framework, while other factors traverse all the levels. An interesting observation from the semiotic perspective is that the behavioural effects of UM are best captured at the social level, where the interpretation of signs based on (informal) norms, beliefs and expectations lead to social effects such as actions.

This opens up the analysis to another related concept in organisational semiotics, namely that of 'information field' (Stamper, 2000). An information field can be seen as a set of norms that are shared by a group of people or a society, which binds them with a shared sense of values and meanings. The information field paradigm was originally introduced as a reaction to the traditional information flow paradigm that assumed that systems can be built solely based on the assumption that information is uniquely interpreted. Table 1 summarises the comparison between these two paradigms as suggested by de Moor and Weigand (2002). As the table suggests, in the information field paradigm, to perceive, understand, value, and act are the objectives in assessing information, rather than to control it. This viewpoint better suits the type of investigation carried out in the context of this paper. The concept of information field can be effectively applied to explain the ways in which UM can be interpreted in different ways by the user, and the differences in norms may lead to unexpected or unintended behaviours.

Table 1. Comparison of information flow and information field (adapted from de Moor & Weigand, 2002, p. 277)

	Information Flow	**Information Field**
Design process	Representation	Interpretation
Objective	Control	Perceive, understand, value, act
Control logic	Rule	Norms

Stamper et al. (2000) further categorise norms into the following categories: *perceptual norms, evaluative norms, cognitive norms,* and *behavioural norms*. Among these categories of norms, evaluative norms, which are related to the system of values associated with signs. This would include how signs are evaluated as a norm, e.g., how a monitoring device is considered. This relates to the factors of justification and perhaps trust through cognitive norms. Evaluative norms also affect how boundaries, social and physical, are perceived (Stamper et al., 2000). Behavioural norms would influence how users would behave based on the behavioural intention formed through the assessment of factors. In this way a further analysis of the behaviour model proposed can be carried out as further work.

CONCLUSION

A model for understanding and predicting the effects of Ubiquitous Monitoring on human behaviour has been presented in this paper. The model consists of a series of factors, identified in the monitoring and pervasive literature, believed to influence behaviour and includes: trust, awareness, control, boundaries, context, justification and intrusion. The model is augmented by the TPB, which provides a theoretical link between the factors and human behaviour. Even though research has been carried out on existing monitoring technologies, the pervasive nature of UM means certain issues will not have been examined. The undesirable impacts of UM technologies have been given little consideration by developers and if these issues are not considered when designing and implementing IPSs, there are likely to be undesirable side-effects. Our model is envisaged to give a designer the ability to improve existing and future IPS and ubiquitous monitoring systems.

This paper has also framed the UM problem using a socio-technical framework describing ubiquitous computing environments and attempted to unify this with the semiotic framework. Adopting the semiotic viewpoint has contributed to a more structured understanding of UM and further analysis should yield an improvement of the model. This includes the interpretation of the problem based on the information field paradigm, in particular through associating the behaviour influencing factors with the social psychological taxonomy of norms.

Organisational semiotics offers a rich set of methods and framework to analyse UM, and the work presented in this paper is a first attempt to apply this approach to UM. Future work should expand the application of semiotic analysis to enhance the behaviour model, which in turn would contribute to the further understanding of the problem domain of UM.

ACKNOWLEDGMENT

The authors would like to thank the anonymous reviewers for their valuable comments on the earlier version of this paper.

REFERENCES

Ajzen, I. (1991). The theory of planned behavior. *Organizational Behavior and Human Decision Processes*, *50*, 179–211. doi:10.1016/0749-5978(91)90020-T

Ajzen, I., & Fishbein, M. (1980). *Understanding attitudes and predicting social behavior*. Upper Saddle River, NJ: Prentice Hall.

Albrechtslund, A. (2007). House 2.0: Towards an ethics for surveillance in intelligent living and working environments. In *Proceedings of the Computer Ethics Philosophical Enquiry*.

Beaudin, J., Intille, S., & Tapia, E. M. (2004). Lessons learned using ubiquitous sensors for data collection in real homes. In *Proceedings of the CHI Extended Abstracts on Human Factors in Computing*.

Botan, C., & Vorvoreanu, M. (2005). What do employees think about electronic surveillance at work? In Weckert, J. (Ed.), *Electronic monitoring in the workplace: Controversies and solutions* (pp. 123–144). Hershey, PA: IGI Global.

Clements-Croome, D., Noy, P., & Liu, K. (2006). *Occupant behaviour analysis*. Retrieved from http://www.serg.soton.ac.uk/idcop/outcomes/ID-COP_Scientific_Report_OccupantBehaviour.pdf

Davis, F. D. (1989). Perceived usefulness, perceived ease of use, and user acceptance of information technology. *Management Information Systems Quarterly*, *13*(3), 319–340. doi:10.2307/249008

Davis, G. B. (2002). Anytime/anyplace computing and the future of knowledge work. *Communications of the ACM*, *45*(12), 67–73. doi:10.1145/585597.585617

de Moor, A., & Weigand, H. (2002). Towards a semiotic communication quality model. In Liu, K., Clarke, R. J., Andersen, P. B., & Stamper, R. K. (Eds.), *Orgnizational semiotics: Evolving a science of information systems* (pp. 275–285). Amsterdam, The Netherlands: Kluwer Academic.

Hartshorne, C., & Weiss, P. (1960). *Collected papers of C. S. Peirce (1931-1935)*. Cambridge, MA: Harvard University Press.

Hayes, G. R., Poole, E. S., Iachello, G., Patel, S. N., Grimes, A., & Abowd, G. D. (2007). Physical, social, and experiential knowledge in pervasive computing environments. *Pervasive Computing, 6*(4), 56–63. doi:10.1109/MPRV.2007.82

Intille, S. S., Larson, K., Beaudin, J., Tapia, E. M., Kaushik, P., Nawyn, J., et al. (2005). The PlaceLab: A live-in laboratory for pervasive computing research [Video]. In *Proceedings of the Pervasive Video Program.*

Intille, S. S., Tapia, E. M., Rondoni, J., Beaudin, J., Kukla, C., Agarwal, S., et al. (2003). Tools for studying behavior and technology in natural settings. In *Proceedings of the Conference on Ubiquitous Computing.*

Jonsson, K. (2006). The embedded panopticon: Visibility issues of remote diagnostics surveillance. *Scandinavian Journal of Information Systems, 18*(2), 7–28.

Konomi, S., & Roussos, G. (2007). Ubiquitous computing in the real world: Lessons learnt from large scale RFID deployments. *Personal and Ubiquitous Computing, 11*(7), 507–521. doi:10.1007/s00779-006-0116-1

Langheinrich, M., Coroamă, V., Bohn, J., & Mattern, F. (2005). Living in a smart environment – Implications for the coming ubiquitous information society. *Telecommunications Review, 15*(1).

Liu, K. (2000). *Semiotics in information systems engineering*. Cambridge, UK: Cambridge University Press. doi:10.1017/CBO9780511543364

Nguyen, D. H., & Mynatt, E. D. (2002). *Privacy mirrors: Understanding and shaping socio-technical ubiquitous computing systems*. Atlanta, GA: Georgia Institute of Technology.

Roethlisberger, F. J., & Dickson, W. J. (1939). *Management and the worker*. Cambridge, MA: Harvard University Press.

Rule, J., & Brantley, P. (1992). Computerized surveillance in the workplace: Forms and distributions. *Sociological Forum, 7*(3), 405–423. doi:10.1007/BF01117554

Sheng, H., Nah, F., & Siau, K. (2006). An experimental study on u-commerce adoption: Impact of personalization and privacy concerns. In *Proceedings of the Fifth Annual Workshop on HCI Research in MIS.*

Stamper, R. K. (2000). New directions for systems analysis and design. In Filipe, J. (Ed.), *Enterprise information systems* (pp. 14–39). Amsterdam, The Netherlands: Kluwer Academic.

Stamper, R. K., Althaus, K., & Backhouse, J. (1988). MEASUR: Method for eliciting, analyzing and specifying user requirements. In *Computerized assistance during the information systems life cycle*. Amsterdam, The Netherlands: Elsevier Science.

Stamper, R. K., Liu, K., Hafkamp, M., & Ades, Y. (2000). Understanding the roles of signs and norms in organizations - A semiotic approach to information systems design. *Behaviour & Information Technology, 19*(1), 15–27. doi:10.1080/014492900118768

Stanton, J. M. (2000). Reactions to employee performance monitoring: Framework, review and research directions. *Human Performance, 13*(1), 85–113. doi:10.1207/S15327043HUP1301_4

Vorvoreanu, M., & Botan, C. H. (2000). Examining electronic surveillance in the workplace: A review of theoretical perspectives and research findings. In *Proceedings of the Conference of the International Communication Association.*

Weiser, M. (1991). The computer for the 21st century. *Scientific American, 265*(3), 94–104. doi:10.1038/scientificamerican0991-94

Zweig, D. (2005). Beyond privacy and fairness concerns: Examining psychological boundary violations as a consequence of electronic performance monitoring. In Weckert, J. (Ed.), *Electronic monitoring in the workplace: Controversies and solutions*. Hershey, PA: IGI Global.

Zweig, D., & Webster, J. (2002). Where is the line between benign and invasive? An examination of psychological barriers to the acceptance of awareness monitoring systems. *Journal of Organizational Behavior*, *23*(5), 605–633. doi:10.1002/job.157

This work was previously published in the International Journal of Information Systems and Social Change, Volume 2, Issue 3, edited by John Wang, pp. 36-47, copyright 2011 by IGI Publishing (an imprint of IGI Global).

Chapter 13

Innovation or Imitation:
Some Economic Performance and Social Welfare Policy Perspectives

Soheil Ghili
Sharif University of Technology, Iran

Hengameh Shams
Sharif University of Technology, Iran

Madjid Tavana
La Salle University, USA

ABSTRACT

This paper develops a mathematical model of innovation in technology with two main characteristics. First, it discusses the endogenously made decision on not only how much to innovate, but also, how much to imitate. Second, it demonstrates that the decision to innovate or imitate are not mutually exclusive and a firm can innovate and imitate simultaneously. A mathematical model is presented, and the authors explain the barriers to innovation development and diffusion. The model is further used to investigate the effectiveness of two technology innovation and imitation policies. It is shown that an intellectual property right (IPR) policy will better function if the price of innovation is set to a level lower than the cost of innovation. The concept "superfluous innovation" (innovations whose costs are higher than their benefits) is also proposed and developed through investigating the policy of levying subsidies on innovation.

INTRODUCTION

The tradeoff between innovation development and innovation diffusion has been widely studied in the technology change and industry performance literature. This tradeoff arises from the fact that

what we do to prevent free availability of existing innovative discoveries to all producers, although beneficial from an ex-post efficiency standpoint, will fail to provide the ex-ante incentives for further innovation (Arrow, 1962; Cohen & Levin, 1989; Cohen & Levinthal, 1990). Such a dilemma is called the appropriability problem.

DOI: 10.4018/978-1-4666-2922-6.ch013

In spite of its long history, the problem of how firms decide on innovation-related issues is still of much interest and importance. In its simplest form, the appropriability problem is concerned about a firm deciding on whether or not to innovate - or how much to innovate - based on the extent to which the innovation is appropriable (Arrow, 1962; Quirmbach, 1986; Ireland & Stoneman, 1986; Levin, 1988; Saracho, 1996; Sakakibara, 2002).

A newly emerging theme, mostly seen in Boldrin and Levin's (2009a, 2009b, 2009c, 2009d) recent works is divisibility according to which, the knowledge embodied in goods or people is not costlessly available. Therefore, if one firm is to imitate another, it should devote some time, money and energy into buying some copies of the goods in which the knowledge is embodied (Boldrin & Levine, 2008, 2009a). In other words, there is no free and unpriced "spillover." Therefore, the matter of interest, here, is not only the innovator deciding how much to innovate due to the degree of appropriability, but also, how the imitator decides on how much to imitate due to the degree of divisibility (which we will refer to as imitability). Boldrin and Levin (2009b, 2009c, 2009d) have studied the appropriability and divisibility decisions and their effect on market size and structure.

In this paper, we study the interdependency between innovation and imitation and show that a firm can be an innovator and an imitator at the same time. While the two-fold role of a firm has been addressed sporadically in the field literature (such as Spence, 1984), none of these studies have considered the interdependency between innovation and imitation. Not all industries can be partitioned into innovators (i.e. north) and adopters (i.e. south) as argued by Helpman (1993) and Akiyama and Furukawa (2009). This gap in the current literature might have led to some misdefinitions of the problem. For instance, it is always postulated that a firm has less incentive for investing on easily imitable innovations only because it may then be imitated by others (Arrow,

1962; Quirmbach, 1986; Ireland & Stoneman, 1986; Levin, 1988; Saracho, 1996; Sakakibara, 2002; Spence, 1984).

In this paper we construct a model to address the question of how much to imitate in the context of the two-fold role of a firm (i.e. innovator and imitator) in a strictly competitive game setting. We will investigate how firms decide on how much to innovate and imitate and how their decisions affect social welfare under different conditions of imitability. We will also investigate the effects of two widely-noted policies: first, the intellectual property right (IPR) and second, the policy of treating innovation as a public good. We argue that under different conditions of imitability, these policies may have different effects on the outcome of the complex decision structure and hence on social welfare.

The remainder of the paper is organized as follows. In the next section, we provide the main assumptions of the model and then we present the mathematical model. In the following section, we provide the results obtained from solving the mathematical model. We also explain the insights gained from the solution results and compare them to the current appropriability literature. We then investigate the subsidy and IPR policies and examine their effect on the proposed model. The last section is devoted to conclusions and future research directions.

ASSUMPTIONS OF THE MODEL

For the sake of simplicity, we assume that there are two firms producing the same product at the same price in the market. We also assume equal market share but different production costs for the two firms. Since price and market share are assumed to be equal, the firm with lower production cost will make more profits. The market is also assumed competitive, thus, one firm's cost reduction will lessen the price and consequently the profit of the other firm. As we will see in the following sections, *appropriability* is captured in the model by linking production cost of each firm to the profit of the other firm.

Each firm can develop innovation by itself or imitate the innovation developed by the other firm. Developing or imitating one unit of innovation results in reduction in the firm's production cost but it has some costs associated with it. The effects of innovations have been considered solely on production cost, not on the quality of goods. Such an assumption is commonplace in the literature (Cole, 2001).

One unit of innovation developed by a firm and one unit of innovation imitated from the other firm will equally reduce production cost although the cost of developing one unit of innovation may differ from that of imitating it. Therefore, the problem faced by each firm is to optimize the extent of its innovation development and imitation (i.e. to maximize its profit). In other words, the decision-making variables for each firm are the extent of innovation development and imitation and the objective is to maximize profit.

We further assume perfect information in the model which is also a common assumption in innovation diffusion models (Hylton, 2003; Goodwin et al., 2005). This assumption is realistice since each firm can predict how the other firm will decide. If one firm did not know that the other firm will imitate it, appropriability would not have evolved as a major concern in the literature on innovation diffusion and political economy.

In this model, there is no variable named "welfare" but the price of the product in the market is considered as an index of social welfare. Such an assumption is made on one hand for the sake of the simplicity; on the other hand, it accords with the common assumption of welfare economics which perceives more consumption as an index of welfare (*i.e.*, ISEW or the index of sustainable economic welfare) (Nordhaus & Tobin, 1972; Daly & Cobb, 1989). In our model, a lower price is an indicator of a higher consumption rate since the lower price, although not leading to an increase in consumption, leaves more disposable income for people.

MATHEMATICAL MODEL

The mathematical notations and definitions presented in Appendix A are used to derive the models presented in this section. The price of the product is obtained from Equation (1):

$$\text{Price} = P_0 + \frac{Cu_1 + Cu_2}{2} \tag{1}$$

where *Price* is the market price of the product and Cu_1 and Cu_2 are the total costs of producing one unit of the product for firm 1 and firm 2, respectively. The relationship between the price and Cu is based upon the assumption of a competitive market and therefore a reduction in the total cost results in a reduction in the market price. P_0 is a constant indicating the average profit per unit. Although the price could have been formulated as in a Cournot duopoly, we did not do so for two reasons: simplicity and the assumption of strict competition. First, our main objective in this study is to investigate how firms' decisions on innovation and imitation interplay in a competitive environment. Focusing on the form of the competition will only add to the complexity of the model and divert us from the main objective of this study. Second, the competition represented by Equation (1) is strict because the sum of the firms' profits is equal to the constant P_0. Cournot setup could not be used here because of the assumption of strict competition. Although we are not going to use the theorems related to strictly competitive games, we do use its logic when providing explanations for the results.

We find each firm's profit by multiplying its profit per unit times the number of units sold (Q):

$$\text{Pr}\,ofit_1 = \left(\text{Price} - Cu_1\right) \times Q = \left(P_0 + \frac{Cu_2 - Cu_1}{2}\right) \times Q \tag{2}$$

where Q is equal for both firms and Equation 2 can be used similarly to find the profit for firm 1. In addition, the appropriability is captured by these equations since Cu_1 is not the sole determinant of Profit_1. Cu_2 also plays a role in determining the profit for firm 1- the less Cu_2, the less Profit_1. This is the reason why firm 1 decides on less innovation when its innovation is imitated.

The total costs for each firm consists of two components: direct or variable costs and indirect or fixed costs. The direct costs of producing a product can be represented by:

$$Cu_{\text{Direct}} = C_0 \times \left(\frac{m}{i_d + i_i}\right) \tag{3}$$

where C_0 and m are coefficients for regulating the dimension of the equation. The dimension of C_0 is the same as that of Cu_{Direct} (the unit of money over the unit of product) and dimension of m is the same as that of i_d and i_i (the units of innovations and imitations, respectively). $i_{d,1}$ represents the units of innovations developed in firm 1 and $i_{d,2}$ represents the units of innovations developed in firm 2. Similarly, $i_{i,1}$ and $i_{i,2}$ are the units of innovation imitated by firm 1 and firm 2, respectively. Therefore, firm 1 imitates from firm 2 and vice versa. As we mentioned earlier, the effect of one unit of innovation on the production cost has nothing to do with whether the product is developed within the firm or it is imitated. This fact is also captured by Equation (3).

The second component of the firm's total cost is indirect production expenses including costs of developing and imitating innovations:

$$Cu_{\text{Indirect}} = \frac{\int_0^{i_d} C_d d_x + \int_0^{i_i} C_i d_x}{nQ} \tag{4}$$

where C_i is the marginal cost of imitating one unit of innovation and C_d is the marginal cost of developing one unit of innovation. As we will see later, the C_i's are formulated subject to the following constraints: $i_{i_1} \leq i_{d_2}$ and $i_{i_2} \leq i_{d_1}$, because no firm is able to imitate more than the amount of innovation developed by the other firm. n equals the number of years during which the innovation is usable; that is, the time period during which the innovation has an effect on the total costs before becoming obsolete due to substitution. n is assumed exogenous and constant. The reason for including n in the denominator of the formula is that all of the terms in this model represent a period of one year (*i.e.* this is a static one-year model). Consequently, only one out of the n years during which the innovation affects Cu_{Direct} is considered. Therefore, it is necessary to take into consideration $1 / n^{\text{th}}$ of its associated cost. In addition, we include Q in the denominator because Cu_{Direct} is an overhead cost.

The marginal cost of imitating innovation for firm 1 is obtained by Equation (5) in which the aforementioned constraint has been taken into account:

$$C_{i,1} = I_0 \times \frac{i_{i,1}}{i_{d,2} - i_{i,1}} \tag{5}$$

According to this equation, where $i_{i,1} = i_{d,2}$, the cost of imitating one more unit will approach infinity. Therefore, the constraint $i_{i,1} \leq i_{d,2}$ is applied to Equation (5). A similar formula is used for the second firm's imitation cost. I_0 is a coefficient which makes the dimensions of the two sides consistent and is also an index for the imitability of the innovation. That is, the less I_0, the less expensive is the imitating innovation, making it more imitable. If I_0 is negligible or zero, the innovation is easily imitable and if I_0 is large, the innovation is non-imitable. I_0 plays a significant role in this study since the main goal of this paper is to analyze the situation in different states of imitability. I_0 is considered the independent variable in all the analyses and is the variable on the horizontal axis in all of the figures.

The marginal cost of developing innovation is assumed to be equal to the constant D_0:

$$C_{d,1} = C_{d,2} = D_0 \qquad (6)$$

which means that the per-unit expense of developing innovation is D_0. If we substitute Equations (5) and (6) into Equation (4) and integrate the two sides of the resulting equations, C_i and C_d can be eliminated and the equation below will be obtained:

$$Cu_{\text{Indirect},1} = \frac{D_0 \times i_{d,1}}{nQ} - \frac{I_0}{nQ} \times \left[i_{i,1} + i_{d,2} \times \ln\left(\frac{i_{d,2} - i_{i,1}}{i_{d,2}}\right) \right] \qquad (7)$$

The total cost of each firm equals the sum of its direct and indirect costs.

$$Cu = Cu_{\text{Direct}} + Cu_{\text{Indirect}} \qquad (8)$$

We have thus far modeled the physical structure of the problem. Next, we model the firms' decisions which involve four equations; two of them are concerned with how much innovation the firm should develop, and the other two correspond to how much innovation the firm should imitate.

Firms determine the degree of their innovation imitation based on the amount of their innovation development. As mentioned before, the effect that one unit of imitated innovation has on the direct cost of production equals the effect of one unit of developed innovation. Therefore, if $C_i > C_d$, the firm can reduce its indirect costs by developing

one more unit of innovation in exchange for one less unit of imitation, while keeping its direct costs unchanged. Similar conditions exist for the situation where $C_i < C_d$[1]. As a result, the optimum decision concerning imitation is made under the assumption that $C_i = C_d$, and, hence:

$$C_{i,1} = C_{d,1} \qquad (9)$$

$$C_{i,2} = C_{d,2} \qquad (10)$$

and two other equations maximize the firms' profits with respect to developed innovations:

$$\frac{\partial \text{Profit}_1}{\partial i_{d,1}} = 0 \qquad (11)$$

$$\frac{\partial \text{Profit}_2}{\partial i_{d,2}} = 0 \qquad (12)$$

MODEL SIMULATION AND THE ANALYSIS OF THE RESULTS

In order to solve this model, we need to solve Equations (9-12) for four unknown quantities: $i_{d,1}$, $i_{d,2}$, $i_{i,1}$ and $i_{i,2}$. It should be noted that the perfect information assumption is used here. For example, when substituting Equation (10) in Equation (11), we are implicitly assuming that firm 1 is completely aware of the second firm's policy on imitation. In addition, the main objective is to study how the imitability of an innovation impacts upon the simulation results. Therefore, all the parameters are taken into consideration as constants except for I_0 which is used to sketch the figures.

We first equalize the marginal cost of innovation development to that of the innovation imitation (Equation (9) for firm 1 and Equation (10) for firm 2). As a result, the amount of imitation is obtained in terms of the amount of development. By solving Equation (9) for $i_{i,1}$ the amount of imitation for firm 1 is:

$$i_{i,1} = i_{d,2} \times \frac{D_0}{D_0 + I_0} \tag{13}$$

and the amount of imitation for the firm 2 is:

$$i_{i,2} = i_{d,1} \times \frac{D_0}{D_0 + I_0} \tag{14}$$

As Equations. (13) and (14) show, while I_0 is negligible, a major fraction of the developed innovations is imitated. When I_0 is large, the amount of imitation is minimal.

Equations (13) and (14) show how the innovation imitated (i_i) by one firm is related to the innovation developed (i_d) by the other firm. When these equations are substituted in Equations (11) and (12), the i_i s are eliminated and Equation (11) and Equation (12) are turned into a system of two equations with two unknowns. By solving this system, we obtain $i_{d,1}$ and $i_{d,2}$ as seen in Box 1.

As expected, the amount of developed innovation for both firms is equal due to the symmetry assumption in the problem definition. As I_0 approaches infinity, i_d approaches $\sqrt{\frac{mC_0nQ}{D_0}}$. This fact is proved in Appendix B.

The result of the model (Equation (15)) is presented schematically in Figure 1.

$(D_0 = 1, m = 5, C_0 = 1, nQ = 100)$

A further analysis of the solution reveals some significant results. First and foremost, the amount of developed innovations is an increasing function of I_0. In other words, if imitation costs increase, firms will tend to develop more innovations. Second, interestingly, those willing to anticipate the effect of imitation costs on the amount of innovation imitation and development without considering the model may conclude that rising imitation costs have a positive effect on the innovation development (resulting in more innovation) and have a negative effect on the innovation imitation (resulting in less imitation). However, as the results show, this is not the case with every level of I_0. Increasing the imitation costs may also increase imitation. It is not difficult to justify this seemingly surprising result based upon the model relationships. When imitation costs rise, a smaller part of the developed innovations is imitated (*i.e.* the spillover fraction falls). Equation (16) can be derived from Equations (13) and (14), introducing the spillover fraction as a decreasing function of I_0 (see Figure 2).

$$\frac{i_{i,1}}{i_{d,2}} = \frac{i_{i,2}}{i_{d,1}} = \frac{D_0}{D_0 + I_0} \tag{16}$$

On the other hand, since the total amount of the developed innovation has increased, the amount of imitation might also rise in spite of the reduction in the spillover fraction. We consider price as an index of welfare due to our interest in studying the effect of imitation costs on social

Box 1.

$$i_{d,1} = i_{d,2} = \frac{D_0 + I_0}{2D_0 + I_0} \times \sqrt{\frac{mC_0I_0nQ}{(D_0 + I_0) \times \left[D_0 + I_0 \times \left(\frac{D_0}{D_0 + I_0} + \ln\frac{I_0}{D_0 + I_0}\right)\right]}} \tag{15}$$

Figure 1. The number of units of developed, imitated and total innovation in accordance with I_0

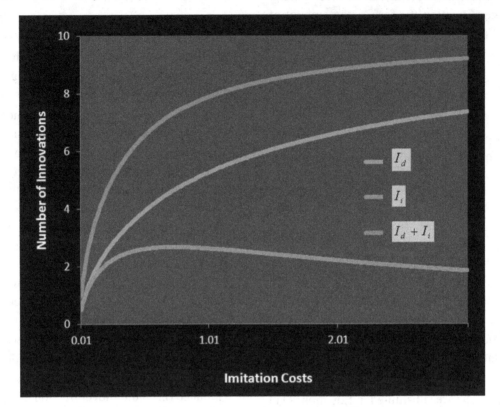

welfare. Figure 3, shows the price variations for different I_0s.

As shown in this figure, the price is highest when I_0 has its lowest value. The reason is that no innovation will be developed because of the appropriability problem. Thus, there will be no imitation either (see Figure 1).

Finally, it is noted that our supposition of the existence of only two firms, weakens the importance of imitation. In a market with, for example, five firms, each firm can imitate from four other firms. When the total amount of the developed innovation is calculated, imitated innovations will gain a coefficient of 4 and developed innovations will gain a coefficient of 1.

Figure 4 demonstrates the widely mentioned inverted U-shaped relationship between the amount of imitability of an innovation and social welfare (O'Donoghue & Zweimuller, 2004; Furukawa,

2007). Although the importance of imitation in reality might be more than what our model shows, analyzing a situation in which there are numerous firms is beyond the scope of this paper.

EXAMINING THE POLICIES ON THE MODEL

In this section, we study two policies that a government can consider towards innovation imitation between firms. We also show the effects of these policies on social welfare. We do not compare the performance of these policies and identify priorities. We intend to show *how* different policies function and provide some insights into this inherently complex problem rather than defining outputs for decision making as suggested by Lyneis (1999). While analyzing each policy, two

Figure 2. The ratio of the imitated to developed innovation in accordance with I_0

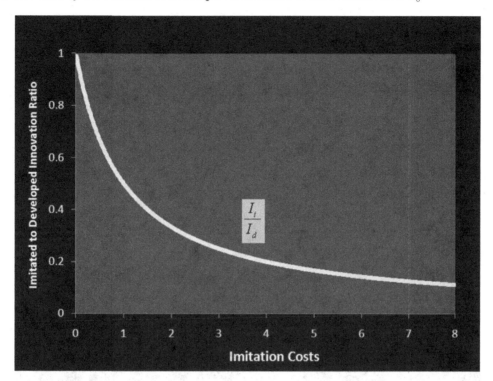

points should be considered: First, how the policy is modeled and entered into the model equations. That is, how manipulating that policy changes the model parameters and equations. Second, how these changes in the equations affect the solution results.

Establishing an IPR for Innovation Policy

One of the prevalent policies for preventing innovation from free imitation is an IPR for innovation policy (Gallini & Scotchmer, 2001). The IPR for innovation policy can resolve the appropriability problem by making the innovation available to other firms in the market in exchange for a predetermined amount of money to the innovation developer firm. This policy changes two parts of the model. First, imitation cost will increase as much as the price of buying the innovation:

$$C_{i,1} = I_0 \times \frac{i_{i,1}}{i_{d,2} - i_{i,1}} + P_i \qquad (17)$$

where P_i is the price of buying one unit of innovation. By substituting Equation (17) in Equation (9) and solving for $i_{i,1}$, the amount of imitated innovation will be obtained:

$$i_{i,1} = i_{d,2} \times \frac{(D_0 - P_i)}{(D_0 - P_i) + I_0} \qquad (18)$$

Second, by implementing this policy, the first firm's indirect costs will go down according to the equation below (there is a similar change for the second firm):

$$Cu_{\text{Indirect},1} = \frac{\int_0^{i_{d,1}} D_0 d_x + \int_0^{i_{i,1}} C_{i,1} d_x - \int_0^{i_{i,2}} P_i d_x}{Q} \qquad (19)$$

Figure 3. The product price for each firm in accordance with $I_0(P_0 = 0.2)$

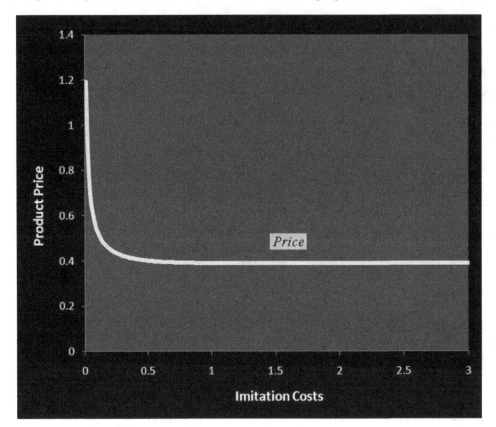

The amount for developed innovation can be obtained by solving this set of equations with the method described earlier (see Box 2).

Figures 5, 6 and 7 show the changes in the amount of developed innovation, innovated imitation and product price in terms of I_0 for both the pre-implementation and post-implementation cases.

As shown in Figure 5, the implementation of the IPR policy stimulates more innovations. Two main reasons can be suggested to understand this phenomenon. First, this policy causes firms to develop more innovations when they know for sure that their innovations will not be imitated easily. Second, the innovator firm knows that if another firm wants to imitate its innovation, it must pay for it. In other words, the production cost of the innovator firm will decrease and the production cost of the imitator firm will increase (which is good for the innovator firm due to the strictly competitive nature of the game). Therefore, the innovator firm can generate some revenues in exchange for their innovation.

Meanwhile, Figure 6 bears another interesting result. The IPR policy, an anti-imitation policy, leads to more imitations. This policy not only lessens the spillover *percentage*, but it also causes more innovations to be developed. As a result, more *absolute* value of the innovations will be imitated.

Taking Figures 5 and 6 into consideration, we can see that an IPR policy is not effective for large I_0s. In other words, for large I_0s, innovation development, innovation imitation and product prices are the same whether or not this policy is implemented. The reason is because this policy

Figure 4. The total units of innovation in a market of 5 firms

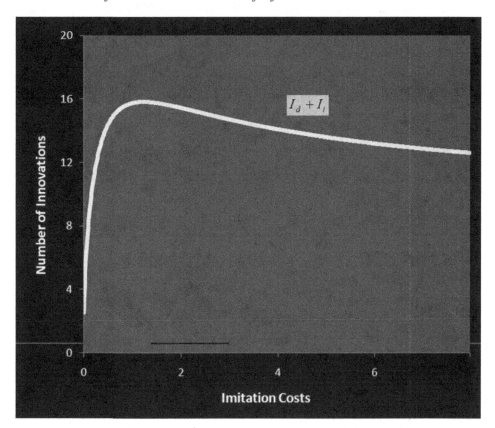

increases imitation costs and it does not work for cases where the imitation costs are already high.

Another issue of concern is how to set P_i to better handle the IPR policy. Note that if the amount of P_i is chosen to be more than D_0, there will be no reason for the firms to imitate from each other because the marginal cost of innovation development will always be less than that of innovation imitation $(C_d < C_i)$. In other words, when P_i is large, developing an innovation is always more economical than imitating the innovation and no imitation will occur.

This important result gives us an interesting insight into adopting suitable policies for innovation prices. As shown in Figure 7 (product price versus I_0 for $P_i = D_0 / 2$), the product prices for low I_0s is less than those for infinite I_0s. In other words, lower prices are reached when P_i is set to quantities smaller than D_0. Therefore, our model proposes setting innovation selling prices

Box 2.

$$i_{d,1} = i_{d,2} = \frac{(D_0 - P_i) + I_0}{2(D_0 - P_i) + I_0} \times \sqrt{\frac{mC_0 I_0 nQ}{[(D_0 - P_i) + I_0] \times \left\{ D_0 + I_0 \times \left[\frac{(D_0 - P_i)}{(D_0 - P_i) + I_0} + \ln \frac{I_0}{(D_0 - P_i) + I_0} \right] - \frac{2P_i(D_0 - P_i)}{(D_0 - P_i) + I_0} \right\}}} \quad (20)$$

Figure 5. The total units of developed innovation (with and without IPR policy $(\frac{P_i}{D_0} = 0.5)$)

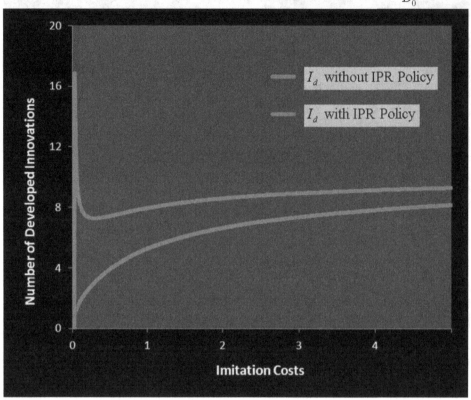

smaller than innovation development costs so that this policy functions more effectively.

This analysis suggests the importance of making a distinction between the two purposes of the IPR policy mentioned above. The first purpose is to motivate the firms by guaranteeing a protection against imitation; and, the second purpose is to motivate them with the profit from selling their innovations. If P_i is very large, it will certainly protect firms against spillover. But, this very protection can also discourage imitation and therefore may result in a loss of earnings for the firm since its innovation is not imitated by others. Hence, a large P_i may reduce the incentives to develop innovations.

Establishing an Imitable Innovation as Public Goods (Tax-Subsidy Policy)

There are two elements in the definition of public goods. First, using public goods by someone is not an impediment for others to use it. Second, it is very costly to prevent someone from using public goods (Varian, 1992). It is obvious that public goods can't be produced by the market because of the free riding problem (Waldman, 1987).

Innovation entails the first characteristic of public goods. When an innovation is used by one firm, it does not prevent other firms from doing so. Meanwhile, if the innovation is imitable, it will also have the second characteristic. It can be stated that the appropriability problem is nothing but the free riding problem in a specific case that the innovation of the product is imitable.

Figure 6. The total units of imitated innovation (with and without IPR policy $(\frac{P_i}{D_0} = 0.5)$)

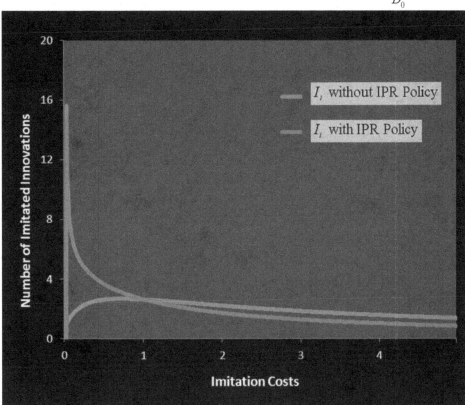

Consequently, it seems that introducing a tax system for compensating firms' innovation costs could be an appropriate policy for encouraging firms to develop more innovations that are easier to imitate. Governments can collect a fraction of the total innovation costs from both firms in the form of taxes. That is, if the innovation development costs for firms 1 and 2 are $C_{D,1}$ and $C_{D,2}$, respectively, the government will levy a tax equal to $\alpha(C_{D,1} + C_{D,2})/2$ on each firm. Then, the government can use these collected taxes to pay $\alpha C_{D,1}$ to firm 1 and $\alpha C_{D,2}$ to firm 2 $(0 < \alpha < 1)$. Accordingly, the government makes firms participate in each other's innovation development costs. This policy changes the equations as follows (similar modifications should be applied to the equations concerning Firm 2):

$$Cu_{\text{Indirect},1} = \frac{(1-\alpha)\int_0^{i_{d,1}} D_0 d_x + \alpha \int_0^{i_{d,2}} C_{d,2} d_x - \int_0^{i_{i,1}} C_{i,1} d_x}{nQ}$$

(21)

$$i_{i,1} = i_{d,2} \times \frac{D_0(1-\alpha)}{D_0(1-\alpha) + I_0}$$

(22)

α is a coefficient which shows what fraction of firms' innovation costs is subject to the tax-subsidy policy. In fact, α is an index of how strictly the government enforces the tax-subsidy policy. If α equals zero, it means that the government collects no taxes at all. In other words, Equations (21) and (22) are the same as Equations. (4) and (13), representing the no tax condition in the model.

215

Figure 7. The product price (with and without IPR policy $(P_0 = 0.2, \frac{P_i}{D_0} = 0.5)$)

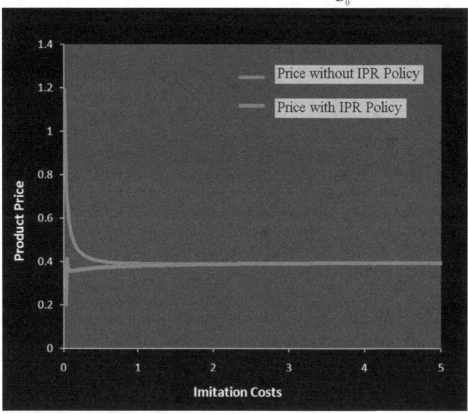

The amount of developed innovation can be obtained by solving this set of equations with the method described earlier (see Box 3).

Figures 8 and 9 show the amount of developed and imitated innovations versus I_0 (Figure 8 schematically represents Equation (23)). Product prices are also shown versus I_0 in Figure 10 (in order to study welfare).

As shown in Figure 8, implementing a tax-subsidy policy stimulates firms to develop more innovations for all quantities of I_0 because firm 1 knows that firm 2 will bear a part of firm 1's innovation costs (a similar story for firm 2). The benefits to the firm are twofold. First, this could result in a reduction of the firm's innovation development costs and ultimately a reduction in the total costs. Second, it will lead to an increase in the rival firm's costs which is beneficial for the

innovating firm due to the strictly competitive nature of the game.

As shown in from Figure 10, the tax-subsidy policy decreases the total production cost for relatively low I_0s. This might be due to the fact that innovation development occurs less when I_0 is low and this policy induces firms to develop more innovations. Figure 10 also shows that when I_0 is large, the tax-subsidy policy not only does not reduce the total production cost, but rather increases it. When analyzing this apparently strange result, it should be taken into consideration that more innovations do not necessarily lead to a higher level of welfare. More innovations lead to more welfare only when the benefits of innovation changes exceed its associated costs. As we already mentioned, while this policy is implemented, a percentage of each firm's innovation

Box 3.

$$i_{d,1} = i_{d,2} = \frac{D_0(1-\alpha) + I_0}{2D_0(1-\alpha) + I_0} \times \sqrt{\frac{mC_0I_0nQ}{[D_0(1-\alpha) + I_0] \times \left\{ D_0(1-\alpha) + I_0 \times \left[\frac{D_0(1-\alpha)}{D_0(1-\alpha) + I_0} + \ln \frac{I_0}{D_0(1-\alpha) + I_0} \right] \right\}}} \quad (23)$$

development costs will be imposed on the other firm. Therefore, firm 1 starts to develop innovations in a large scale because it pays less for its innovation development in comparison with the situation in which there is a no tax-subsidy policy. That is, with a tax- subsidy policy, firm 1 imposes a part of its innovation development costs on firm 2 by means of the tax collected from both of them. Firm 2 acts exactly in the same way and a portion of firm 2's innovation development costs will be imposed on firm 1. In other words, although both firms pay less for developing the innovations, they inadvertently suffer from each other's costs, so they will develop *"superfluous innovations"* (innovations whose costs are higher than their benefits) which can lead to higher total costs and hence higher prices.[2]

In summary, the tax-subsidy policy is only appropriate for the conditions in which the innovation is imitable, that is, it has low imitation cost (*i.e.* low I_0). In other conditions, this policy leads to *superfluous innovations* and increases production costs.

This result further validates our model since we have already stated that any innovation can be considered as public goods where I_0 is low. Therefore, when I_0 is high, innovation cannot be considered as public goods, implying that the free riding problem will not occur. Each firm tends to develop as many innovations as it needs while tax-subsidy policy results in more developing innovations than are needed.

Note that one of the model's assumptions was that market demand is inelastic, that is, the sales quantity (Q) is fixed and is not affected by the price. Such an assumption is not necessarily true for all kind of markets since demand is generally elastic in realistic circumstances. When demand is elastic, the tax-subsidy policy will not create as much negative effect as our model shows. If an increase in innovation development leads to higher prices (having superfluous innovation), higher prices will lead to less demand and less Q, and less Q leads to less profit for the firms. In the real world, firms make decisions on the basis of their profits, so a decrease in profits will guide them to give up developing innovations in a way that is unprofitable.

The superfluous innovation effect, however, is not the only damaging effect of the tax-subsidy policy on the performance of the system. Another effect is that the tax-subsidy policy decreases the marginal cost of innovation development, with the marginal cost of imitation remaining unchanged. Therefore, as shown in Figure 9, we should expect the spillover fraction, and maybe even the absolute value of imitation to decrease [3] (see Equations 16 and 22). We should also note that this effect could not be captured with a model not endogenizing the imitation decisions (*e.g.* Spence, 1984).

Figure 8. The total units of developed innovations (with and without tax-subsidy policy)

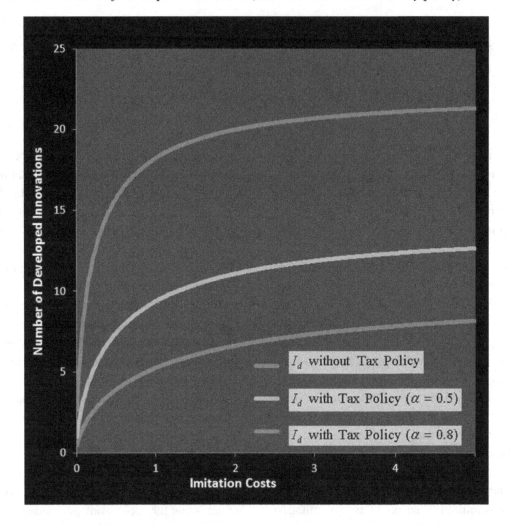

CONCLUSION AND FUTURE RESEARCH DIRECTIONS

The model introduced in this paper explained how the appropriability problem might arise from the interactions between firms. The above-mentioned model differs from the present models in the literature in that it considers two firms which both can simultaneously develop and imitate innovation; therefore, two key points about the firms' decisions are considered simultaneously in this model. First, a firm's decision on how much innovation to develop influences its decision on how much to imitate and vice-versa. Second, a firm's decision about how much to innovate and imitate is dependent on rival firms' decision. We have already examined the effect of imitability of innovation on social welfare through a simple framework which is also close to the real world. In this analysis it was indicated that the imitability of an innovation might not only reduce the extent of creating innovation (known as appropriability) but also reduces the degree of imitating innovations.

We used the model to analyze two policies using a descriptive rather than a prescriptive approach. That is, our purpose was mainly to give insights into the problem, and not to prescribe a solution. We illustrated that IPR policy leads to

Figure 9. The total units of imitated innovations (with and without tax-subsidy policy)

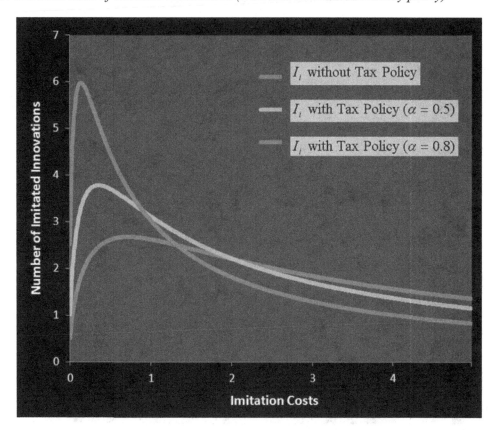

more innovation development in two ways. First, it guarantees that the firm's innovations will not be easily imitated (it appropriates innovation). Secondly, it guarantees that if the firm's innovations are imitated, it will earn a profit from selling that innovation and also an equal cost will be incurred to the imitating competitor. It was also shown that the latter function is not fulfilled when the intellectual property is priced high.

The results have also shown that imitable innovation (having low I_0) can be considered as public goods. By simulating the model, we observed that in this case, tax-subsidy policy might result in more innovation development and more social welfare. Meanwhile, it was noted that this policy does not function properly in the case of non-imitable innovations (having high I_0). Meanwhile, we showed that as I_0 increases, the

effect of the tax-subsidy policy gradually becomes less effective and even harmful to welfare since it causes a large amount of superfluous innovation to be developed. While the aforementioned insight is quite interesting, the main contribution of investigating the effects of a tax-subsidy policy is the proposition of "superfluous innovation".

The output of this work includes not only the insights driven from the model, but also the model itself. This model can be used as an instrument for analyzing how different factors affect the appropriability problem and its recommended policies. Some of these important factors are: asymmetry in production costs or imitation costs between two firms, asymmetry of firms' sizes (which heightens the importance of economies of scale), and innovation obsolescence time, etc. For example, some factors such as cost structure or market share of the two firms can be considered

Figure 10. The product price (with and without tax-subsidy policy ($P_0 = 0.2$))

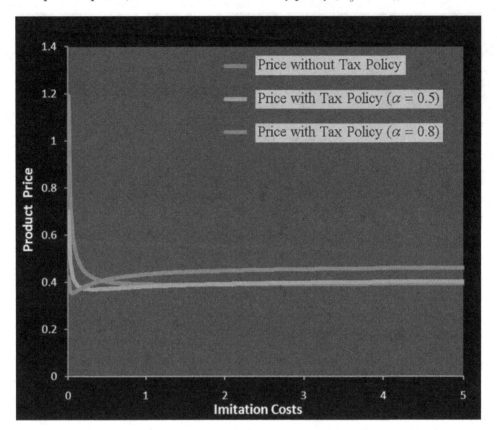

as asymmetric and the results could be solved and interpreted for an asymmetric case.

Furthermore, the boundary of this model can be extended in some aspects. For example, one way might be adding uncertainty effect to the model and then studying interactions between appropriability and uncertainty by omitting the perfect information assumption. Such an omission could lead to the transformation of our static model to a dynamic structure because when firms do not know, they should learn and learning is a dynamic phenomenon. This might be the reason why Arrow (1962) has used some dynamic terms such as "pre-invention" and "post-invention".

Another interesting extension of the model is to consider innovation's obsolescence time as endogenous rather than external and exogenous since some innovations are substituted for others.

The rate of the innovation development might affect its life time (*i.e.,* I_d may affect *n*) and ultimately impact the results and policies extracted from the model. In addition, it would be interesting to investigate the effects of possible cooperation between firms on the amount of innovation development and imitation. In order to do this, one should replace the simple strictly competitive market structure of our model with a Cournot structure - or some other standard structure which allows for the possibility of cooperation and examine the effects of collusions on innovation related issues. Finally, it would be interesting to study our model in the context of the theory of international political economy and the rapidly changing policies and economic conditions of the international community. More precisely, the interlink between economics and politics and the strategy that the state develops to ensure economic stability (Evaghorou, 2010). The systems thinking approach suggested by Kamppinen et al. (2008) could also be useful in understanding and managing the multitudes of conditions and effects that constitute economic stability.

ACKNOWLEDGMENT

The authors would like to thank the anonymous reviewers and the editor for their insightful comments and suggestions.

REFERENCES

Akiyama, T., & Furukawa, Y. (2009). Intellectual property rights and appropriability of innovation. *Economics Letters*, *103*(3), 138–141. doi:10.1016/j.econlet.2009.03.006

Arrow, K. J. (1962). Economic welfare and the allocation of resources for invention. In Nelson, R. R. (Ed.), *The rate and direction of inventive activity: Economic and social factors* (pp. 609–625). Princeton, NJ: Princeton University Press.

Boldrin, M., & Levine, D. K. (2008). Perfectly competitive innovation. *Journal of Monetary Economics*, *55*(3), 435–453. doi:10.1016/j.jmoneco.2008.01.008

Boldrin, M., & Levine, D. K. (2009a). A model of discovery. *The American Economic Review*, *99*(2), 337–342. doi:10.1257/aer.99.2.337

Boldrin, M., & Levine, D. K. (2009b). Does intellectual monopoly help innovation? *Review of Law & Economics. Berkeley Electronic Press*, *5*(3), 1211–1219.

Boldrin, M., & Levine, D. K. (2009c). Market structure and property rights in open source industries. *Washington University Journal of Law and Policy*, *30*, 325–363.

Boldrin, M., & Levine, D. K. (2009d). Market size and intellectual property protection. *International Economic Review*, *50*(3), 855–881. doi:10.1111/j.1468-2354.2009.00551.x

Cohen, W. M., & Levin, R. C. (1989). Empirical studies of innovation and market structure. In Schmalensee, R., & Willig, R. D. (Eds.), *Handbook of industrial organization* (*Vol. 2*, pp. 1059–1107). Amsterdam, The Netherlands: Elsevier.

Cohen, W. M., & Levinthal, D. A. (1990). Absorptive capacity: A new perspective on learning and innovation. *Administrative Science Quarterly*, *35*(1), 128–152. doi:10.2307/2393553

Cole, J. H. (2001). Patents and copyrights: Do the benefits exceed the costs? *The Journal of Libertarian Studies*, *15*(4), 79–105.

Daly, H. E., & Cobb, J. B. (1989). *For the common good*. Boston, MA: Beacon Press.

Evaghorou, E. L. (2010). The state strategy in today's global economy: A reset position in the theory of international political economy. *International Journal of Society Systems Science*, *2*(3), 297–311. doi:10.1504/IJSSS.2010.033496

Furukawa, Y. (2007). The protection of intellectual property rights and endogenous growth: Is stronger always better? *Journal of Economic Dynamics & Control*, *31*(11), 3644–3670. doi:10.1016/j.jedc.2007.01.011

Gallini, N., & Scotchmer, S. (2001). Intellectual property: When is it the best incentive system? In Jaffe, A., Lerner, J., & Stern, S. (Eds.), *Innovation policy and the economy* (*Vol. 2*). Cambridge, MA: MIT Press.

Goodwin, N., Nelson, J. A., Ackerman, F., & Weisskopf, T. (2005). *Microeconomics in context*. Boston, MA: Houghton Mifflin.

Helpman, E. (1993). Innovation, imitation, and intellectual property rights. *Econometrica*, *61*(6), 1247–1280. doi:10.2307/2951642

Hylton, K. N. (2003). *Antitrust law: Economic theory and common law evolution*. Cambridge, UK: Cambridge University Press. doi:10.1017/CBO9780511610158

Ireland, N., & Stoneman, P. (1986). Technological diffusion, expectations and welfare. *Oxford Economic Papers*, *38*(2), 283–304.

Kamppinen, M., Vihervaara, P., & Aarras, N. (2008). Corporate responsibility and systems thinking - tools for balanced risk management. *International Journal of Sustainable Society*, *1*(2), 158–171. doi:10.1504/IJSSOC.2008.022572

Levin, R. C. (1988). Appropriability, R&D spending, and technological performance. *The American Economic Review*, *78*(2), 424–428.

Lyneis, J. M. (1999). System dynamics for business strategy: A phased approach. *System Dynamics Review*, *15*(1), 37–70. doi:10.1002/(SICI)1099-1727(199921)15:1<37::AID-SDR158>3.0.CO;2-Z

Nordhaus, W., & Tobin, J. (1972). *Is growth obsolete?* New York, NY: Columbia University Press.

O'Donoghue, T., & Zweimuller, L. (2004). Patents in a model of endogenous growth. *Journal of Economic Growth*, *9*(1), 81–123. doi:10.1023/B:JOEG.0000023017.42109.c2

Quirmbach, H. C. (1986). The diffusion of new technology and the market for an innovation. *The Rand Journal of Economics*, *17*(1), 33–47. doi:10.2307/2555626

Sakakibara, M. (2002). Formation of R&D consortia: Industry and company effects. *Strategic Management Journal*, *23*(11), 1033–1050. doi:10.1002/smj.272

Saracho, A. I. (1996). The diffusion of a durable embodied capital innovation. *Economics Letters*, *54*(1), 45–50. doi:10.1016/S0165-1765(97)00003-7

Spence, M. (1984). Cost reduction, competition, and industry performance. *Econometrica*, *52*(1), 101–121. doi:10.2307/1911463

Varian, H. R. (1992). *Microeconomic analysis* (3rd ed.). New York, NY: W.W. Norton and Company.

Waldman, M. (1987). Noncooperative entrance, uncertainty, and the free rider problem. *The Review of Economic Studies*, *54*(2), 301–310. doi:10.2307/2297519

ENDNOTES

[1]   Such a formulation captures the fact that a low rate of imitation can occur not only when the imitation process is costly, but also when the innovation development is not difficult to implement.

[2]   Now, we can question the recommendation that all policies should be designed in such a way to encourage more innovation.

[3]   This effect could be greater even when the number of firms is greater than 2.

APPENDIX A

Table A1. Mathematical notations and definitions

$C_{d,1}$	Marginal cost of developing 1 unit of innovation in firm 1
$C_{d,2}$	Marginal cost of developing 1 unit of innovation in firm 2
$C_{D,1}$	Total cost of developing innovation in firm 1
$C_{D,2}$	Total cost of developing innovation in firm 2
$C_{i,1}$	Marginal cost of imitating 1 unit of innovation in firm 1
$C_{i,2}$	Marginal cost of imitating 1 unit of innovation in firm 2
$Cu_{Direct,1}$	Direct production costs for firm 1
$Cu_{Direct,2}$	Direct production costs for firm 2
$Cu_{Indirect,1}$	Indirect production costs for firm 1
$Cu_{Indirect,2}$	Indirect production costs for firm 2
Cu_1	Total costs for firm 1
Cu_2	Total costs for firm 2
D_0	Cost of developing 1 unit of innovation
$i_{d,1}$	Total units of innovation developed by firm 1
$i_{d,2}$	Total units of innovation developed by firm 2
$i_{i,1}$	Total units of innovation imitated by firm 1
$i_{i,2}$	Total units of innovation imitated by firm 2
I_0	A parameter representing innovation imitability
n	Life time of innovation (years)
Price	Price per unit

continued on following page

Table A1. Continued

$Profit_1$	Profit of firm 1
$Profit_2$	Profit of firm 2
P_i	Cost of buying 1 unit of innovation under IPR
P_0	Average profit per unit of the good
Q	Sales volume for each firm
α	A fraction of the firms' innovation costs which is subject to tax- subsidy policy

APPENDIX B

We stated that as I_0 approaches infinity, i_d approaches $\sqrt{\dfrac{mC_0nQ}{D_0}}$. This fact is proved here.

The amount of developed innovations can be obtained through the equation below:

$$i_{d,1} = i_{d,2} = \frac{D_0 + I_0}{2D_0 + I_0} \times \sqrt{\frac{mC_0I_0nQ}{(D_0 + I_0) \times \left[D_0 + I_0 \times \left(\dfrac{D_0}{D_0 + I_0} + \ln \dfrac{I_0}{D_0 + I_0}\right)\right]}}$$

Using the following equation for the natural logarithm:

$$\ln\left(1 + x\right)_{x \to 0} \sim x - \frac{x^2}{2} + \ldots$$

We see that as I_0 (imitation cost) approaches infinity, the amount of developed innovations approaches:

$$\lim_{I_0 \to \infty} I_d = \sqrt{\frac{mC_0I_0nQ}{I_0 \times \left[D_0 - I_0 \times \dfrac{1}{2} \times \left(\dfrac{D_0}{D_0 + I_0}\right)^2\right]}} = \sqrt{\frac{mC_0nQ}{D_0}}$$

This work was previously published in the International Journal of Information Systems and Social Change, Volume 2, Issue #, edited by John Wang, pp. 48-66, copyright 2011 by IGI Publishing (an imprint of IGI Global).

Section 4
Technology Trends and Critical Social Challenges

Chapter 14
Technology Acceptance Model and Determinants of Technology Rejection

Melih Kirlidog
Marmara University, Turkey & North-West University, South Africa

Aygul Kaynak
Marmara University, Turkey

ABSTRACT

Technology Acceptance Model (TAM) is an important tool to understand the dynamics of acceptance of Information Systems in an organization. The model posits that perceived ease of use and perceived usefulness are key factors in the adoption. This study extends TAM for investigating the user rejection of technology by reversing the two key factors into perceived difficulty of use and perceived uselessness. The study was conducted by surveying the customers of an e-banking application in Turkey who disuse the system. The results reveal important hints for the organization that wants to get an insight into the causes of the system disuse.

INTRODUCTION

There is a wide body of research about information systems success and adoption in organizations. There have been several attempts in theorizing the adoption and success of IS. The two most commonly used models for these tasks are Technology Acceptance Model (TAM) and DeLone

DOI: 10.4018/978-1-4666-2922-6.ch014

and McLean's (1992) IS success model where "system quality" and "information quality" have an effect on "user satisfaction" and "actual usage" and these, in turn, have impact on individuals and on the organization. Ten years after the introduction of the model the authors have updated it by adopting the service quality as the third input construct along with system quality and service quality (DeLone & McLean, 2003). The updated model also has the "net benefits" as the final construct replacing "individual impact" and "organizational impact." TAM has been introduced by Davis (1986) in his PhD dissertation and was further developed by the same author (Davis, 1989). TAM is derived from Theory of Reasoned Action (TRA) of Fishbein and Ajzen (1975). The model postulates that "perceived usefulness" and "perceived ease of use" determine the attitudes toward using the system which in turn determines the intention to use. This intention is assumed to lead to the actual use of the system.

An important pillar of the marketing research is consumers' behavior for the technological products or services offered. Electronic banking, a service that has become widespread in the last decade, is widely researched in the context of consumer behavior (Howcroft et al., 2002; Kolodinsky et al., 2004). E-banking systems are desirable for the convenience they offer to the users. They are also desirable for the banks, because a typical e-banking transaction costs about one cent whereas an ATM transaction costs 27 cents and a teller-window transaction costs 107 cents (Dandapani, 2004). An important aspect of e-banking research is the consumers' decision to accept or reject the service along with the technology that facilitates it. Since TAM has the capacity to offer an insight into that decision process, it is widely used in e-banking research (Pikkarainen et al., 2004; Wang et al., 2003).

Sole empiricism and anecdotal approach to the events are indicators of infancy of an academic field. Webster and Watson (2002) argue that, like other management fields, the maturity process of the IS field is directly related with theory development. Theories and models enable us to deduce and induce along the spectrum of specific to general. Theories in social sciences are not static and they can be extended to cover new areas with new constructs. The development of TAM from TRA shows that theories can also breed new theories. Like most other models in IS TAM has been extended by several researchers including its developer. Venkatesh and Davis (2000) extended the model to incorporate social influence processes and cognitive instrumental processes. They tested the extended model longitudinally and called it TAM2. In addition to the constructs that existed in the first version of TAM, the extended model contains constructs such as subjective norm, image, job relevance, output quality, and result demonstrability.

The essence of TAM is for adoption of the technology and the original model's constructs such as usefulness and ease of use have a positive overtone. By reversing these constructs to uselessness and difficulty of use the newly formed model can be a valuable tool for predicting the negative outcome of a possible rejection of information systems by its users. This research attempts to realize this approach by using data from an e-banking application in Turkey. Although it is a common practice to extend theoretical models in IS, this research is a unique approach that reverses the constructs of the model along with the model itself. The resulting model has the capacity to offer an insight to the reasons for disuse of the computerized information systems. By determining the possible areas of user dissatisfaction and analyzing their relative importance to the users the model can be a useful tool for preventing possible failures in future.

PRIOR RESEARCH

TAM is the most commonly used theoretical model in IS research (Lee et al., 2003) and it has been applied and empirically tested in various environments. Liao and Landry (2000) applied the model in analyzing technology adoption in the banking sector and Wang et al. (2003) used TAM to analyze customers' adoption behavior for e-banking. Benamati and Rajkumar (2002) analyzed the decision process for IS development outsourcing through TAM. Vijayasarathy (2004) and Shih (2004) analyzed on-line shopping behavior with TAM.

The results of previous research show that the model was generally found to sufficiently explain the mechanisms of a new ICT adoption in organizations. Lee et al. (2003) analyzed the TAM literature between 1986 and 2003. They chronologically classified TAM-related articles into four periods, namely model introduction, model validation, model extension, and model elaboration. These periods are not distinct; for example, model extension period started in 1994 by Straub's article and it was still in progress in 2003 when the article was published. The model elaboration period started in 2000 by the article of Venkatesh and Davis and this period was also in progress in 2003. Lee et al. also listed the extensions to the model. This list has a range of variables such as accessibility, anxiety, attitude, compatibility, complexity, and usability.

Most of the research about TAM empirically and quantitatively verified the validity of the model. However, recently there were more cautious research findings about its validity in all contexts and calls for integrating it with other frameworks. For example, Legris et al. (2003) reviewed 28 TAM research findings in the 22 articles published from the 1980s to 2001. In this meta-analysis they found that the research findings were generally heterogeneous and they called for integrating TAM to a broader model that includes organizational and social factors. In their cross-cultural study, Straub et al. (1997) have found that TAM holds for the US and Switzerland, but not for Japan. Although TAM seems to be an important theoretical tool for predicting the user acceptance of the ICT systems, such cautions imply that there is still room not only for testing and extending the original model, but also radically altering it such as reversing its constructs.

Straub et al. (1997) have argued that TAM has been widely used and tested in North America but less in other environments and developing countries. Yet, most of them after that article was published, TAM was tested in the conditions of some countries other than the US (Elbeltagi et al.,

2005; Huang et al., 2003; De Vreede et al., 1999; Phillips et al., 1994; Lim, 2003; Rose & Straub, 1998) and was generally found to have satisfactory descriptive power for other cultures as well. Recently, McCoy et al. (2005) examined TAM in the conditions of the USA and Uruguay. Their research was based on Hofstede's cultural dimensions of power distance, uncertainty avoidance, individualism and collectivism, masculinity and femininity indices and they found that TAM functions well in cross-cultural boundaries. Elbeltagi et al. (2005) examined the applicability of TAM in Egypt. They also report that perceived ease of use and perceived usefulness has a significant direct effect on DSS usage.

Proliferation of TAM-based research and e-banking in the world leaded some researchers to examining user acceptance of e-banking through TAM in developing countries. Kamel and Hassan (2003) investigated e-banking acceptance through TAM in Egypt and found that it can be useful in predicting actual usage.

TAM FOR ANALYZING DISUSE

The wide body of literature about TAM has a common trait: As can be expected, all articles use the model from its user acceptance dimension. In other words, the positive notions of perceived usefulness and perceived ease of use are investigated for their effect on "attitude to use" and "behavioral intention to use" the system. Gefen (2003) argues that TAM was originally geared narrowly toward the acceptance of a new system. Although this approach is understandable and it can be regarded as the "native" manner of the model, TAM can also be used at the opposite ends of the continuums of the usefulness and ease of use. As stated above, the model has widely been extended and its robust nature allowed many extensions to fit to the main body smoothly. Hence, it is also plausible to reverse its original constructs without losing the essence of the main structure so that the model will seek to analyze the opposite concept of its original. In other words, the model can be extended to gauge Perceived Uselessness (PU) and Perceived Difficulty of Use (PDU) which have the potential to lead to the disuse of the system. PU and PDU are important reasons for IS failure along with some other reasons such as organizational politics, lack of top management support, and lack of user involvement in the development process. Hence, an approach that originates in analyzing the PU and PDU would be particularly beneficial for understanding the failure of IS projects which are reported to be in alarming proportions (Legris et al., 2003; Heeks & Bhatnagar, 1999).

There is also a subtle issue in the research about IS failure. An overwhelming majority of literature about such failure is either about the failure of in-house system development project (Pan, 2005; Brooks, 1995) or implementation of a purchased software package (Gauld, 2007; Bartis & Mitev, 2008). Yet, it is reasonable to believe that PU and PDU may lead to the suboptimal use or to the total discard of the system some time after its implementation. In such cases approaching TAM from the "negative" side will be helpful in determining the user-perceived problems gauged with quantitative data. Once the most important problems perceived by the users are pinpointed this way, some further effort can fix them, clearly a better option than discarding the system completely. In order to make use of TAM in this way,

it must be extended to capture the weighted importance of the causes of perceived uselessness and perceived difficulty of use. Might be called "Technology Rejection Model", Figure 1 shows this version of the TAM where actual disuse is the dependent variable while PDU and PU are independent variables.

An overwhelming majority of research about TAM is quantitative (some exceptions are Benamati & Rajkumar, 2002; Briggs et al., 1999; De Vreede et al., 1999). Due to the nature of the model, the survey instruments developed aim to measure the extent to which the user perceives the system useful and easy to use. Measuring the reverse requires a different organization of the survey instrument. In other words, the survey instrument must be organized in such a way that the measures of perceived uselessness and perceived difficulty of use should stem from the "traditional" variables of perceived usefulness and perceived ease of use. Since the new model aims to measure the reverse of the original variables the new variables must designate the opposite ends of the relevant spectrum. Referring to the survey instrument used by Davis (1989) a "traditional" TAM survey instrument measures the following for perceived usefulness:

- Accomplishing tasks more quickly,
- Improving job performance,
- Increase productivity,

Figure 1. Technology rejection model

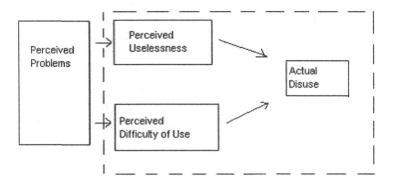

- Enhance effectiveness,
- Making the job easier.

The same instrument measures the extent for the following for perceived ease of use:

- Learning the system is easy,
- It is easy to get the system to do the task,
- Interaction with the system is clear and understandable,
- Interaction with the system is flexible,
- It is easy to become skilful at using the system.

Most of these traits are suitable for measuring the spectrums of perceived usefulness to uselessness as well as the opposite ends of the spectrums. However, to get a better insight for perceived uselessness and usage difficulty, the survey instrument must accurately measure the extent to which particular problems play role in the actual disuse. These particular problems are easy to determine by having a series of pre-interviews with the users. All areas of user complaints are possible sources for disuse and the model should contain these complaints as variables.

The Case Study for Analyzing Disuse

The data for this study was collected from the customers of the Turkish branch of a multinational bank who ceased using its e-banking application. Although not formally performed, user complaints were determined by the bank employees by continuously querying the customers about their satisfaction with the system. Hence, possible problem areas were well-known but their levels of effect on the intension to disuse were unknown to the bank. The present research aimed to clarify those effects and the questions in the survey instrument shown in Table 1 were prepared for this purpose. The instrument had 7-point Likert scale (1-strongly agree, 7-strongly disagree).

The survey instrument contains questions about the independent variables PDU and PU. The items in the constructs that have been derived as a result of the unstructured interviews proved to be the reverse of some of the items in the above-mentioned "traditional" TAM measures (Davis, 1989). There are seven items in the PDU construct which measure the extent to which the system is perceived to be difficult. There are mainly two groups of items related with the perceived difficulty. The first group is related with the internal character-

Table 1. Survey questions and descriptive statistics

		N	Mean	St. dev.
PDU1	Navigation to find the desired function is too long and complicated.	78	2.50	1.10
PDU2	There is too much detail that makes the system difficult to use.	78	2.26	0.99
PDU3	Usage of the system is difficult because of the Internet connectivity problem.	78	3.79	1.44
PDU4	Usage of the system is difficult because of the hardware token.	78	3.85	1.30
PDU5	Customer support unit is inadequate and responds late.	78	4.82	1.32
PDU6	Usage of Internet Banking systems of other banks is easier.	78	1.92	0.88
PDU7	Generally, usage of the system is difficult and complex.	78	2.17	1.16
PU1	The response time of the system is too slow.	78	2.74	1.10
PU2	The system does not have absolutely necessary functions for my job.	78	2.09	0.94
PU3	Generally, the system is useless for me.	78	2.73	1.11

PDU → Perceived difficulty of use

PU → Perceived uselessness

istics of the system where navigation is regarded to be difficult and the system is perceived to have too many details. In the second group there are items related with the external functions of the system such as connection problems, existence of the token, inadequate customer support, and the negative perception about the system in comparison with the other systems in the market. The items for PU are the perception about the speed and lack of some functionalities of the system. The lack of functionalities is an internal attribute of the system. The speed problem can both be regarded as external and internal depending on the source of the problem.

METHODOLOGY AND DATA COLLECTION

Research Setting

In Turkey usage of e-banking is quite common and almost all major locally-owned and multinational banks offer services to their corporate and individual customers through the Internet. Since the e-banking services offered to corporate and individual customers are quite different from each other, the bank involved in this study has two separate departments, each specializing for addressing the requirements of those two segments. This study investigates both types of the customers.

The system for corporate customers is available globally. It has customers in over 90 countries and the users can choose one of the 26 languages to communicate with the system. Since security is an important concern for e-banking, the system uses an additional security layer beyond traditional static password. This is accomplished by a hardware token, a credit card-sized device that generates a different password at each use. A separate password is required to invoke the token card itself. This way of authentication prevents phishing, the most important instrument of e-banking

fraud in Turkey. Further, for corporate banking, it inhibits the use of the system by unauthorized persons within the organization even if they know the "previous" password. Although several of the banks in Turkey use such tokens in their systems on a voluntary basis, it is compulsory for the system investigated. Traditional security measures such as 128 bit Secure Sockets Layer (SSL) encryption is available in the system and all customer activities are monitored round-the-clock by a unit which is responsible for detecting suspicious behavior and intervening if necessary.

As can be expected, some customers discontinue their usage of the system after some period of use, and some others never start using it even if they get their user id, password, and token. In the bank's terminology the e-banking customers are classified in the following groups:

- **Activated:** Client definitions are created in the system and the customers have been given user-id, password, and token. However, they have not actually started to use the system yet.
- **Active:** The customers who had some sort of activity in the system in the last two months.
- **Inactive:** The customer who did not have any activity in the system in the last two months.
- **Deleted:** The customers who have declared that they will not use the system any further after some period of use or who have never started to use it after a specific time after activation.

An ever-increasing share of banking activities are accomplished through the Internet due to the convenience and time-saving attributes of this medium. This results in dramatic decrease in transaction costs not only for the bank customers, but also for the bank itself. Thus, it is important to understand the underlying factor(s) for the shift in status both for preventing and fixing the perceived problems of the customers and convincing them to re-use the system.

The bank's experience shows that once the customers are in inactive status it is very difficult for convincing them to use the system again. This sometimes results in the total loss of the customer because of the increasing importance of the e-banking transactions in banking activities. Thus, it is important to get an insight of the problems encountered by the e-banking customers not only for the e-banking system alone, but also for the entire bank activities.

PRE-INTERVIEWS

As an organizational research method Bryman (1989) differentiates interviews as structured and unstructured. Structured interviews are "collections of specific and precisely formulated questions" whereas unstructured interviews are conducted in a very informal way, sometimes without a pre-formulated series of questions. The unstructured interviews may prove to be an ideal mechanism to get an insight for the problems they face when using the system. Yet, although this method is good for identifying the problems, it offers little value about the relative importance of them. Quantitatively determining the relative importance of the problems can be done with the questionnaires where the items are formed by the unstructured interviews.

Understanding the perceived relative importance of these particular problems and concentrating on solving a few of the most important ones can revitalize the system which is doomed for failure. For example, the effect of poor response time in an application leading to the disuse of the system can only be investigated by such an approach. Such an implementation of extended TAM can be useful not only for general acceptance of IT applications within the organization, but also for determining the underlying causes of customers' disuse of an e-commerce system such as e-banking.

Kelly et al. (2008) argue that open questions allow respondents to report their own feelings, beliefs and impressions. Thus, in order to get a better insight of the perceived problems, questions in the survey instrument also allowed respondents to enter their thoughts and impressions. Such an approach with open-ended questions proved to be useful for capturing the underlying problems for not using the system.

DATA COLLECTION

The data were collected from the deleted and inactive customers of an Istanbul branch of the bank. The branch had 251 activated e-banking customers during the research was conducted. 19 of them were in deleted and 76 of them were in inactive status. All inactive customers declared that they will not use the system in future. Since some corporate customers had more than one user, a total of 107 users were identified. The questionnaire was sent to those 107 users through e-mail with an accompanying message indicating the purpose of the study. The users were requested to fill in the questionnaire and send it back through e-mail. 64 usable questionnaires were collected this way. The customers who failed to respond were contacted again and were reminded about the study. During this second round of contacts the users were offered to fill in the questionnaire together on the phone if they wished. Some users agreed and 14 more questionnaires were filled this way. Thus, at the end of the process a total of 78 usable responses were collected. This makes 73% response rate.

SURVEY INSTRUMENT DESIGN

The survey instrument contains the customer complaints about the difficulty and uselessness of the system. Since a thorough analysis of these complaints has never been made before, the effect of these individual arguments on the disuse of the system was not known. There are seven questions related with the PDU and three questions related with the PU and all are independent variables of the model. The dependent variable is the actual disuse of the system which is the case for all respondents.

The first two questions in the PDU group are for determining complaints about navigation on the screen and the details which could be regarded by the users as unnecessary. The ever-increasing sophistication of electronic banking systems bring increasing functionality which in turn results in escalated number of screens and details. The resulting complexity in navigation might be irritating for many users who are interested in the basic use of the system. On the other hand, sophistication and enhanced functionalities are important for users who are interested in the advanced tasks of the system. So, like all information systems, a trade-off must be made between functionality and simplicity. In this case it seems that the bank's preference is inclined towards increased functionality, in conformity with the general trend.

The symmetry of the PDU2 is gauged by PU2 which is about the lack of necessary details in the system. Oftentimes such details are critically important for advanced users and their deficiency results in perceived uselessness of the system.

The PDU3 is about the usage difficulty resulting from the Internet connectivity problems, possibly attempting to connect dial-up instead of a broadband connection. In such a case the bank does not have much to do other than offering the customer a free broadband connection. The perceived difficulty of use by connection difficulties is associated with the more serious problem of perceiving the system as useless because of slow response time that is gauged in PU1. Oftentimes, the former tends to progress into the latter.

The PDU4 is about the difficulty of using the hardware token and the bank's options are limited here, too. Due to the widespread phishing and e-bank fraud activity in Turkey additional security measures are required in order to prevent serious losses.

An important problem that is closely associated with usage difficulty is the perceived inadequacy of the Customer Support Unit which is the subject the PDU5. Possible problems in this issue can be solved by the bank and it must act promptly to retain the current users who might decide to disuse the system in future because of the inadequacy of this unit.

Table 2. Pearson's correlation matrix for PDU and PU

	PDU1	**PDU2**	**PDU3**	**PDU4**	**PDU5**	**PDU6**	**PDU7**	**PU1**	**PU2**
PDU1									
PDU2	.371**								
PDU3	.115	.111							
PDU4	.063	.051	.066						
PDU5	.403**	.326**	.090	.120					
PDU6	.483**	.368**	.162	.285**	.336**				
PDU7	.362**	.144	.021	.190*	.097	.422**			
PU1	.547**	.469**	.230*	.227*	.264**	.490**	.280**		
PU2	.532**	.506**	.052	.361**	.170	.541**	.535**	.575**	
PU3	.620**	.348**	.046	.160	.348**	.337	.419**	.570**	.494**

N=78 * .05 significance level ** .01 significance level

The PDU6 question is about the competitors' e-banking systems and since it is a comparison with other systems, it is an important measure of the difficulty of use from the users' point of view.

ANALYSIS AND DISCUSSION

Internal consistency test revealed that Cronbach's alpha values were 0.63 for PDU and 0.78 for PU. The values are over acceptable value of 0.6 set by Nunnally (1967).

Although the disuse of the system is affected by both PDU and PU, the latter is more important for users in deciding to disuse. This is partly related with the fact that a difficult-to-use system can still be used if it is useful, but there is no point to continue to use or adopt a useless system. There is also a progressive relationship between the two. If the perceived difficulty of use exceeds a certain threshold, the users deem the system as useless. This relationship between PDU and PU is shown in Figure 1 and the progressing of perceived connectivity difficulties into perceived uselessness because of the "slowness" of the system that has been discussed before is an example to this fact. This resembles the causal relationship in the original TAM between Perceived Ease of Use and Perceived Usefulness where the former leads to the latter.

Table 1 shows that the inadequacy and unresponsiveness of the Customer Support Unit is a common source of complaint with the highest point of mean (4.82) among the items. This indicates that the bank has to move swiftly and decisively in order to prevent possible disuse in future. Unlike external items which are beyond the control of the bank, it can easily enhance the services of this unit. The other two problematic PDU items are the difficulty of use of the hardware token and the connectivity problem. Elimination of the token is too risky in the high-fraud environment but the bank can shift to some other easier-to-use

Table 3. Rotated factor loadings

	Factor 1	Factor 2	Factor 3
PDU1	.79		
PU3	.70		
PDU2	.68		
PU1	.66		
PDU5	.65		
PDU4		.75	
PU2	.49	.70	
PDU7		.68	
PDU6	.49	.55	
PDU3			.87
Percentage of variance	30.1	21.6	10.7
Percentage of variance - cumulative	30.1	51.7	62.4

hardware. The connectivity is also an external problem to the bank, but it can offer free or cheap Internet connection to some or all of the Internet customers. The low mean figure attributed to the easiness of other banks' systems indicates that this is not a very important problem in disuse.

The PU item about slowness of the system must be further investigated by the bank. It may either stem from the inadequacy of the hardware and software of the bank or may be related with some other connection problems. In the former case the bank can consider shifting to more powerful hardware due to the relative difficulty of enhancement of the sophisticated software. The comparatively low mean (2.09) of the next PU item indicate that the software is usually considered to have the necessary functions required in an e-banking system.

Pearson's correlations for PDU and PU are shown in Table 2.

Principal component analysis was utilized with varimax rotation with Kaiser normalization. The items less than 0.4 were suppressed. The analysis produced three factors which are shown in Table 3. The factors explain 62.4% of the variance.

IMPLICATIONS AND CONCLUSION

Although e-banking is a very desirable service both for consumers and providers some users may choose not to use it. In most cases the disuse of the system stems from shifting to another provider. There may be several reasons for this shift some of which are based on the inadequacies of the system. This article shows that TAM, a model that can explain the motives and mechanisms for technology acceptance can be reversed to understand the relative importance of the motives and mechanisms of technology rejection. Although there have been several extensions of the TAM, to the best of our knowledge, such an extension of the model by reversing it is unique.

Unlike traditional information systems that have been developed for internal purposes such as accounting, e-banking is outward-oriented and potential users have absolute freedom to accept or reject the system. The TAM literature some of which has been discussed in this article covers both internal and external systems. In other words, TAM is applicable not only for the acceptance of the systems by internal users, but also for external users. Since an external system is an extension of an organization's services to the customers, acceptance or rejection of those services have more important implications that internal systems. Like the original model, Technology Rejection Model discussed in this article can be used to gain an insight to the consumer behavior which has the determining factor in an organization's success.

There can be several causes of system failures such as budget and project time overruns, poor project management practices, and inexperienced system development staff. A large majority of the literature about IS failures is about the failure or suboptimal use in software development or initial implementation phases. Yet, the rejection of the system by some or all of the users after some implementation time is also an important problem and little is known about this fact in the IS research (Poulymenakou & Holmes, 1996). The "inversed"

TAM can be an important tool in understanding the user rejection of a software system not only after some time of implementation, but also in the initial phases of the implementation as well.

There is a wide body research about TAM and an important portion of this contains the extensions of the model to particular situations. Drawing from TAM, it is proposed that the most important reason for the user rejection is perceived difficulty of use and perceived uselessness. Once the particular causes of these two areas of complaints and their relative perceived importance are determined, the model can be used to fix problems and avoid total disuse.

REFERENCES

Bartis, E., & Mitev, N. (2008). A multiple narrative approach to information systems failure: A successful system that failed. *European Journal of Information Systems*, *17*(2), 112–124. doi:10.1057/ejis.2008.3doi:10.1057/ejis.2008.3

Benamati, J., & Rajkumar, T. M. (2002). The application development outsourcing decision: An application of the technology acceptance model. *Journal of Computer Information Systems*, *42*(4), 35–43.

Briggs, R. O., Adkins, M., Mittleman, D., Kruse, J., Miller, S., & Nunamaker, J. F. (1999). A technology transition model derived from field investigation of GSS use aboard the U.S.S. Coronado. *Journal of Management Information Systems*, *15*(3), 151–195.

Brooks, F. P. (1995). *The mythical man-month* (Anniversary ed.). Reading, MA: Addison-Wesley.

Bryman, A. (1989). *Research methods and organization studies*. London, UK: Unwin Hyman. doi:10.4324/9780203359648doi:10.4324/9780203359648

Dandapani, K. (2004). Success and failure in web-based financial services. *Communications of the ACM*, *47*(5), 31–33. doi:10.1145/986213.986233doi:10.1145/986213.986233

Davis, F. D. (1986). *Technology acceptance model for empirically testing new end-user information systems theory and results*. Unpublished doctoral dissertation, MIT, Cambridge, MA.

Davis, F. D. (1989). Perceived usefulness, perceived ease of use, and user acceptance of information technology. *Management Information Systems Quarterly*, *13*(3), 319–340. doi:10.2307/249008doi:10.2307/249008

De Vreede, G., Jones, N., & Mgaya, R. J. (1999). Exploring the application and acceptance of group support systems in Africa. *Journal of Management Information Systems*, *15*(3), 197–234.

DeLone, W. H., & McLean, E. R. (1992). Information systems success: The quest for the dependent variable. *Information Systems Research*, *3*(1), 60–95. doi:10.1287/isre.3.1.60doi:10.1287/isre.3.1.60

DeLone, W. H., & McLean, E. R. (2003). The DeLone and McLean model of information systems success: A ten-year update. *Journal of Management Information Systems*, *19*(4), 9–30.

Elbeltagi, I., McBride, N., & Hardaker, G. (2005). Evaluating the factors affecting DSS usage by senior managers in local authorities in Egypt. *Journal of Global Information Management*, *13*(2), 42–65. doi:10.4018/jgim.2005040103doi:10.4018/jgim.2005040103

Fishbein, M., & Ajzen, I. (1975). *Belief, attitude, intention and behavior: An introduction to theory and research*. Reading, MA: Addison-Wesley.

Gauld, R. (2007). Public sector information system project failures: Lessons from a New Zealand hospital organization. *Government Information Quarterly*, *24*(1), 102–114. doi:10.1016/j.giq.2006.02.010doi:10.1016/j.giq.2006.02.010

Gefen, D. (2003). TAM or just plain habit: A look at experienced online shoppers. *Journal of End User Computing*, *15*(3), 1–13. doi:10.4018/joeuc.2003070101doi:10.4018/joeuc.2003070101

Heeks, R., & Bhatnagar, S. (1999). Understanding success and failure in information age reform. In R. Heeks (Ed.), *Reinventing government in the information age* (pp. 49–74). London, UK: Routledge. doi:10.4324/9780203204962doi:10.4324/9780203204962

Hofstede, G. (2001). *Culture's consequences* (2nd ed.). Thousand Oaks, CA: Sage.

Howcroft, B., Hamilton, R., & Hewer, P. (2002). Consumer attitude and the usage and adoption of home-based banking in the United Kingdom. *International Journal of Bank Marketing*, *20*(2-3), 111–121. doi:10.1108/02652320210424205doi:10.1108/02652320210424205

Huang, L., Lu, M., & Wong, B. K. (2003). The impact of power distance on email acceptance: Evidence from the PRC. *Journal of Computer Information Systems*, *44*(1), 93–101.

Kamel, S., & Hassan, A. (2003). Assessing the introduction of electronic banking in Egypt using the technology acceptance model. *Annals of Cases on Information Technology*, *5*, 1–25.

Kelly, D., Harper, D. J., & Landau, B. (2008). Questionnaire mode effects in interactive information retrieval experiments. *Information Processing & Management*, *44*(1), 122–141. doi:10.1016/j.ipm.2007.02.007doi:10.1016/j.ipm.2007.02.007

Kolodinsky, J. M., Hogarth, J. M., & Hilgert, M. A. (2004). The adoption of electronic banking technologies by US consumers. *International Journal of Bank Marketing*, *22*(4-5), 238–259. doi:10.1108/02652320410542536doi:10.1108/02652320410542536

Lee, Y., Kozar, A. K., & Larsen, K. R. T. (2003). The technology acceptance model: Past present, and future. *Communications for the AIS, 12,* 752–780.

Legris, P., Ingham, J., & Collerette, P. (2003). Why people use information technology? A critical review of the technology acceptance model. *Information & Management, 40,* 191–204. doi:10.1016/S0378-7206(01)00143-4doi:10.1016/S0378-7206(01)00143-4

Liao, Z., & Landry, R. (2000). An empirical study on organizational acceptance of new information systems in a commercial bank environment. In *Proceedings of the 33ʳᵈ Hawaii International Conference on System Sciences* (pp. 2021-2030).

Lim, J. (2003). A conceptual framework on the adoption of negotiation support systems. *Information and Software Technology, 45,* 469–477.

McCoy, S., Everard, A., & Jones, B. M. (2005). An examination of the technology acceptance model in Uruguay and the US: A focus on culture. *Journal of Global Information Technology Management, 8*(1), 27–45.

Nunnally, J. (1967). *Psychometric theory.* New York, NY: McGraw-Hill.

Pan, G. S. C. (2005). Information systems project abandonment: A stakeholder analysis. *International Journal of Information Management, 25*(2), 173–184. doi:10.1016/j.ijinfomgt.2004.12.003doi:10.1016/j.ijinfomgt.2004.12.003

Phillips, L. A., Calantone, R., & Lee, M. (1994). International technology adoption: behavior, structure, demand certainty and culture. *Journal of Business and Industrial Marketing, 9*(2), 16–28. doi:10.1108/08858629410059762doi:10.1108/08858629410059762

Pikkarainen, T., Pikkarainen, K., Karjaluoto, H., & Pahnila, S. (2004). Consumer acceptance of online banking: An extension of the technology acceptance model. *Internet Research, 14*(3), 224–235. doi:10.1108/10662240410542652doi:10.1108/10662240410542652

Rose, G., & Straub, D. (1998). Predicting general IT use: Applying TAM to the Arabic world. *Journal of Global Information Management, 6*(3), 39–46.

Shih, H. (2004). An empirical study on predicting user acceptance of e-shopping on the Web. *Information & Management, 41*(3), 351–368. doi:10.1016/S0378-7206(03)00079-Xdoi:10.1016/S0378-7206(03)00079-X

Straub, D. (1994). The effect of culture on IT diffusion e-mail and fax in Japan and the U.S. *Information Systems Research, 5*(1), 23–47. doi:10.1287/isre.5.1.23doi:10.1287/isre.5.1.23

Straub, D., Keil, M., & Brenner, W. (1997). Testing the technology acceptance model across cultures: A three country study. *Information & Management, 33,* 1–11. doi:10.1016/S0378-7206(97)00026-8doi:10.1016/S0378-7206(97)00026-8

Venkatesh, V., & Davis, F. (2000). A theoretical extension of the technology acceptance model: Four longitudinal field studies. *Management Science, 46*(2), 186–204. doi:10.1287/mnsc.46.2.186.11926doi:10.1287/mnsc.46.2.186.11926

Venkatesh, V., & Morris, M. G. (2000). Why don't men ever stop to ask for directions? Gender, social influence, and their role in technology acceptance and usage behavior. *Management Information Systems Quarterly, 24*(1), 115–139. doi:10.2307/3250981doi:10.2307/3250981

Vijayasarathy, L. R. (2004). Predicting consumer intentions to use on-line shopping: The case for an augmented technology acceptance model. *Information & Management, 41*(6), 747–762. doi:10.1016/j.im.2003.08.011doi:10.1016/j.im.2003.08.011

Wang, Y. S., Wang, Y. M., Lin, H. H., & Tang, T. I. (2003). Determinants of user acceptance of Internet banking: An empirical study. *International Journal of Service Industry Management, 14*(5), 501–519. doi:10.1108/09564230310500192doi:10.1108/09564230310500192

Webster, J., & Watson, Watson, R. T. (2002). Analyzing the past to prepare for the future: Writing a literature review. *Management Information Systems Quarterly, 26*(2), xiii–xxiii.

This work was previously published in the International Journal of Information Systems and Social Change, Volume 2, Issue 4, edited by John Wang, pp. 1-12, copyright 2011 by IGI Publishing (an imprint of IGI Global).

Chapter 15
Scale Economies in Indian Commercial Banking Sector:
Evidence from DEA and Translog Estimates

Biresh K. Sahoo
Xavier Institute of Management, Bhubaneswar, India

Dieter Gstach
Vienna University of Economics & Business Administration, Austria

ABSTRACT

Two alternative estimation models, i.e., a translog cost function and data envelopment analysis (DEA) based on a cost model are compared and contrasted in revealing scale economies in the Indian commercial banking sector. The empirical results indicate that while the translog cost model exhibits increasing returns to scale for all the ownership groups, the DEA model reveals economies of scale only for foreign banks, diseconomies of scale for nationalized banks, and both economies and diseconomies of scale for private banks. The divergence of the results obtained from these two estimation models should concern model builders. From an empirical perspective the definition of scale economies through a constant input mix is very restrictive. The DEA cost model is much more flexible in this respect: It neither requires the restrictive assumptions that the unit factor prices are always available with certainty, nor that these prices are exogenous to the firms. However, the very volatile nature of the banking industry might question the validity of the empirical estimates in this deterministic setting. Therefore, further research is required to examine the bank performance behavior using both SFA and chance constrained DEA for the comparison in a stochastic setting.

DOI: 10.4018/978-1-4666-2922-6.ch015

INTRODUCTION

There is a widely held general belief that competition, a driving force behind numerous important policy changes, exerts downward pressure on costs, reduces slacks, provides incentives for the efficient organization of production, and even drives innovation forward. To analyze the performance of firms, the concept of *productivity growth* has been widely used in the literature, and the sources of this growth are largely due to contributions from either *scale economies (returns to scale)*[1] or *technical change* or both[2].

The returns to scale (RTS) property of a production function is regularly used to describe the relationship between scale and efficiency. As for terminology: Constant RTS are said to prevail at a point on the production frontier if an increase of all inputs by 1% leads to an increase of all outputs by 1%. Decreasing RTS are present if outputs increase by less than 1%, while increasing RTS exist if they increase by more than 1%.

An appropriate estimation strategy for the underlying production (technology) structure is essential in understanding RTS characteristics of firms. We find in the literature (Färe et al., 1988) that there are two approaches to the estimation of RTS: the neoclassical approach and the axiomatic approach. The former (usually estimated with some parametric econometric technique) gives one a *quantitative* measure of RTS, whereas the latter approach (to be estimated in a nonparametric fashion via data envelopment analysis (DEA))[3] yields both *qualitative* and *quantitative* information about RTS. The latter is been researched in many studies (Førsund, 1996; Banker et al., 1996; Sueyoshi, 1997; Fukuyama, 2000, 2001, 2003; Førsund & Hjalmarsson, 2004; Tone & Sahoo, 2004, 2005, 2006; Hadjicostas & Soteriou, 2006; Førsund et al., 2007; Sueyoshi & Sekitani, 2007a, 2007b; Podinovski et al., 2008)[4]. However, both methods have become important analytical tools in the empirical evaluation of RTS.

Note that all standard methods of determining RTS proceed by examining tangential planes to the frontier that can be drawn through a given point. This is done either by looking at the constant term (the variable u_0 originally introduced in the literature by Banker et al., 1984) that represents the intercept of that plane with the plane in which all inputs are set to zero or, by observing the weights of the corner points of the facet of the frontier associated with that plane. This determination, however, may be difficult because the plane need not be unique. In this study we will, therefore, deal with both the lower and upper bounds of RTS.

However, the difference between the econometric and DEA approaches lies in the construction of the efficiency frontier and the calculation of a measure for scale economies along the frontier. The advantage of the econometric approach is that it allows for a formal statistical testing of hypotheses and construction of confidence intervals (Hjalmarsson et al., 1996). However, the problem with this approach is that it is parametric and can confound the effects of *misspecification* of functional form with scale economies; and further, *flexible* functional forms are susceptible to multicollinearity, and theoretical restrictions may be violated (Reinhard et al., 2000). DEA, however, has the advantage of both, being non-parametric in nature which means *less* susceptible to specification error; and being able to accommodate multiple inputs and multiple outputs. We will be critically analyzing in this paper the nature of scale economies properties from both *nonparametric* cost DEA and a *parametric* translog cost function on the common premise that they are both *deterministic* in nature.

For empirical illustration, we will be analyzing in this paper the RTS characteristics of banking in India, looking particularly at the impact of competition on scale performance of banks with respect to ownership[5]. For a growing economy like India, the faster growth of any sector hinges upon an efficiently and reliably operating financial system, and hence the Indian banking sector is selected.

The Indian financial sector, which had been operating in a closed and regulated environment, underwent a radical change during the nineties. To promote efficiency and competition, therefore, the Reserve Bank of India (RBI) initiated in 1992 a number of reforms such as entry deregulation, branch delicensing, deregulation of interest rates, and allowing public sector banks to raise up to 49% of their equity in the capital market. All of this gave rise to the heightened competitive pressure in the banking industry. These changes came in the form of greater use of automatic teller machines and internet banking, huge increases in housing and consumer credit, stronger and more transparent balance sheets and product diversification. A major intent of these policies is to have a radical transformation in the operating landscape of the Indian banks. In this scenario we believe that banks are in the pursuit of enlarging their size using available *scale economies* in order to enhance their asset base and profit so as to meet global standards.

It should be noted, as pointed out by an anonymous referee, that along with some beneficial innovations, one disastrous innovation introduced into the industry during this period is the creation of new exotic lending instruments and the willingness of local and international banks to assume risk levels that earlier generations of bankers would never had dreamed of. These were eventually to wreak havoc with the entire international financial system, thus leading to world-wide recession.

In studying the Indian banking sector it is important to examine the banks' RTS behaviors across the entire spectrum of ownership types, which might yield valuable information concerning performance differentials. This will enable us to investigate the economic linkage between ownership and performance in the light of the property right hypothesis (Alchian, 1965; de Alessi, 1980) and public choice theory (Niskanen, 1971;

Levy, 1987). According to the property right hypothesis, private enterprises should perform more efficiently and profitably than public enterprises, because of a supposed strong link between the markets for corporate control and the efficiency of private enterprises. While this argument may apply more to developed countries, testing for performance differentials across ownership types in the banking industry of a developing country like India can yield some insights in the success of the reform process.

The remainder of this paper proceeds as follows. First we provide a theoretical discussion examining relative strengths and weaknesses of *nonparametric* cost DEA and *parametric* translog cost function methods for measuring scale economies. We deal with the selection of input and output data of the Indian commercial banking sector. Results are presented and discussed, followed by some concluding remarks.

METHODOLOGY

We will be discussing here the evaluation of scale economies, first, based on a nonparametric cost-based DEA model, and then, based on a parametric translog cost function. We deal with n banking firms each using m inputs to produce s outputs. For each firm h ($h = 1,2,\ldots, n$), we denote, respectively, the physical input and output vectors by $x_h \in R^m$ and $y_h \in R^s$. Let the unit input price vector be $w_h \in R^m$. The *observed* cost of firm h can be obtained as $c_h = \sum_{i=1}^{m} w_{ih} x_{ih}$ ($h = 1,2,\ldots,n$).

In the primal framework the returns to scale (RTS) or scale elasticity (SE), $\varepsilon(x, y)$ or the *Passus Coefficient* (as in the terminology of Frisch, 1965), is defined as the ratio of the maximum proportional (β) expansion of outputs (y) for a given proportional (α) expansion of inputs (x). So, differentiating the transformation function $F(y, x) = 0$ with respect to scaling factor α, and then equating it to zero yields the following local scale elasticity measure[6]:

$$\varepsilon(x,y) \equiv -\sum_{i=1}^{m} x_i \frac{\partial F(\cdot)}{\partial x_i} \Big/ \sum_{r=1}^{s} y_r \frac{\partial F(\cdot)}{\partial y_r} \tag{1}$$

In the dual framework $\varepsilon(x,y)$ can be defined as

$$\varepsilon(y;w) \equiv C(y;w) \Big/ \sum_{r=1}^{s} y_r \frac{\partial C(y;w)}{\partial y_r} \tag{2}$$

Here $C(y;w) \equiv \min_x \{wx : x \in L(y)\}$ represents the minimum cost of producing output vector y when the input price vector is w where $L(y)$ is the input set representing the set of input vectors that all yield at least the output vector y.

Scale Economies in Nonparametric Approach

To describe the scale elasticity evaluation procedure in the nonparametric DEA approach, let us first consider a simple technology involving single input, i.e., total cost (c) and multiple outputs. Then the following describes the efficiency determination with the DEA model[7] for firm h (see Box 1).

The cost efficiency model (3) is based on a cost-based technology set (T^C) defined in Box 2.

For an efficient firm h, $\theta^* = 1$, and $F(y_h, c_h) = 0$ holds, i.e.,

$$F(y_h, c_h) \equiv \sum_{r=1}^{s} u_r^* y_{rh} - v^* c_h + u_0^* = 0 \tag{5}$$

where u_r, v and u_o are the dual multipliers obtained from the following Lagrange function (see Box 3.)

The cost efficiency frontier for firm h is then derived from (5) as

$$c(y_h) = c_h^* = \beta_0 + \sum_{r=1}^{s} \beta_r y_{rh} \tag{7}$$

where $\beta_0 = u_0^* / v^*$ and $\beta_r = u_r^* / v^*$.

The scale elasticity of firm h, $\varepsilon(y_h)$ can be derived from (7) as

$$\varepsilon(y_h) = \frac{c_h^*}{\sum_{r=1}^{s} (\partial c_h^* / \partial y_{rh}) y_{rh}} = \frac{c_h^*}{\sum_{r=1}^{s} n_r y_{rh}} = \frac{1}{1 - u_0^*} \tag{8}$$

Box 1.

$$\text{Min } \theta \quad \text{s.t.} \quad \sum_{j=1}^{n} c_j \lambda_j \le \theta c_h, \sum_{j=1}^{n} y_{rj} \lambda_j \ge y_{rh} \ (\forall \ r), \sum_{j=1}^{n} \lambda_j = 1, \lambda_j \ge 0 \tag{3}$$

Box 2.

$$T^C \equiv \left\{ (c,y) \Big| \sum_{j=1}^{n} c_j \lambda_j \le c, \sum_{j=1}^{n} y_{rj} \lambda_j \ge y_r \ (\forall \ r), \sum_{j=1}^{n} \lambda_j = 1, \lambda_j \ge 0 \right\} \tag{4}$$

Firm h experiences (local) economies of scale if $\varepsilon(y_h) > 1$ $(u_0^* > 0)$, (local) diseconomies of scale if $\varepsilon(y_h) < 1$ $(u_0^* < 0)$; and no (local) economies or diseconomies of scale if $\varepsilon(y_h) = 1$ $(u_0^* = 0)$. In case of multiple optimal solutions in u_0^*, one can find out the lower and upper bounds of scale economies by computing the lower and upper bounds of u_0^* in the spirit of Tone and Sahoo (2006).

However, following Sueyoshi (1997), scale economies, $n_h(y_h; w_h)$ can also be obtained from the dual of the following cost DEA model:

$$c(y_h; w_h) = \min_{x,n} \sum_{i=1}^{m} w_{ih} x_i$$

$$\text{s.t.} \sum_{j=1}^{n} x_{ij} n_j \leq x_i (\forall\ i),\ \sum_{j=1}^{n} y_{rj} n_j \geq y_{rh} (\forall\ r),$$

$$\sum_{j=1}^{n} n_j = 1,\ n_j \geq 0 \tag{9}$$

as $\varepsilon(y, w) \equiv \dfrac{1}{1 - [n_h^* / c(y_h; w_h)]} \tag{10}$

where n_h^* is the dual variable corresponding to $\sum_{j=1}^{n} n_j = 1$.

The DEA model (9) is based on a factor-based technology set $\left(T^F\right)$, which can be defined as seen in Box 4.

Note that if one computes the scale elasticity in under the assumptions of the BCC model of Banker et al. (1984), which is based on technology set T^F in (10), then it is same as the scale elasticity measure in the cost environment (10) (Cooper et al., 1995; Tone & Sahoo, 2005).

The DEA cost model (9) is based on a number of simplifying assumptions: a) input prices are exogenously given, b) input prices are measured and known with certainty by the firms, and c) the underlying factor-based technology set (T^F) is convex. However, in many real-world situations, prices are not exogenous, but vary according to the actions by the firms (Chamberlin, 1933; Robinson, 1933). Also, firms often face *ex ante* price uncertainty while making production decisions (Sandmo, 1971; McCall, 1967). And, the maintained axiom of a convex, factor-based technology set (T^F) [8] assumes away some economically important technological features such as *indivisible* production activities, *economies of scale* (= increasing RTS) and *economies of specialization* (diseconomies of scope), which all result from *concavities* in production functions, as pointed out by Farrell (1959).

Box 3.

$$L = -n + \sum_{r=1}^{s} u_r^* \left(\sum_{j=1}^{n} y_{rj} \lambda_j - y_{rh} \right) + v^* \left(\theta c_h - \sum_{j=1}^{n} c_j \lambda_j \right) + u_o^* \left(\sum_{j=1}^{n} n_j - 1 \right) \tag{6}$$

Box 4.

$$T^F = \left\{ (x, y) \middle| \sum_{j=1}^{n} x_{ij} n_j \leq x_i,\ \sum_{j=1}^{n} y_{rj} n_j \geq y_r, \sum_{j=1}^{n} n_j = 1, n_j \geq 0 \right\} \tag{11}$$

In fact, looking at the DEA-related economic literature (Afriat, 1972; Hanoch & Rothschild, 1972; Varian, 1984) on production analysis, it appears that convexity properties are motivated from the perspective of economic objectives (e.g., cost minimization), but are not an inherent feature of a production technology. However, looking at the structures of real-world production leads one to suspect the harmless character of the convexity postulate. For details see Cherchye et al. (2000a, 2000b) and Kuosmanen (2003) who provide empirical as well as theoretical arguments, less in favor of, but, mostly against the convexity postulate in DEA. Therefore, the DEA efficiency model (1), which is based on a cost-based technology set (T^C) scores well over DEA model (7) based on a factor-based technology set (T^F) in exhibiting realistic scale economies behavior of firms.

The other fundamental shortcoming associated with DEA cost model (9) in measuring scale economies of a real-world firm is that in this measure the output is related to the inputs only by defining the input-mix in a special way, e.g., as a replication measure, as a size measure, or as a long-run measure of only one input such as plant and machinery or capital. The replication measure is purely statistical in nature, often, used in statistical theory of design of experiments. But this lacks any economic meaning as the techniques and inputs used at a higher scale are very different from those used at a lower scale of production[9]. The size measure in terms of inputs is not unequivocal. In agricultural economics it is natural to measure size by acres, but in industrial manufacturing there is no such natural physical measure due to heterogeneity of capital. Note, however, that if the current input mix can be represented by a size measure, then the 'size' elasticity of output is a good measure of RTS. This can be accomplished for example for plant and machinery. Note also that the empirical implementation of scale elasticity measure (10) requires the availability of level

data on all unit input prices, which are rarely available at the firm level, while costs are often readily available and well measured.

However, in contrast, the DEA cost model (3) is flexible because it does not require the input-mix (or input cost mix) to be constant to estimate the scale elasticity of firms. As we see in real life, when production is expanded, firms often *change* the organization of their processes or the characteristics of their inputs, which are economically more attractive than the replicated alternatives. We therefore argue that the concept of production technology should be broad enough to capture the fundamental interactions between increases in output and increases in the specialization of functions such as changes in techniques, organization, characteristics of inputs, input factor proportions, etc. And these fundamental interactions are inevitable *when the economic theory of production is being revitalized in response to urgent questions raised by advances in technology, shortages of various natural resources, and intensified competition in international markets* (The italics are from Gold, 1981, p. 6). Therefore, in the light of the aforementioned problems, we argue that the cost model in (9) is very restrictive to capture scale economies behaviors of firms.

Scale Economies in Parametric Approach

Let us now consider the following cost model where each firm *h* is assumed to minimize its cost $c(y_h, w_h)$ to produce its output vector y_h by choosing an input vector x_h at the market price vector w_h subject to a production technology constraint $F(y_h, x_h, t)$ in period t as follows:

$$\min \quad c\,(y_h, w_h, t) = \sum_{i=1}^{m} w_{ih} x_i(y, w) \qquad (12)$$

subject to $F(y_h, x_h, t) = 0$

Consistent with most of the firm-efficiency literature, we employ the following variable returns to scale multi-product translog cost function $c\ (y,w,t)$ [10] of a firm in period t (see Box 5).

In order for the cost function $c\ (y,w,t)$ defined in (13) to be well behaved, it must be *homogeneous of degree one*, *monotonic*, and *concave in prices*, in which case the restrictions: (13a-g) should hold [11].

A number of additional parametric restrictions can be imposed to reflect a priori knowledge about the specific structure of the underlying production technology. For example, if the technology structure is *homothetic*, then $c\ (y,w,t)$ has to additionally satisfy the following restrictions:

$$\gamma_{ir} = 0\ (\forall\ i,r);\ \delta_i = 0\ (\forall\ i);\ \text{and}\ \eta_r = 0\ (\forall\ r) \tag{14}$$

Further, if $c\ (y,w,t)$ is *homogeneous* in outputs, the following restrictions should apply (see Box 6).

The production structure exhibits constant returns to scale (CRS) if $c\ (y,w,t)$ satisfies (15) along with

$$\sum_{r=1}^{s} \beta_r = 1 \tag{16}$$

Finally, if the underlying technology is CRS Cobb-Douglas, then along with the previous restrictions, $c\ (y,w,t)$ should additionally satisfy

Box 5.

$$\ln c\ (y,w,t) = \alpha_0 + \sum_{i=1}^{m} \alpha_i \ln w_i + \sum_{r=1}^{s} \beta_r \ln y_r + \frac{1}{2} \sum_{i=1}^{m} \sum_{i'=1}^{m} \alpha_{ii'}(\ln w_i)(\ln w_{i'})$$
$$+\ \frac{1}{2} \sum_{r=1}^{s} \sum_{r'=1}^{s} \beta_{rr'}(\ln y_r)(\ln y_{r'}) + \sum_{i=1}^{m} \sum_{r=1}^{s} \gamma_{ir}(\ln w_i)(\ln y_r) \tag{13}$$
$$+\ \sum_{i=1}^{m} \delta_{it}(\ln w_i)t + \sum_{r=1}^{s} \eta_{rt}(\ln y_r)t - \tau\ t - \frac{1}{2} \upsilon\ t^2$$

to estimate the cost-output relationship with the following restrictions:

$$\sum_{i=1}^{m} \alpha_i = 1;\text{(13a)}$$

$$\sum_{i'=1}^{m} \alpha_{ii'} = 0\ (\forall\ i);\text{(13b)}$$

$$\alpha_{ii'} = \alpha_{i'i}\ (\forall\ i,\ i');\text{(13c)}$$
$$\beta_{rr'} = \beta_{r'r}\ (\forall\ r,\ r');\text{(13d)}$$

$$\sum_{i=1}^{m} \gamma_{ir} = 0\ (\forall\ r);\text{(10e)}$$

$$\sum_{i=1}^{m} \delta_{it} = 0;\text{(13f)}$$

$$\sum_{r=1}^{s} \eta_{rt} = 0\text{(13g)}$$

$$\alpha_{ii'} = 0 \; (\forall \; i) \tag{17}$$

Using Shephard's Lemma: $\dfrac{\partial c(\cdot)}{\partial w_i} = x_i(y, w, t)$, the cost-share equation for input x_i, i.e., s_i can be obtained from the cost function (13) as seen in Box 7.

There are two alternative approaches to estimate the relevant parameters of the cost function $c\;(\cdot)$. One way is to directly estimate (13) by ordinary least squares subject to the restrictions: 13a-g implied by symmetry and linear homogeneity in input prices. However, as argued by Christensen and Green (1976), efficiency in the estimates of parameters can be improved if the cost function (13) with restrictions (13a-g) is jointly estimated with its (m-1) factor share equations (18) by the seemingly unrelated regression (SUR) method. The latter also takes contemporaneous correlation in the disturbances of cost function and its factor-share equations into account.

But, as argued by Guilkey and Lovell (1980), one might not gain much by SUR estimation when it is applied to the translog approximation rather than to the true underlying functional form of the cost function. Truncating the approximation at second order terms forces one to commit a type of specification error such that the system disturbances are no longer well behaved leaving little hope for improvement in estimation efficiency. They cited one study by Summers (1965) who found in an extensive Monte Carlo analysis of the various simultaneous equation models that the single equation models frequently outperformed simultaneous equation models in the presence of specification error, and therefore conjectured similar results to hold in case of specification error arising out of truncation of approximation at second order terms. See also Dixton et al. (1987) where it is argued that the cross-equation parameter equality is rarely found to be satisfied, thus justifying the use of single equation OLS with standard restrictions.

The local returns to scale[12] of firm h (ε_h) is then computed by Equation 19, as seen in Box 8.

In the parametric literature the *ad hoc* functional forms chosen for the production technology without knowing the underlying relationship between inputs and outputs, often cast doubt on the results. To cite one such study here: For all functional forms which Hasenkamp (1976) considered, his findings suggest economies of scale whereas for flexible functional forms, his results reveal economies of specialization (i.e., violation of *convexity*). In spite of massive empirical evi-

Box 6.

$$\gamma_{ir} = 0 \; (\forall \; i, r); \; \beta_{rr'} = 0 \; (\forall \; r, r'); \; \delta_i = 0 \; (\forall \; i); \; \text{and} \; \eta_r = 0 \; (\forall \; r) \tag{15}$$

in which case the degree of homogeneity is $1 \Big/ \sum_{r=1}^{s} \beta_r$.

Box 7.

$$s_i = \frac{\partial \ln c(\cdot)}{\partial \ln w_i} = \frac{w_i}{c(\cdot)} \frac{\partial c(\cdot)}{\partial w_i} = \frac{w_i x_i(\cdot)}{c(\cdot)} = \alpha_i + \sum_{i'=1}^{m} \alpha_{ii'} \ln w_{i'} + \sum_{r=1}^{s} \gamma_{ir} \ln y_r + \delta_i t \quad (\forall \; i) \tag{18}$$

where $\sum_{i'=1}^{m} w_i x_i(\cdot) = c(\cdot)$ and $\sum_{i'=1}^{m} s_i = 1.$

Box 8.

$$\varepsilon_h = \left[\sum_{r=1}^{s} \frac{\partial \ln c(w,y)}{\partial \ln y_{rh}}\right]^{-1} = \left[\sum_{r=1}^{s} \beta_r + \sum_{r=1}^{s}\sum_{r'=1}^{s} \beta_{rr'} \ln y_{rh} + \sum_{r=1}^{s}\sum_{i=1}^{m} \gamma_{ir} \ln w_{ih} + \sum_{r=1}^{s} \eta_{rt} t\right]^{-1} \quad (19)$$

dence against the convexity axiom for factor-based technologies, the parametric approach imposes such restrictions on the parameters of the cost function to bring in *convexity*.

The Data

The Indian banking sector is taken as an empirical illustration of the two methods in estimating scale economies. In the literature there are two approaches to measure bank efficiency[13] - production approach and intermediation approach. In the former banks use capital, labor and other non-financial inputs to provide deposits and advances (Ferrier & Lovell, 1990). In the latter approach, however, a bank is treated as a producer of intermediation services - by transforming risk and maturity profile of funds received from depositors to investment or loan portfolios of different risk and maturity profiles. Banks also provide services for which specific charges are levied and monetary value of non-interest income is sometimes considered as another important output variable.

Keeping in mind that banks, besides being profit driven, are also forced to take up economic and social responsibilities, this study has adopted the intermediation approach by considering three outputs - investments (y_1), performing loan assets (y_2) and non-interest income (y_3), and three inputs - fixed assets (x_1), borrowed funds (x_2) and labor (x_3), which seem most relevant in analyzing management's success in controlling cost and generating revenues. All the monetary values of inputs and outputs that have been deflated using the wholesale price index

with base 1993-94 and are expressed in crores of Indian rupees (Rs.) (Note that Rs.1 crore = Rs.10 million).

Concerning the prices of inputs, the unit prices of 'fixed assets' (w_1), 'borrowed funds' (w_2) and 'labor' (w_3) are taken, respectively, as the 'non-labor operational cost per rupee amount of fixed asset', 'average interest paid per rupee of borrowed funds' and 'average staff cost'. As regards the output prices, unit prices of 'investments', 'performing loan assets' and 'non-interest income' are, respectively, taken as 'average interest earned on per rupee unit of investment', 'average interest earned on per rupee unit of performing loan assets' and 'non-interest fee-based income on per rupee of working funds.' The input and output data as well as their prices have been taken from the various sections of 'Statistical Tables Relating to Banks in India', Reserve bank of India and from Indian Banking Association publications[14].

This study is conducted on 70 banks (26 nationalized banks (NB), 26 private banks (PB) and 18 foreign banks (FB)) for the eight-year period 1997-98 to 2004-05. So we have in total 560 (= 70 x 8) commercial banks in our sample period to construct a single overall frontier, which provides a benchmark against which to calculate the scale elasticity of each bank in each year. This approach has some advantage in terms of yielding information on trends in performance, which would not have been available, had we used DEA to calculate annual frontiers since the benchmarks would likely change from year to year.

Empirical Results and Discussion

Translog Cost Function Estimates

The translog (TL) cost model (equation: 10) for the Indian commercial banking sector is estimated in five ways: first with no restrictions (translog, TL(UR)), second with normal restrictions (13a-g) as assumed by theory (non-homothetic, TL(NH)), third with both (13a-g) and (14) (homothetic, TL(H)), fourth with (13a-g), (14) and (15) (homogeneous) and finally with (13a-g), (14), (15), and (16) (CRS). The DEA cost structure is estimated using model (3). Table 1 exhibits the complete set of parameter estimates with the five different models in translog scheme.

To test whether the underlying cost structure is homothetic, homogeneous or homogeneous of degree one (CRS), we employed the likelihood ratio test statistic (LR $= -2(n_R - n_{UR})$) where n_R and n_{UR} are the log likelihood values for the restricted and the unrestricted model (13), respectively. The LR test statistic is distributed asymptotically as a χ^2 random variable, with the number of independent restrictions being imposed as the degrees of freedom. Figure 1 reports these LR test statistics for different technology structures. It can be seen that all the imposed technology structures, i.e., homotheticity, homogeneity and CRS, are comfortably rejected, thus suggesting that nonhomotheticity better represents the production structure of the Indian commercial banking industry. Therefore, we henceforth do not consider the possibility of a homothetic production in further analysis.

Comparison of DEA and Translog SE Estimates

We now turn to examine the average trend scale elasticity (SE) of the commercial banking industry in India obtained from the DEA cost model [Model (3)] on one hand, and the unconstrained translog and constrained nonhomothetic models on the other hand to see whether there is any divergence between the two. The comparative SE estimates are exhibited in Figure 2. As expected, TL(UR) exhibits lower average estimates of around 1.07 as compared to its constrained non-homothetic counterpart, i.e., TL(NH) of around 1.18. While TL(UR) shows a slightly declining trend from 1.09 to 1.06, TL(NH) exhibit a monotonically increasing trend from 1.1 to 1.21. However, in contrast, DEA exhibits a declining trend starting with economies of scale estimate of 1.32 and ending with diseconomies of scale estimate of 0.98.

Translog SE Estimates w.r.t. Ownership

To understand the distribution of scale economies in the translog model with respect to ownership, we display the average annual SE trends in both unconstrained and constrained non-homothetic environments in Figure 2 and Figure 3 respectively. In the unconstrained model (see Figure 3) we found both foreign and private banks exhibiting higher SE (IRS) compared to the average, and the nationalized banks performing below the average.

However, the picture is completely different in the constrained non-homothetic case. One can see in Figure 4 that all three bank types show an improvement in their SE behavior. Foreign banks, though started with relatively low SE estimates, have shown consistent and drastic improvement in their performance over years, and even surpassed both private and nationalized banks in the end (last two years) of our investigation period.

Cost DEA SE Estimates w.r.t Ownership

Turning now to cost-based DEA SE estimates (based on model (4)), we see in Figure 5 that in contrast to the translog case, both nationalized and private banks show very low SE estimates, with the former exhibiting diseconomies of scale

Figure 1. Cost function estimation for all models

Table 1: Cost function estimation for all models

	Translog	Nonhomothetic	Homothetic	Homogeneity	CRS
α_0	-3.182** (1.223)	2.696* (1.356)	2.150** (0.601)	1.852** (0.345)	1.786** (0.362)
α_1	-0.0080 (0.100)	0.774** (0.123)	0.629** (0.076)	0.699** (0.071)	0.571** (0.073)
α_2	0.137 (0.141)	-0.181 (0.144)	-0.054 (0.079)	-0.066 (0.071)	-0.015 (0.074)
α_3	-0.228 (0.150)	0.406** (0.122)	0.424** (0.060)	0.367** (0.058)	0.444** (0.060)
β_1	0.543 (0.440)	0.850 (0.546)	-0.407 (0.313)	0.452** (0.078)	0.440** (0.081)
β_2	0.516 (0.552)	-0.781 (0.710)	0.971* (0.472)	0.406** (0.088)	0.543** (0.090)
β_3	-0.207 (0.214)	0.361 (0.339)	0.214 (0.208)	-0.026 (0.039)	0.017 (0.040)
α_{11}	-0.037** (0.009)	-0.071** (0.014)	-0.073** (0.012)	-0.081** (0.012)	-0.046** (0.011)
α_{22}	0.015 (0.022)	-0.001 (0.018)	0.017 (0.014)	0.017 (0.014)	0.047** (0.014)
α_{33}	-0.020 (0.011)	0.003 (0.008)	0.016* (0.007)	0.011 (0.007)	0.021** (0.007)
α_{12}	0.021 (0.011)	0.037** (0.013)	0.036** (0.011)	0.037** (0.011)	0.010 (0.011)
α_{13}	0.003 (0.008)	0.034** (0.008)	0.037** (0.007)	0.044** (0.007)	0.036** (0.008)
α_{23}	0.025* (0.011)	-0.036** (0.010)	-0.053** (0.008)	-0.055** (0.008)	-0.057** (0.008)
β_{11}	-0.141 (0.101)	-0.190 (0.166)	-0.268* (0.130)		
β_{22}	0.009 (0.130)	0.161 (0.209)	-0.380** (0.134)		
β_{33}	-0.017 (0.025)	0.053 (0.044)	0.051 (0.039)		
β_{12}	0.084 (0.097)	0.095 (0.153)	0.380** (0.090)		
β_{13}	-0.041 (0.055)	-0.058 (0.094)	-0.059 (0.090)		
β_{23}	0.041 (0.064)	-0.044 (0.111)	-0.024 (0.105)		

	Translog	Nonhomothetic	Homothetic	Homogeneity	CRS
γ_{11}	-0.004 (0.029)	0.064 (0.041)			
γ_{21}	-0.143** (0.034)	-0.178** (0.046)			
γ_{31}	0.0589* (0.029)	0.114** (0.030)			
γ_{12}	0.033 (0.032)	-0.076 (0.045)			
γ_{22}	0.144** (0.040)	0.207** (0.048)			
γ_{32}	-0.026 (0.031)	-0.131** (0.031)			
γ_{13}	-0.010 (0.013)	0.005 (0.019)			
γ_{23}	0.013 (0.019)	-0.017 (0.018)			
γ_{33}	-0.013 (0.013)	0.012 (0.017)			
δ_{1t}	-0.006 (0.005)	-0.011 (0.007)			
δ_{2t}	0.003 (0.008)	-0.008 (0.009)			
δ_{3t}	-0.011 (0.007)	-0.019** (0.008)			
η_{1t}	-0.004 (0.027)	0.033 (0.043)			
η_{2t}	0.020 (0.0293)	-0.025 (0.045)			
η_{3t}	-0.020* (0.009)	-0.008 (0.016)			
τ	-0.203* (0.104)	-0.149 (0.153)	0.100 (0.062)	0.104 (0.063)	0.109 (0.066)
υ	0.006 (0.008)	0.006 (0.014)	-0.009 (0.013)	-0.011 (0.013)	-0.008 (0.014)
Rests.	0	9	19	25	26
R^2	0.9522	0.8416	0.8270	0.8188	0.8002
Adj R^2	0.9490	0.83391	0.8219	0.8155	0.7970
LR test statistic (χ^2):		670.298	719.784	745.829	800.320

Note: Standard errors are in parentheses. *: $p < 0.05$, **: $p < 0.01$

throughout the whole period and the latter from 2001-02 onwards. Foreign banks are the clear winner, exhibiting economies of scale consistently over all the years of our study period.

The finding of clear economies of scale for foreign banks, and clear diseconomies of scale for nationalized banks is understandable because of RBI's branching policy. As Bhattacharyya et al. (1997) pointed out, under the RBI's branching policy, Indian (nationalized) banks are required to open branches but are not allowed to close unprofitable branches, and this policy prevents banks from optimizing their resources across the branch network because banks have neither control over the location of the branches nor the ability to close the loss making branches. How-

Figure 2. Comparative DEA and translog SE estimates over years

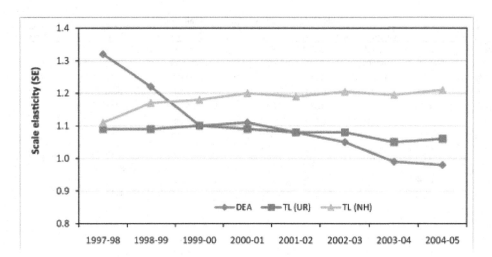

ever, foreign banks are mostly young, and tend to have smaller branch networks, since they have not yet fully expanded their business and have not been forced by regulators to expand branch networks beyond their optimal size. However, the evidence on the SE nature of private banks is mixed, showing both economies and diseconomies of scale. This might be due to the fact that Indian private banks are of two types: One follows the tradition of nationalized banks (old type) whereas the other go with foreign banks (new type).

Note that the conflicting signals concerning scale economies of banks from both translog and

DEA approaches might be due to their inherently different theoretical set up. First, in the translog approach the cost model employed is based on two assumptions[15]: a) input prices are exogenous to firms[16], and b) the underlying technology structure is convex where the scale elasticity estimates obtained from both primal (production) and dual (cost) environments appear to be the same, thus giving the illusion that *returns to scale* and *economies of scale* are one and the same[17]. However, the cost model employed in DEA (Model 3) does not employ any of these two assumptions. The factor prices can be endogenous when there are pecuni-

Figure 3. Unrestricted TL SE estimates w.r.t. ownership

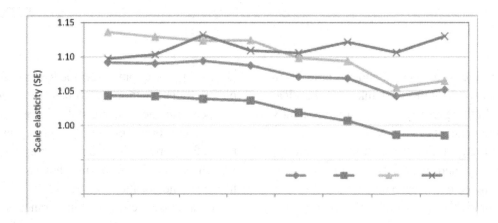

Figure 4. Nonhomothetic TL SE estimates w.r.t. ownsership

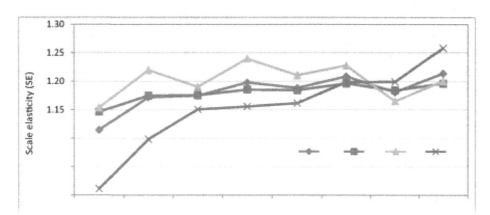

ary effects associated with the firm's employment of factors and these might be strong enough to cause economies (or diseconomies) of scale even though firm's technology exhibits decreasing (or increasing) returns to scale. Therefore, one can find firms in this setup not only controlling over the mix and quantities of inputs used but also exercising control over input prices. Furthermore, the underlying factor-based production technology can be concave because of many important technological and economic features such as *indivisibilities*, *economies of scale* and *scope*, which all arise due to *concavities* in production.

Second, the weights implied by DEA and translog methods most likely differ because of different assumptions regarding disposability of inputs and outputs. The translog method permits weak disposability in both inputs and outputs while the DEA method employed in this paper imposes strong disposability in both cost and outputs[18]. From the translog estimates one finds that the input weight of borrowed funds (x_2) and the output weight of performing loan assets (y_2)

Figure 5. DEA SE estimates w.r.t. ownership [cost model]

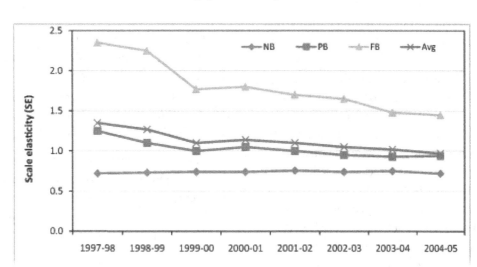

to be negative while DEA cost and output weights are always positive but near zero in some cases.

Third, the other probable reason behind the difference in estimates between translong and DEA is the deterministic nature of the structure of technology employed in both schemes. As one anonymous referee pointed out, during our study period there were new classes of loans, and new classes of debt obligations introduced which did not even exist before, and which were eventually to wreak havoc with the entire international financial system. Since the key feature of the international banking at the end of this period was the blatant neglect of risk, with banks of all stripes and color loading up on high-risk assets, conducting research on bank performance in any country in a deterministic setting might be questionable. Therefore, further research is desired using both chance constrained DEA as presented in Land et al. (1993) and stochastic frontier analysis (SFA).

CONCLUDING REMARKS

This paper applies and compares two alternative approaches to estimate scale economies to study the performance of the Indian commercial banking sector: a translog cost function and a DEA cost model. The common premise underlying both approaches is the *deterministic* nature of the observations. Our empirical results indicate that while the translog cost model exhibits increasing returns to scale for all three ownership types, the DEA model reveals economies of scale only for foreign banks, diseconomies of scale for nationalized banks, and both economies and diseconomies of scale for private banks. The divergence of the results obtained from these two schemes might arise from two possible sources: one is the model set up and the other is the assumed deterministic nature of production employed in both these schemes. As argued earlier, though the DEA cost model is more flexible than the translog scheme, the very volatile nature of the banking industry

during our study period might question the validity of the empirical estimates in deterministic setting. Therefore, further research on bank performance behavior is needed, using both SFA and chance constrained DEA for comparison in a stochastic setting.

ACKNOWLEDGMENT

We are sincerely thankful to the two anonymous referees and John Wang (Editor-in-Chief) for their constructive comments and suggestions that proved most useful in improving the first draft of this paper. The first author wishes to thank the Board of Austrian Science Fund (FWF), for financial support in the form of a *Lise Meitner* scholarship at Vienna University of Economics and Business Administration (WU-Wien). The remaining errors, if any, are the responsibility of the authors.

REFERENCES

Afriat, S. (1972). Efficiency estimation of production functions. *International Economic Review*, *13*, 568–598. doi:10.2307/2525845

Alchian, A. A. (1965). Some economics of property rights. *Il Politico*, *30*, 816–829.

Banker, R. D., Chang, H., & Cooper, W. W. (1996). Equivalence and implementation of alternative methods for determining returns to scale in data envelopment analysis. *European Journal of Operational Research*, *89*, 473–481. doi:10.1016/0377-2217(95)00044-5

Banker, R. D., Charnes, A., & Cooper, W. W. (1984). Some models for estimating technical and scale inefficiencies in data envelopment analysis. *Management Science*, *30*, 1078–1092. doi:10.1287/mnsc.30.9.1078

Basu, S., & Fernald, J. G. (1997). Returns to scale in US production: Estimates and implications. *The Journal of Political Economy, 105,* 249–283. doi:10.1086/262073

Baumol, W. J., Panzar, J. C., & Willig, R. D. (1998). *Contestable markets and the theory of industry structure.* New York, NY: Harcourt Brace Jovanovich.

Berger, A. N., & Humphrey, D. (1992). Measurement and efficiency issues in commercial banking. In Griliches, Z. (Ed.), *Output measurement in the service sector.* Chicago, IL: Chicago University Press.

Berger, A. N., & Mester, J. L. (1997). Beyond the black box: What explains differences in the efficiencies of financial institutions? *Journal of Banking & Finance, 21,* 87–98. doi:10.1016/S0378-4266(97)00010-1

Bhattacharyya, A., Lovell, C. A. K., & Sahay, P. (1997). The impact of liberalization on the productive efficiency of Indian commercial banks. *European Journal of Operational Research, 98,* 332–345. doi:10.1016/S0377-2217(96)00351-7

Borland, J., & Yang, X. (1995). Specialization, product development, evolution of the institution of the firm, and economic growth. *Journal of Evolutionary Economics, 5,* 19–42. doi:10.1007/BF01199668

Bouhnik, S., Golany, B., Passy, S., Hackman, S. T., & Vlatsa, D. A. (2001). Lower bound restrictions on intensities in data envelopment analysis. *Journal of Productivity Analysis, 16,* 241–261. doi:10.1023/A:1012510605812

Briec, W., Kerstens, K., & Vanden Eeckaut, P. (2004). Non-convex technologies and cost functions: Definitions, duality and nonparametric tests of convexity. *Journal of Economics, 81,* 155–192. doi:10.1007/s00712-003-0620-y

Burgress, D. F. (1974). A cost minimization approach to import demand equations. *The Review of Economics and Statistics, 56,* 224–234.

Chamberlin, E. H. (1933). *The theory of monopolistic competition: A reorientation of the theory of value.* Cambridge, MA: Harvard University Press.

Cherchye, L., Kuosmanen, T., & Post, T. (2000a). What is the economic meaning of FDH? A reply to Thrall. *Journal of Productivity Analysis, 13,* 259–263. doi:10.1023/A:1007827126369

Cherchye, L., Kuosmanen, T., & Post, T. (2000b). *Why convexify? An assessment of convexity axioms in DEA.* Helsinki, Finland: Helsinki School of Economics and Business Administration.

Christensen, L. R., & Green, W. H. (1976). Economies of scale in U.S. electric power generation. *The Journal of Political Economy, 84,* 655–676. doi:10.1086/260470

Christensen, L. R., Jorgenson, D. W., & Lau, L. J. (1971). Congugate duality and the transcendental logarithmic production function. *Econometrica, 39,* 255–256.

Christensen, L. R., Jorgenson, D. W., & Lau, L. J. (1973). Transcendental logarithmic production frontiers. *The Review of Economics and Statistics, 55,* 28–45. doi:10.2307/1927992

Cooper, W. W., Thompson, R. G., & Thrall, R. M. (1996). Introduction: Extensions and new developments in DEA. *Annals of Operations Research, 66,* 3–46. doi:10.1007/BF02125451

De Alessi, L. (1980). The economics of property rights: A review of the evidence. In Richard, O. Z. (Ed.), *Research in law and economics: A research annual (Vol. 2,* pp. 1–47). Greenwich, CT: JAI Press.

Deprins, D., Simar, L., & Tulkens, H. (1984). Measuring labor efficiency in post offices. In Marchand, M., Pestieau, P., & Tulkens, H. (Eds.), *The performance of public enterprises: Concepts and measurements*. Amsterdam, The Netherlands: North-Holland.

Despotis, D. K. (2005). A reassessment of the human development index via data envelopment analysis. *The Journal of the Operational Research Society*, *56*, 969–980. doi:10.1057/palgrave. jors.2601927

Devereux, M. B., Head, A. C., & Lapham, B. J. (1996). Monopolistic competition, increasing returns, and the effects of government spending. *Journal of Money, Credit and Banking*, *28*, 233–254. doi:10.2307/2078025

Diewert, W. E., & Fox, K. J. (2008). On the estimation of returns to scale, technical progress and monopolistic markups. *Journal of Economics*, *145*, 174–193.

Dixton, B., Garcia, P., & Anderson, M. (1987). Usefulness of pretests for estimating underlying technologies using dual profit functions. *International Economic Review*, *28*, 623–633. doi:10.2307/2526570

Emrouznejad, A., Parker, B. R., & Tavares, G. (2008). Evaluation of research in efficiency and productivity: A survey and analysis of the first 30 years of scholarly literature in DEA. *Socio-Economic Planning Sciences*, *42*, 151–157. doi:10.1016/j.seps.2007.07.002

Färe, R., & Grosskopf, S. (1985). A nonparametric cost approach to scale efficiency. *The Scandinavian Journal of Economics*, *87*, 594–604. doi:10.2307/3439974

Färe, R., Grosskopf, S., & Lovell, C. A. K. (1985). *The measurement of efficiency of production*. Boston, MA: Kluwer Academic.

Färe, R., Grosskopf, S., & Lovell, C. A. K. (1988). Scale elasticity and scale efficiency. *Journal of Institutional and Theoretical Economics*, *144*, 721–729.

Farrell, M. J. (1959). Convexity assumptions in theory of competitive markets. *The Journal of Political Economy*, *67*, 377–391. doi:10.1086/258197

Ferrier, G., & Lovell, C. A. K. (1990). Measuring cost efficiency in banking: Econometric and linear programming evidence. *Journal of Econometrics*, *46*, 229–245. doi:10.1016/0304-4076(90)90057-Z

Førsund, F. R. (1996). On the calculation of scale elasticity in DEA models. *Journal of Productivity Analysis*, *7*, 283–302. doi:10.1007/BF00157045

Førsund, F. R., & Hjalmarsson, L. (2004). Calculating scale elasticity in DEA models. *The Journal of the Operational Research Society*, *55*, 1023–1038. doi:10.1057/palgrave.jors.2601741

Førsund, F. R., Hjalmarsson, L., Krivonozhko, V., & Utkin, O. B. (2007). Calculation of scale elasticities in DEA models: Dsirect and indirect approaches. *Journal of Productivity Analysis*, *28*, 45–56. doi:10.1007/s11123-007-0047-5

Frisch, R. (1965). *Theory of production*. Dordrecht, The Netherlands: D. Reidel Publishing.

Fukuyama, H. (2000). Returns to scale and scale elasticity in data envelopment analysis. *European Journal of Operational Research*, *125*, 93–112. doi:10.1016/S0377-2217(99)00200-3

Fukuyama, H. (2001). Returns to scale and scale elasticity in Farrell, Russell and additive models. *Journal of Productivity Analysis*, *16*, 225–239. doi:10.1023/A:1012558521742

Fukuyama, H. (2003). Scale characterizations in a DEA directional technology distance function framework. *European Journal of Operational Research*, *144*, 108–127.

Griliches, Z., & Ringstad, V. (1971). *Economies of scale and the form of the production function.* Amsterdam, The Netherlands: North-Holland.

Guilkey, D. K., & Lovell, C. A. K. (1980). On the flexibility of the translog approximation. *International Economic Review, 21,* 137–147. doi:10.2307/2526244

Hadjicostas, P., & Soteriou, A. C. (2006). One-sided elasticities and technical efficiency in multi-output production: A theoretical framework. *European Journal of Operational Research, 168,* 425–449. doi:10.1016/j.ejor.2004.05.008

Hanoch, G. (1970). Homotheticity in joint production. *Journal of Economic Theory, 2,* 423–426. doi:10.1016/0022-0531(70)90024-4

Hanoch, G., & Rothschild, M. (1972). Testing assumptions in production theory: A nonparametric approach. *The Journal of Political Economy, 80,* 256–275. doi:10.1086/259881

Hasencamp, G. (1976). A study of multiple-output production functions. *Journal of Econometrics, 4,* 253–262. doi:10.1016/0304-4076(76)90036-1

Hjalmarsson, L., Kumbhakar, S. C., & Heshmati, A. (1996). DEA, DFA and SFA: A comparison. *Journal of Productivity Analysis, 7,* 303–328. doi:10.1007/BF00157046

Jones, C. I. (2004). *Growth and ideas.* Cambridge, MA: National Bureau of Economic Research.

Kuosmanen, T. (2003). Duality theory of non-convex technologies. *Journal of Productivity Analysis, 20,* 273–304. doi:10.1023/A:1027399700108

Land, K. C., Lovell, C. A. K., & Thore, S. (1993). Chance-constrained data envelopment analysis. *Managerial and Decision Economics, 14,* 541–554. doi:10.1002/mde.4090140607

Levy, B. (1987). A theory of public enterprise behavior. *Journal of Economic Behavior & Organization, 8,* 75–96. doi:10.1016/0167-2681(87)90022-9

Lozano, S., & Gutierrez, E. (2008). Data envelopment analysis of the human development index. *International Journal of Society Systems Science, 1,* 132–150. doi:10.1504/IJSSS.2008.021916

McCall, J. J. (1967). Competitive production for constant risk utility functions. *The Review of Economic Studies, 34,* 417–420. doi:10.2307/2296560

Morrison, C. J. (1992). Unraveling the productivity growth slowdown in the United States, Canada and Japan: The effects of subequilibrium, scale economies and markups. *The Review of Economics and Statistics, 74,* 381–393. doi:10.2307/2109482

Niskanen, W. (1971). Bureaucrats and politicians. *The Journal of Law & Economics, 18,* 617–643. doi:10.1086/466829

Panzar, J. C., & Willig, R. D. (1977). Economies of scale in multioutput production. *The Quarterly Journal of Economics, 41,* 481–493. doi:10.2307/1885979

Podinovski, V. V., Førsund, F. R., & Krivonozhko, V. E. (2009). A simple derivation of scale elasticity in data envelopment analysis. *European Journal of Operational Research, 197,* 149–153. doi:10.1016/j.ejor.2008.06.015

Reinhard, S. R., Lovell, C. A. K., & Thijssen, G. J. (2000). Environmental efficiency with multiple environmentally detrimental variables: Estimated with SFA and DEA. *European Journal of Operational Research, 121,* 287–303. doi:10.1016/S0377-2217(99)00218-0

Robinson, J. (1933). *The economics of imperfect competition.* London, UK: Macmillan.

Rosenberg, N. (1963). Technological change in the machine tool industry, 1840-1910. *The Journal of Economic History, 23,* 414–443.

Rosenberg, N. (1981). Why in America? In Mayr, O., & Post, R. C. (Eds.), *Yankee enterprise, the rise of the American system of manufactures.* Washington, DC: Smithsonian Institution Press.

Saen, R. E., & Azadi, M. (2009). The use of super-efficiency analysis for strategy ranking. *International Journal of Society Systems Science, 1*, 281–292. doi:10.1504/IJSSS.2009.022819

Sahoo, B. K., Mohapatra, P. K. J., & Trivedi, M. L. (1999). A comparative application of data envelopment analysis and frontier translog production function for estimating returns to scale and efficiencies. *International Journal of Systems Science, 30*, 379–394. doi:10.1080/002077299292335

Sahoo, B. K., Sengupta, J. K., & Mandal, A. (2007). Productive performance evaluation of the banking sector in India using data envelopment analysis. *International Journal of Operations Research, 4*, 1–17.

Sahoo, B. K., & Tone, K. (2009a). Decomposing capacity utilization in data envelopment analysis: An application to banks in India. *European Journal of Operational Research, 195*, 575–594. doi:10.1016/j.ejor.2008.02.017

Sahoo, B. K., & Tone, K. (2009b). Radial and non-radial decompositions of profit change: With an application to Indian banking. *European Journal of Operational Research, 196*, 1130–1146. doi:10.1016/j.ejor.2008.04.036

Sandmo, A. (1971). On the theory of the competitive firm under price uncertainty. *The American Economic Review, 61*, 65–73.

Scarf, H. E. (1981a). Production sets with indivisibilities part I: Generalities. *Econometrica, 49*, 1–32. doi:10.2307/1911124

Scarf, H. E. (1981b). Production sets with indivisibilities part II: The case of two activities. *Econometrica, 49*, 395–423. doi:10.2307/1913318

Scarf, H. E. (1986). Neighborhood systems for production sets with indivisibilities. *Econometrica, 54*, 507–532. doi:10.2307/1911306

Scarf, H. E. (1994). The allocation of resources in the presence of indivisibilities. *The Journal of Economic Perspectives, 8*, 111–128.

Sengupta, J. K., & Sahoo, B. K. (2006). *Efficiency models in data envelopment analysis: Techniques of evaluation of productivity of firms in a growing economy*. London, UK: Palgrave Macmillan.

Shi, H., & Yang, X. (1995). A new theory of industrialization. *Journal of Comparative Economics, 20*, 171–189. doi:10.1006/jcec.1995.1008

Sokoloff, K. L. (1988). Inventive activity in early industrial America: Evidence from patent records. *The Journal of Economic History, 48*, 813–850. doi:10.1017/S002205070000663X

Starrett, D. A. (1977). Measuring returns to scale in the aggregate and the scale effect of public goods. *Econometrica, 45*, 1439–1455. doi:10.2307/1912310

Sueyoshi, T. (1997). Measuring efficiencies and returns to scale in Nippon telegraph and telephone in production and cost analysis. *Management Science, 43*, 779–796. doi:10.1287/mnsc.43.6.779

Sueyoshi, T., & Sekitani, K. (2007a). Measurement of returns to scale using non-radial DEA model: A range-adjusted measure approach. *European Journal of Operational Research, 176*, 1918–1946. doi:10.1016/j.ejor.2005.10.043

Sueyoshi, T., & Sekitani, K. (2007b). The measurement of returns to scale under a simultaneous occurrence of multiple solutions in a reference set and a supporting hyperplane. *European Journal of Operational Research, 181*, 549–570. doi:10.1016/j.ejor.2006.05.042

Summers, R. (1965). A capital intensive approach to the small sample properties of various simultaneous equation estimates. *Econometrica, 33*, 1–41. doi:10.2307/1911887

Thrall, R. M. (1999). What is the economic meaning of FDH? *Journal of Productivity Analysis, 11,* 243–250. doi:10.1023/A:1007742104524

Tone, K., & Sahoo, B. K. (2003). Scale, indivisibilities and production function in data envelopment analysis. *International Journal of Production Economics, 84,* 165–192. doi:10.1016/S0925-5273(02)00412-7

Tone, K., & Sahoo, B. K. (2004). Degree of scale economies and congestion: A unified DEA approach. *European Journal of Operational Research, 158,* 755–772. doi:10.1016/S0377-2217(03)00370-9

Tone, K., & Sahoo, B. K. (2005). Evaluating cost efficiency and returns to scale in the life insurance corporation of India using data envelopment analysis. *Socio-Economic Planning Sciences, 39,* 261–285. doi:10.1016/j.seps.2004.06.001

Tone, K., & Sahoo, B. K. (2006). Re-examining scale elasticity in DEA. *Annals of Operations Research, 145,* 69–87. doi:10.1007/s10479-006-0027-6

Tulkens, H. (1993). On FDH analysis: Some methodological issues and applications to retail banking, courts and urban transit. *Journal of Productivity Analysis, 4,* 183–210. doi:10.1007/BF01073473

Tulkens, H., & Vanden Eeckaut, P. (1995). Non-parametric efficiency, progress and regress measures for panel data: methodological aspects. *European Journal of Operational Research, 80,* 474–499. doi:10.1016/0377-2217(94)00132-V

Varian, H. R. (1984). The nonparametric approach to production analysis. *Econometrica, 52,* 279–297. doi:10.2307/1913466

Yang, X. (1994). Endogeneous vs. exogeneous comparative advantage and economics of specialization vs. economies of scale. *Journal of Economics, 60,* 29–54. doi:10.1007/BF01228024

Yang, X., & Ng, Y. K. (1993). *Specialization and economic organization.* Amsterdam, The Netherlands: Elsevier Science.

Yang, X., & Rice, R. (1994). An equilibrium model endogenizing the emergence of a dual structure between the urban and rural sectors. *Journal of Urban Economics, 35,* 346–368. doi:10.1006/juec.1994.1020

ENDNOTES

[1] In the literature the concepts of either 'returns to scale' (RTS) or 'economies of scale' (EOS) are generally taken both to be satisfactory ways to treat *scale economies* without bothering much about whether or not they actually refer to the same causal factors. Much of earlier literature on production economics, in fact, treats the concept of RTS as synonymous with that of EOS, which is only true under two stringent conditions: 1) a homothetic production structure and 2) input prices exogenous to firms.

[2] The sources of economic growth in the U.S. economy have been due to large increasing returns to scale and less due to technical progress (Rosenberg, 1963, 1981; Sokoloff, 1988; Morrison, 1992; Devereux et al., 1996; Basu & Fernald, 1997; Jones, 2004).

[3] DEA has also been widely used in various application areas, viz., human development index (Despotis, 2005; Lozano & Gutierrez, 2008), strategy (Saen & Azadi, 2009), etc. See the recent paper by Emrouznejad et al. (2008) for an excellent survey on the scholarly literature on the evaluation of research in efficiency and productivity.

[4] Barring Tone and Sahoo (2005, 2006), the literature cited here deals precisely with the estimation of scale economies in terms of *returns to scale* but not *economies of scale.* These two concepts have distinctive caus-

ative factors that do not permit them to be used interchangeably. See Sahoo et al. (1999) and Tone and Sahoo (2003) for a discussion on the historical evolution of the concept of scale since the days of the Classicists.

[5] Various other aspects of productive performance in the Indian commercial banking sector have been studied in the literature. Most recent studies in the Indian banking literature include Sengupta and Sahoo (2006) and Sahoo et al. (2007) who have examined various production and economy measures of efficiency; Sahoo and Tone (2008) who have decomposed profit change into its six mutually exclusive drivers: *two* drivers of total factor productivity change - pure technical efficiency change and technical change, *three* drivers of activity change - scale change, resource-mix change and product-mix change, and finally, price change; and Sahoo and Tone (2009) who have decomposed technological capacity utilization (CU) into its various meaningful components such as technical inefficiency, economic CU and optimal capacity idleness.

[6] See Hanoch (1970), Starrett (1977), Panzar and Willig (1977) and Baumol et al. (1982) for a detailed discussion on this issue.

[7] Note that Färe and Grosskopf (1985) are the first to introduce the above cost frontier approach (3) as a way of estimating *scale efficiency*. However, they assumed constant unit factor prices which each firm in the market would face alike.

[8] Surprisingly, even though convexity postulate is under severe attack for assuming away some important technological features, it has rarely been exposed to empirical tests in DEA (a noticeable exception is Briec et al., 2004). Barring a few authors like Thrall (1999), recent literature favoring the dropping of convexity axiom include, among others, Deprins et al. (1984), Scraf (1981a, 1981b, 1986, 1994), Tulkens (1993), Tulkens

and Vanden Eeckaut (1995), Bouhnik et al. (2001), Tone and Sahoo (2003), and Kuosmanen (2003); whereas the literature on some exciting economic analysis arising from violation of convexity include Yang and Ng (1993), Yang (1994), Yang and Rice (1994), Borland and Yang (1995) and Shi and Yang (1995). The only argument favoring the convexity postulate is that it enables reduction of small sample errors. But this comes at the cost of possible specification errors, which are less of a problem with large samples.

[9] For replication any firm can be replicated by a new firm. But this has no practical and economic meaning, because it is not a controlled experiment. It is usually very hard to judge the economic relevance (i.e., effect) of a replicated firm of a given size (i.e., treatment) in the real world unless it is comparable to an actual firm of that size. The issue involved here is how representative a replicated firm is for an actual firm to which the investigator would like to project. It can be argued that a replicated firm may not properly represent a real-world firm because of 'indivisibilities' associated not only with technology, but also with unique attributes like geographic location, innovative mangers, etc. If there are no such indivisibilities, there is no economic justification for the existence of large firms, i.e., there is nothing to be gained by organizing economic activity in large, durable and complex units such as assembly lines, bridges, transportation and communication networks, giant presses, and complex manufacturing plants, which are available in specific discrete sizes, and whose economic usefulness manifests itself only when the scale of operation is large.

[10] The translog functional forms for production were first introduced by Grilliches and Ringstad (1971) and Christensen et al. (1971, 1973). However, Burgress (1974) was the

first to introduce this translog form for a cost function. This function is assumed to be a second-order Taylor's series approximation in logarithms of outputs and input prices to an arbitrary cost function. The more general specification of the translog cost function imposes no prior restrictions on the production structure, that is, it does not impose, *ex ante, neutrality, homotheticity, homogeneity, constant returns to scale*, or *unitary elasticities of substitutions*; in fact it allows testing these alternative configurations.

[11] However, in the case of a translog production function the parameters concerning inputs and cross-products of inputs are unrestricted, which is precisely why the cost function represents only the economically relevant subset of the production technology with the given input prices.

[12] Diewert and Fox (2008) have shown that the reciprocal of returns to scale $(\sum_{r=1}^{s} \partial \ln c(w, y) / \partial \ln y_r)$ equals the degree of homogeneity defined as sum of β_r, i.e., $\sum_{r=1}^{s} \beta_r$ when the following two additional sets of constraints are satisfied: $\sum_{r'=1}^{s} \beta_{rr'} = 0 \ (\forall \ r)$ and $\sum_{r=1}^{s} \gamma_{ir} = 0 \ (\forall \ i)$ along with (13a-g).

[13] See Berger and Mester (1997) for a comprehensive discussion of these two approaches.

However, the distinction between these two approaches seems to be dubious since the approaches that have been actually used in banking studies - the 'asset approach' and user-cost approach' - are often viewed as variants of the intermediation approach. See Berger and Humhprey (1992) for the detailed discussion.

[14] See Sahoo and Tone (2009a, 2009b) for the details on the selection of inputs and outputs and choice of study period.

[15] Besides these two assumptions, one needs to implicitly assume that the data on unit market prices of factor inputs are *always* available with certainty for this model to be applied in practice.

[17] See Baumol et al. (1988, pp. 63-64) where it is shown that any differentiable cost function, whatever the number of outputs involved, and whether or not it is derived from a homogeneous production process, has a local degree of homogeneity, which is reciprocal of the homogeneity parameter of the production process.

[18] One could, however, also accommodate a weak disposability assumption in DEA. See, for example, among others Färe et al. (1985) and Tone and Sahoo (2004, 2006).

This work was previously published in the International Journal of Information Systems and Social Change, Volume 2, Issue 4, edited by John Wang, pp. 13-30, copyright 2011 by IGI Publishing (an imprint of IGI Global).

Chapter 16
Stigmergic Hyperlink:
A New Social Web Object

Artur Sancho Marques
CEG-IST, ESGTS /Polytechnic Institute of Santarém, Portugal

José Figueiredo
CEG-IST /DEG Technical University of Lisbon, Portugal

ABSTRACT

Inspired by patterns of behavior generated in social networks, a prototype of a new object was designed and developed for the World Wide Web – the stigmergic hyperlink or "stigh". In a system of stighs, like a Web page, the objects that users do use grow "healthier", while the unused "weaken", eventually to the extreme of their "death", being autopoieticaly replaced by new destinations. At the single Web page scale, these systems perform like recommendation systems and embody an "ecological" treatment to unappreciated links. On the much wider scale of generalized usage, because each stigh has a method to retrieve information about its destination, Web agents in general and search engines in particular, would have the option to delegate the crawling and/or the parsing of the destination. This would be an interesting social change: after becoming not only consumers, but also content producers, Web users would, just by hosting (automatic) stighs, become information service providers too.

INTRODUCTION

Regular Web hyperlinks have limitations like unidirectional linkage, unverified destination, and issues related to relevance, reputation and trust (Leuf, 2006), some addressed by technologies like XLink (W3C, 2001). The stigmergic hyperlinks

DOI: 10.4018/978-1-4666-2922-6.ch016

(stighs) we designed embody an alternative for some specific applications and propose some interesting paradigmatic changes.

"Stigmergy" means the mark of the work (Grassé, 1959), and it is a form of indirect communication, effective for some distributed control problems (Marco Dorigo, 2004).

Stighs are "stigmergic" because they communicate indirectly and, as collective, display emergent behaviors (that are not the result of centralized control mechanisms).

Stighs are also "hyperlinks" because they can look and feel like regular hypertext hyperlinks (Wardrip-Fruin, 2004). The main difference is that stighs have a "life" attribute that increases when they are used and decays at a natural pace, eventually down to a "death" level. This dynamic drives interesting emergent systemic behaviors.

We elaborate on the stigh's architecture in the context of pervasive stigmergy, also in human-human relations and use other author's taxonomy (Parunak, 2005) to frame our proposed object and look at it from a social epistemology perspective (Marsh & Onof, 2007), that includes Word Wide Web cases.

We discuss applications for stighs at two scale levels: the single Web page and the "generalized usage" scenario. For the Web page scale applications, we provide a demo at http:/stigh.org that can exhibit the behaviors that support decentralizing the finding of useful Web resources across a community of users and the automatic replacement of undesired destinations. These applications have similarities with recommendation systems (Linden, Smith, & York, 2003).

Regarding the generalized usage scenario, we discuss decentralizing search and stighs and the Deep Web. Decentralizing search would increase the "calculative capacity" (Callon & Muniesa, 2003) of Web authors and that would be a noticeable social change – this could be achieved if search engines could, at least partially, outsource some of their tasks, like crawling and/or parsing (Brin & Page, 1997), delegating on stighs or equivalent objects. The same delegation could be an approach to the Deep Web – the Web that the search engines ignore (Bergman, 2001) – because it is hard to generalize how to crawl it. Instead of generalizing, specialized stighs could handle particular cases.

The structure of this paper is as follows: first we explain the stigmergic hyperlink from a high level perspective, including the architecture of the individual agent and its environment, one possible taxonomical classification, and a social epistemology perspective of stigmergy and stighs. We then discuss possible applications from the single Web page scale to the generalized usage scenario. In order to better understand the internals of stigmergic hyperlinks, there is a detailed engineering section. After one illustrated example of stighs in action, we state some technological considerations and, finally, future work.

STIGMERGIC HYPERLINK(S)

The Stigh

Social insects are one inspiration for this new type of hypertext object we call the *stigmergic hyperlink* – stigh, for short.

Social insects, such as ants and termites are capable of complex behaviors like traveling long paths to/from food, and building structures. As individuals they wouldn't be able to do it but as a collective body they indirectly communicate and forms of organization emerge. Stigmergy is this indirect communication via modifications on the environment and it can be a solution for distributed control problems (Marco Dorigo, 2004).

Stigmergy – from the Greek *stigma* (mark) and *ergon* (work) – means "the mark of work", in the sense that agents/workers acting/working on their local environment "mark" it in a way that probabilistically other agents/workers will acknowledge. This will reinforce the work process without the need for direct communication. The expression was introduced by the French biologist Pierre-Paul Grassé (Grassé, 1959)

Grassé observed that termites, when building a nest, modify their local environment by aggregating mud balls, marked with pheromones. These balls are more likely to be placed where other pheromone-marked mud balls already are, than elsewhere – the nest is built from this continuous depositing process, having arches as its fundamental building block (Grassé, 1959).

Stighs have a "life" attribute that reflects (marks) what the users of the Web page where they "live in" have been doing (working) with them. Web page visitors perform like ants or termites, leaving a digital pheromone that reinforces a stigmergic hyperlink's life, whenever they click it. On the other hand, a neglected (not used) stigh will slowly wither, until its life level eventually zeroes, which would represent its death.

Although "under the hood" stighs are not regular hypertext hyperlinks, they can be displayed by Internet browsers exactly like if they were – see (Wardrip-Fruin, 2004) on Hypertext – and they will behave accordingly: when someone clicks a stigh the browser loads the corresponding destination resource.

Regular hyperlinks have limitations like unidirectional linkage, unverified destination, and issues related to relevance, reputation and trust (Leuf, 2006). More recent technologies, like W3C's XLink (W3C, 2001), try to address some of the limitations. Stighs are not a replacement for the regular hyperlinks – they build on them to represent an alternative adequate for specific and pertinent purposes.

While an individual stigh is a very simple object – a hyperlink with a floating energy level, reflecting its relative usage, and some actions/ methods – what emerges from a system of stighs, and/or from its collaboration with other Web entities, is more complex and can have very interesting applications.

AGENT AND ENVIRONMENT ARCHITECTURE

(Parunak, 2005) explores human-human Stigmergy and its ubiquitousness. He also presents an architecture and a taxonomy for Stigmergy. Stighs fit in both frames, as described below.

Regarding the architecture of Stigmergy there are two fundamental components: a population of agents and an environment. A collection of stigmergic hyperlinks living in the same Web page is a population of agents. A Web page containing stighs is their environment.

Each *agent* has:

- An *internal state*, "usually not directly visible to other agents".

In the case of stighs, all attributes are protected and only indirectly exposed via "properties". These "attributes" and "properties" are meant in the context of the computer programming language C#, which was our choice for the current prototype. The main attribute is the hyperlink's life, which represents its energy level. Other attributes include a clicks counter and the destination URL.

- *Sensors,* "give access to some environment's state variables".

Stighs' sensors read values that are shared via the Web page/environment, like a "trigger" value, that sets the "natural" decay pace for any stigh's life. These read-only shared values act as the system's configuration, imposing the decay rhythm, the maximal and the minimal values for the life attribute.

- *Actuators*, that "change some of the environment's state variables".

When a stigmergic hyperlink is clicked, its OnClick actuator/method directly increases the object's own clicks counter and can, indirectly, update a global clicks counter.

- A *program* that "governs the evolution of its state over time".

For stighs, this program is their methods collection, namely getStronger and getWeaker, which, respectively, increase and decrease, the object's life, in response to click events or to the natural decay.

The *environment* has:

- A *state*, with "certain aspects generally visible to the agents".

Since stighs are part of a Web page's structure, they explicitly compose the environment where they work on – hence the environment's state is also composed by the states of all embedded stighs.

- A *program* "that governs the evolution of its state over time".

This program could be seen as the pairs of {events, event handlers}, which the page supports. For example, for the click event, stighs will respond with their OnClick method as the event handler – usually, stighs will only respond to clicks. But, on a higher level, the page's program/behavior is just an emerging function from the embodied objects/agents, namely the stighs.

Agent Behavior Taxonomy

One possible classification of the type of stigmergy distinguishes between "marker" and "structure" (sematectonic) based action (Wilson, 1975).

Stigmergic hyperlinks have a mark that reflects their relative usage - their life attribute. This mark is reinforced when Web users use them. Eventually, because a "healthy" stigh will probabilistically remain longer in the system, it might be more used. In this positive feedback sense, a stigmergic hyperlinks system could be considered "marker" based.

But stighs are themselves part of the environment's structure and their destination resource should be a reason to follow them, or not. In this sematectonic perspective, one structural element is the reason for the users' actions, so structure

also provides signals for the work that will, or not, happen.

Moreover, Web content authors can decide the stigh's appearance, for example allowing the life attribute to be rendered / shown, as it is by default, in the current demo version, available at http://stigh.org/.

This life value can signal the same information that the number of people bookmarking a resource does in social bookmarking systems like http://del.icio.us/ – that is, people will use/work the resource based on the work of previous users.

The same way people might browse to some del.icio.us destination, because it has already been visited by "many", as made explicit by the number of its visitors, a stigmergic hyperlink destination might be visited, just because it is more "lively" than others, as made explicit by its shown life attribute.

So, a stigmergic hyperlinks system is simultaneously a marker and a sematectonic case of stigmergy. While the structural destination attribute is determinant for the browsing endpoint, the life mark captures and signals what a community of agents has been doing, and that information can be relevant for users.

A Social Epistemology Perspective of Stigmergy and Stighs

In Marsh and Onof (2007) the authors consider that the subject of social epistemology is "the formation, acquisition, mediation, transmission and dissemination of knowledge in complex communities", and "recommend a stigmergic framework for social epistemology". Because stigh systems capture the relative browsing preferences of a community of Web users and convey that acquired information following a stigmergic approach, they are another manifestation of how pervasive stigmergic systems are in human societies (Parunak, 2005), including the World Wide Web.

Amazon.com's recommendation system and Google's search algorithm are discussed in (Marsh

& Onof, 2007) from a stigmergic perspective and as WWW examples, along atemporal references, like a market place of commodities. In all cases, the loop *agents(s) → environment → agent(s) → environment* is key for the reasoning: Amazon users' behavior might be influenced by others' recommendations and will itself contribute to future recommendations, explicitly (by rating some product, for example), or implicitly (not by rating, but by buying, for example).

In the following section we explain how a stigmergic hyperlinks system can perform as recommendation system and how stighs could contribute to an alternative decentralized approach to the Web search problem.

STIGMERGIC HYPERLINKS APPLICATIONS

In this section we discuss applications on two scale levels: the single Web page and, on the limit, the whole Web.

The current prototype is a single page stigmergic hyperlinks system that can only perform the "Decentralizing the finding of useful Internet Resources" and the "Automatic replacement of undesired destinations" applications.

The other applications that we discuss depend on wider scale usages. Considering the World Wide Web an information market that permanently calculates – in the (Callon & Muniesa, 2003) meaning – its goods and services, structure and content, from the interactions of many heterogeneous agents, that produce and consume information, we look at some "calculations" and try to approach some of their stages using stighs.

DECENTRALIZING THE FINDING OF USEFUL WEB RESOURCES ACROSS A COMMUNITY OF USERS

A system of stighs can locally (at the single Web page scale), automatically (without human intervention) and transparently (requiring no extra user input) capture the navigation preferences of the community of its users.

When clicked/used, a stigmergic hyperlink will see its energy level increase; when not clicked/not used, the stigh's vitality will decrease. In time, appreciated stighs/links will "live" and the neglected will "die" being replaced by new destinations, potentially more useful to the community, chosen from a pool of possibilities that consists of all the hyperlinks contained inside the resources pointed by the surviving links, on the assumption that they probably connect to more attractive destinations than the terminated stigh did.

This all happens automatically without the need for a "Webmaster" to update the local pages, and transparently, without requiring visitors to login or input extra data – users just need to browse and won't distinguish between stighs and regular hyperlinks.

This way, a system of stighs is an approach for the maintenance of a dynamic Web of preferred destinations for a community of users.

AUTOMATIC REPLACEMENT OF UNDESIRED DESTINATIONS

The problem of automating the replacement of undesired destinations, like connections to resources that disappeared (dead-links), is related to the prior: odds are that dead-links and their corresponding http-404-error responses won't captivate anyone so, in time, as users perceive the destination as useless, the stigh with that URL will see its energy level sink to death/replacement. Again, the replacement will happen automatically.

Hence, a system of stighs also works as a solution for Webmasters to publish-and-forget external links (links to destinations outside their control) believing that, in time, if for example an external resource changes location, the corresponding stigh won't remain indefinitely pointing to what is gone, as a regular static hyperlink would.

Because once published any stigh system escapes its creator's control, stigh systems are examples of feral hypertext systems (Walker, 2005).

LIKE A RECOMMENDATION SYSTEM

Amazon.com's recommendation system, via item-to-item collaborative filtering (Linden et al., 2003), was designed to identify and suggest related items, from a catalog of million(s). Since users shop or browse via the Website's pages, those pages are the environment that mediates the (indirect) communication between the human agents. The agents' interaction is stigmergic because it is indirect and every user's buying and/ or ranking behavior might affect everyone else's.

The life attribute of a stigmergic hyperlink can be interpreted as a reading of its relative quality, to the community of users of a mediating Web page: users might decide to (not) browse to a destination, based on its life value alone; when doing so, they will be influencing the signal/recommendation. Thus, the emergent behavior of a stigmergic hyperlink system is comparable to that of a ranking/recommendation system.

DECENTRALIZING SEARCH

Search engines like the ones made available by Google, Yahoo, and others, are precious intermediates between searchers and content. They are also hugely outnumbered: the Internet grows in new data by the minute while new search services are a rare event. This disproportion reflects the dominant centralized approach to the Internet search problem.

Crawling and indexing are major tasks for search engines (Brin & Page, 1997). Stigmergic hyperlinks have a method for the retrieval of information about their destination that might contribute to decentralize crawling and/or indexing. This method processes (crawls to and retrieves data about) the linked resource, returning information about it. If the returned information is of value to some search engine's procedure, for example as complementary data that can guide to higher quality results, or even as a principal data source, there would be an alternative to the current non delegating approach – search engines would be able to outsource some processing and Web agents, in general, would have many more data sources available.

The external information retrieval method is not exposed in the current prototype – it is yet to be decided the default information to return, and how it can be invoked by Web agents. We intend to allow stigh users to be able to replace the default method by their own (or do nothing and accept the default behavior).

In a scenario of generalized usage, the stigh could, transparently, transform Web users, hosting stighs, into information providers about hyperlinks' destinations, eventually paving the road for a change in search engine design, via the externalization and decentralization of tasks. This would increase the "calculative capacity" (Callon & Muniesa, 2003) of Web authors.

Stighs and the Deep Web

The publicly indexable Web is the fraction of the Web that search engines cover, but it is a subset that ignores large amounts of high-quality information (Raghavan & Garcia-Molina, 2000). Many Web contents are only available on very precise requests, some requiring user authentication, and some denying access to bots. These hard to index contents, often stored in databases of undisclosed structure, unreachable to the classical search engines, are popularly known as the Deep Web (Bergman, 2001).

Content producers (authors) and content consumers (readers) can facilitate the finding of relevant Web content, by providing metadata, like tags, from prebuilt closed classification systems (taxonomies) and/or creating their own labels (folksonomies) (Marlow, Naaman, Boyd, & Davis, 2006). The hyperlink alone can represent much information about its destination (Marchiori, 1997), not only via the URL, but also via its anchor text and other data (Brin & Page, 1997). This applies to both the "surface" (publicly indexable Web) and to the Deep Web.

The Deep Web particularity is that it is much harder to generalize and automate how to crawl it.

The stigh's architecture can contribute to an approach to the Deep Web challenge: search engines could abstract the specialized crawling required for deep sites and delegate that on stighs with very specific information retrieval methods, or other objects capable of performing the equivalent task. Those hosting the stighs would be the ones directly addressing the Deep Web issue, for the specific cases they would link to.

ENGINEERING SYSTEM INTERNALS

Every Web page with at least one stigh object is considered a stigh system.

A stigh object has some attributes and actions/ methods: the more relevant attributes are "life" (its remaining energy), "URL" (the resource to where it points), and "text" (the anchor text, usually shown underlined); the more usual methods are "getStronger" (increases life) and "getWeaker" (decreases life).

Whenever a stigh is clicked its "life" increases by a configurable reward. The other stighs in the page are not directly affected, but the system has one "trigger" property that affects all: this integer value imposes an ageing/withering pace for

everyone, measured in page clicks. In the current demo, trigger equals 2, meaning that every two clicks in the page/stigh system, all stigh objects will lose vitality.

In the current demo configuration, the loss of life is half the reward, and the reward is double the trigger, but all this is adjustable. Notice that different values determine different systems.

The system also imposes a maximum value for "life" in order to minimize the long term effects of fake usages, aimed at artificially inflating the hyperlink's energy.

Stighs are high level entities, processed on the Web server side. Because plain regular hyperlinks are processed on the client side and the Web server needs to track the "life" of stighs, we needed a solution to take the hyperlink processing to the server side: in the current prototype stighs are buttons disguised as regular HTML links.

When someone clicks a stigh object, that stigmergic hyperlink's attributes change globally, for all the active browser sessions and not just for who did the click. This way a stigh can be marked/ worked by a global community. One consequence is that what a user sees in his/her browser window, might not correspond to the latest version of the page. In some occasions this can surprise someone who clicked to browse to a destination, but then the browser loaded a different resource... That might have happened because, during the time that lapsed between the page load and the click, other user(s) interacted with the system in a way that caused the replacement of the object that was still being displayed at the out-of-date browser.

The "death" or replacement of a stigmergic hyperlink is performed by a method currently named "extinction". This method is more complex than the trivial sums/subtractions involved in "getStronger" and "getWeaker". Currently when a stigh "dies" what really disappears is the neglected destination, replaced by a new one, drawn from the pages linked by the "survivors". The assumption here is that pages on similar subjects probably interconnect so a living/pleasing stigh eventually connects to other potentially agreeable locations.

The picking of a replacement destination is done by first spinning a roulette wheel with as many slices as the number of survivors. The more "life" a stigh has, the bigger will be its slice in the roulette wheel, therefore increasing, but not guaranteeing, its chances of being chosen as the seed. The seed's destination is then scanned for HTML links and one of them will be randomly picked to set the replacement stigh's URL.

One interesting situation happens when the stigh selected for seeding is "sterile", in the sense of pointing to a page that is a dead-end, linking to no other places. When the selected page can't provide destinations, the "dying" stigh performs a radical maneuver, randomly picking one entry from Google's Zeitgeist service, which lists the top searched expressions, for the running week. The idea is to introduce some mutation, in order to avoid stagnation, allowing some serendipity.

AN EXAMPLE

The following illustrated example is taken from a stigh system, composed by four individual stigmergic hyperlinks, ageing/withering every two clicks ("trigger" = 2 clicks).

As previously stated, in the current demo configuration, the loss of life, due to ageing/withering, is half the reward, and the reward is double the "trigger", so selection reward = 2*trigger = 2*2 = 4 and ageing penalty = reward/2 = 4/2 = 2.

In this example, all stighs were born with "life" = 10, but each could have been declared with a different value. Low initial "life" values facilitate the observation of "death" events.

Every stigh has a unique identifier. In this case, the identifiers are CS1, CS2, CS3 and CS4, standing for "class stigh object" #1, #2, #3 and #4. Currently, stighs can run on any Web server that can serve .NET ASPX pages: the declaration syntax for objects hosted in such pages is behind the scope of this article, but, for stighs, it looks like this:

```
<ccl:cStigh ID="CS1" aLife="10" aUrl="http://www.digg.com" runat="server" aConfirmMessage="Follow the link?" aShowConfirm="true"></ccl:cStigh>
```

Figure 1 illustrates the initial system. The tooltips (text on yellow background) appear only when the mouse goes over the stigmergic hyperlinks, but the image was composed in order to show all the tooltips at the same time. The displayed "pace" value is the system "trigger".

The tooltips' numbers read that all the stighs have a current "life" of 10 and a maximum of 100 (life: 10/100). No stigh has been clicked individually, or on the page, or on the whole hosting Website (#clicks link/page/site: 0/0/0). The page holds 4 stighs and the Web server created a total of 4 stigh objects (#stighs page/abs: 4/4).

Figure 1. Web page with 4 new stighs

If the Web page was reloaded, the absolute number of stighs (abs) would then read 8, because a total of 8 stighs would have been served by the Web server, but no other number would change. If there were other pages with stighs – other stigh systems – then the absolute number of objects created would also account for their presence.

If a visitor clicks the acm.org (CS2) link, then that object's "life" will become 10+4=14 and the click counters will reflect the event. It is important to understand that a regular visitor typically does not see the updated numbers, because, in response to the click, he/she will be visiting the destination URL. One way to see the updated data is to use the browser's "go back" function and reload the page, or run a "debug" version of the system that behaves exactly like the "end-user" version, except that it doesn't load the linked destinations and thus allows watching the objects' state evolution without "back and refresh". See Figure 2.

Assuming that the next click is again on CS2 then CS2's "life" will become 14+4=18, minus 2 (=16) because the system aged and all objects withered. The other stighs life is now 10-2=8.

Assume that the next click is on CS3 (gamasutra.com): its "life" will be 8+4=12; no other "life" attributes will change. Insisting on CS3, its "life" will become 12+4-2=14. CS1 and CS4, still neglected, will weaken to 8-2=6. CS2 will weaken to 16-2=14. See Figure 3.

If the next two clicks are on CS1 (digg.com) then CS1 will strengthen to 6+4=10, and then to 10+4-2=12. CS2 and CS3 will weaken to 14-2=12. CS4 will weaken to 6-2=4, nearing death… See Figure 4.

If the next click is on CS2 and the following on CS3, those stighs' "life" will strengthen to 12+4-2=14. CS1 will weaken to 12-2=10. CS4 will weaken to 4-2=2. See Figure 5.

Assume now that CS1, then CS3, will be clicked. CS1 will strengthen to 10+4-2=12. CS3 will strengthen to 14+4-2=16. CS2 will weaken to 14-2=12. CS4 will weaken to 2-2=0. This will spark its "death".

Figure 2. System, after click #1 on CS2 (acm.org). The page was also reloaded

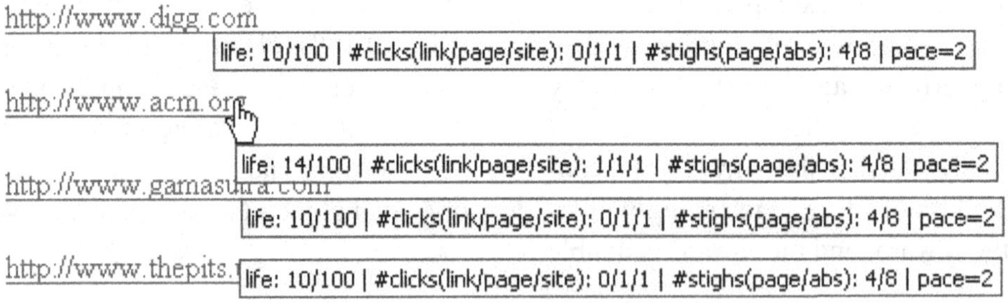

http://www.digg.com
life: 10/100 | #clicks(link/page/site): 0/1/1 | #stighs(page/abs): 4/8 | pace=2

http://www.acm.org
life: 14/100 | #clicks(link/page/site): 1/1/1 | #stighs(page/abs): 4/8 | pace=2

http://www.gamasutra.com
life: 10/100 | #clicks(link/page/site): 0/1/1 | #stighs(page/abs): 4/8 | pace=2

http://www.thepits.
life: 10/100 | #clicks(link/page/site): 0/1/1 | #stighs(page/abs): 4/8 | pace=2

Figure 3. System, after click #2 on CS2 (acm.org) and clicks #3 and #4, both on CS3 (gamasutra.com)

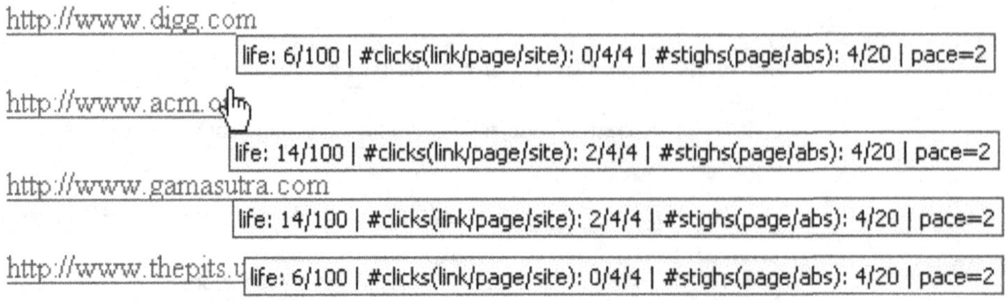

http://www.digg.com
life: 6/100 | #clicks(link/page/site): 0/4/4 | #stighs(page/abs): 4/20 | pace=2

http://www.acm.o
life: 14/100 | #clicks(link/page/site): 2/4/4 | #stighs(page/abs): 4/20 | pace=2

http://www.gamasutra.com
life: 14/100 | #clicks(link/page/site): 2/4/4 | #stighs(page/abs): 4/20 | pace=2

http://www.thepits.
life: 6/100 | #clicks(link/page/site): 0/4/4 | #stighs(page/abs): 4/20 | pace=2

On "death", CS4 will draw a replacement URL from the survivors set = {CS1, CS2, CS3}. Chances are that CS3 will seed the new resource but there is no guarantee. The proportional roulette wheel to be used is computed and shown in Figure 6.

In this example an article from Wikipedia replaced the previous CS4's URL. We could learn that it was extracted from CS1 / digg.com, by looking at the logs that are kept for each stigmergic hyperlink.

The new stigh is healthy, with half the maximum life (100/2=50). Higher values facilitate the survival of newborns; lower values contribute to a certain effervescence of the system.

This 10 clicks example shows all the main events in the life of a system of stighs. As illustrated, the stigmergic hyperlinks can look like regular HTML links and apparently behave like them… but they are running on the server side and keeping track of a "life" attribute, which strengthens on clicks and weakens at a regular

Figure 4. System, after clicks #5 and #6, both on CS1 (digg.com)

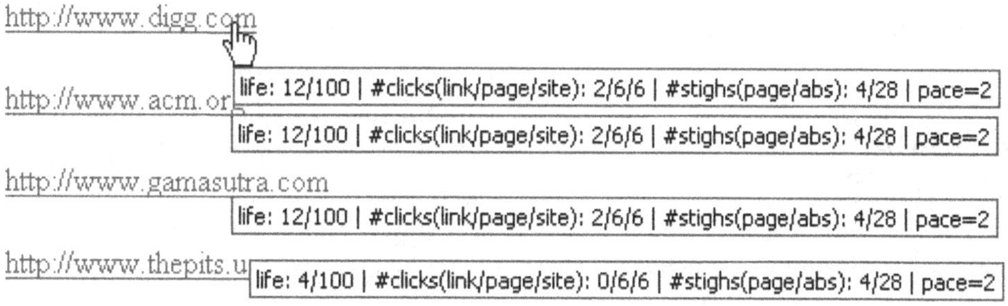

http://www.digg.com
http://www.acm.or life: 12/100 | #clicks(link/page/site): 2/6/6 | #stighs(page/abs): 4/28 | pace=2
life: 12/100 | #clicks(link/page/site): 2/6/6 | #stighs(page/abs): 4/28 | pace=2
http://www.gamasutra.com
life: 12/100 | #clicks(link/page/site): 2/6/6 | #stighs(page/abs): 4/28 | pace=2
http://www.thepits.u life: 4/100 | #clicks(link/page/site): 0/6/6 | #stighs(page/abs): 4/28 | pace=2

Figure 5. System, after clicks #7 (on CS2) and #8 (on CS3)

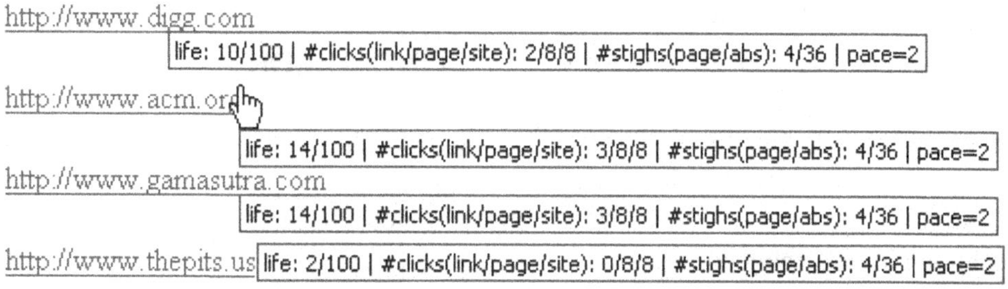

http://www.digg.com
life: 10/100 | #clicks(link/page/site): 2/8/8 | #stighs(page/abs): 4/36 | pace=2
http://www.acm.org
life: 14/100 | #clicks(link/page/site): 3/8/8 | #stighs(page/abs): 4/36 | pace=2
http://www.gamasutra.com
life: 14/100 | #clicks(link/page/site): 3/8/8 | #stighs(page/abs): 4/36 | pace=2
http://www.thepits.us life: 2/100 | #clicks(link/page/site): 0/8/8 | #stighs(page/abs): 4/36 | pace=2

Figure 6. Probabilities in the roulette wheel, for seeding a replacement for CS4, after clicks #9 (on CS1) and #10 (on CS3)

stigh ID	life	probability	P
CS1	12	2/∑ = 12/40 = 3/10	0.3
CS2	12	12/∑ = 12/40 = 3/10	0.3
CS3	16	16/∑ = 16/40 = 2/5	0.4
CS4	0	0	
sum of lives = ∑	40	sum of p =	1

pace – whenever the shared environment signals "withering time".

Stighs can read/"smell" the pheromone/"life" of their siblings in the same immediate page/"territory" and use that information on "death" time in order to select their successors, via a proportional roulette wheel draw (see Figure 7).

The demo page is available at http://stigh.org – if trying it, please remember to "go back and refresh", after any click(s), in order to see the objects' state evolution reflected in your browser, or you'll be just browsing to the chosen destinations. There is also an artificial user, which simulates a specified number of random clicks on stighs, to speed up the observation of extinctions – again, "go back and refresh" after using it.

TECHNOLOGICAL CONSIDERATIONS

Current stigmergic hyperlinks are instances of a class named cStigh, programmed in C#. The class was compiled in "Visual Studio 2005", as a "class library project" – the output is a DLL file (Webstigh.dll).

Equipping a .NET enabled Website with stigmergic hyperlinks just requires the hosting of the Webstigh.dll file at its \bin folder. This class library includes cStigh, allowing the site to serve its instances (stighs). See Figure 8, for a class diagram of cStigh.

Other technologies are being considered.

FUTURE WORK

Currently, each stigh has its own dedicated state URL: for example, for the CS1 object on stigh.org, the URL is http://stigh.org/~_Default. aspx_CS1_STATE.TXT. This logging solution is granular and doesn't provide a systemic view. We will address this with an administrator's console, oriented to the big picture, but also allowing to

Figure 7. Proportional roulette wheel to be used

Proportional roulette

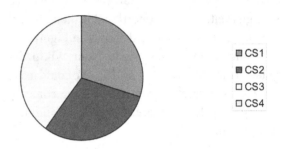

read/set individual variables. The console's main purpose will be to provide a more informative reading about all that has been happening in the stigmergic hyperlinks system.

Also in the "to do" list is exposing the method for the retrieval of external information about the destination resource and decide how Web entities can use it.

Finally, we'll research how the stigmergic hyperlink technology can relate to business models, for example supporting dynamic pricing systems based on relative usage, or supporting new Web relations with the potential for monetization.

CONCLUSION

We introduced the prototype of a new Web object, named "stigmergic hyperlink", or "stigh": a server side hyperlink with a floating energy level and some actions.

Stigmergy was first observed on social insects, but has always been pervasive on human-human interactions. We addressed the concept from its roots to recent Web cases.

We argued that a system of stighs can perform as a recommender system of preferred Web destinations and as an automatic solution for the replacement of undesired links. These applications emerge at the single Web page level and can be observed on the public demo available at http://

Figure 8. cStigh class diagram

stigh.org/. We also provided a 10 clicks illustrated example and detailed the internals of the object.

At the World Wide Web scale, we discussed applications that have disruptive potential: to decentralize the current approach to (Deep) Web search would create a new social Web of information service providers.

We now need better tools for the observation of stighs and to expose the method for the information retrieval about the destination, in order to make the potential clearer.

REFERENCES

Bergman, M. K. (2001). The deep web: Surfacing hidden value. *Journal of Electronic Publishing, 7*(1). doi:10.3998/3336451.0007.104

Brin, S., & Page, L. (1997). *The anatomy of a large-scale hypertextual web search engine.* Retrieved from http://infolab.stanford.edu/~backrub/google.html

Callon, M., & Muniesa, F. (2003). Les marchés économiques comme dispositifs collectifs de calcul. *Reseaux, 122,* 189–233.

Grassé, P. P. (1959). La reconstruction du nid et les coordinations interindividuelles chez bellicositermes natalensis et cubitermes sp. la théorie de la stigmergie: Essai d'interprétation du comportement des termites constructeurs. *Insectes Sociaux, 6*(1). doi:10.1007/BF02223791

Leuf, B. (2006). Enhancing the web. In Leuf, B. (Ed.), *The semantic web: Crafting infrastructure for agency* (pp. 3–30). New York, NY: John Wiley & Sons. doi:10.1002/0470028173.ch1

Linden, G., Smith, B., & York, J. (2003). *Amazon. com recommendations.* Retrieved from http://www.computer.org/portal/web/csdl/doi/10.1109/MIC.2003.1167344

Marchiori, M. (1997). *The quest for correct information on the web: Hyper search engines.* Retrieved from http://www.w3.org/People/Massimo/papers/quest_hypersearch.pdf

Marco Dorigo, T. S. (2004). *From real to artificial ants: Ant colony optimization.* Cambridge, MA: MIT Press.

Marlow, C., Naaman, M., Boyd, D., & Davis, M. (2006). *Tagging paper, taxonomy, flickr, academic article, to read.* Paper presented at the Conference on Hypertext and Hypermedia, Odense, Denmark.

Marsh, L., & Onof, C. (2007). *Stigmergic epistemology, stigmergy cognition.* Retrieved from http://mpra.ub.uni-muenchen.de/10004/

Parunak, H. V. D. (2005). *Expert assessment of human-human stigmergy.* Ann Arbor, MI: Altarum Institute.

Raghavan, S., & Garcia-Molina, H. (2000). *Crawling the hidden web.* Retrieved from http://ilpubs.stanford.edu:8090/456/1/2000-36.pdf

W3C. (2001). *XML linking language (xlink).* Retrieved from http://www.w3.org/TR/xlink/

Walker, J. (2005). *Feral hypertext: When hypertext literature escapes control.* Paper presented at the Conference on Hypertext and Hypermedia, Salzburg, Austria.

Wardrip-Fruin, N. (2004). *What hypertext is.* Paper presented at the Conference on Hypertext and Hypermedia, Santa Cruz, CA, USA.

Wilson, E. O. (1975). *Sociobiology: The new synthesis.* Cambridge, MA: Belknap Press of Harvard University Press.

This work was previously published in the International Journal of Information Systems and Social Change, Volume 2, Issue 4, edited by John Wang, pp. 31-43, copyright 2011 by IGI Publishing (an imprint of IGI Global).

Chapter 17
Social and Cultural Challenges in ERP Implementation:
A Comparative Study Across Countries and Cultures

Sapna Poti
Indian Institute of Technology Madras, India

Sanghamitra Bhattacharyya
Indian Institute of Technology Madras, India

T.J. Kamalanabhan
Indian Institute of Technology Madras, India

ABSTRACT

This paper studies the differential practices of change management in organizations of western origin and compares it with the best practices prevalent in Indian organizations, with special emphasis on social and cultural challenges faced in these countries. Since Enterprise Resource Planning (ERP), as part of an information and communication technology (ICT) initiative, is frequently associated with organization change and transformation in relation to its adaptation, it has been used as the context in this study. The impact of social factors and cultural challenges on change management processes and elements are compared and contrasted using multiple case studies from USA, Canada, European (Western/Eastern) and Indian organizations who have adopted ERP technologies. The conceptual framework highlights cultural and social factors that affect ERP implementation, and offers suggestions to researchers to empirically test these influences using sophisticated analytical methods and develop change strategies and practices in response to these challenges. Further, it also draws attention to the need for a contemporary, result-oriented, quantitatively measurable framework of change management at the individual and enterprise levels. It is expected that such an approach would result in better buy-in from all stakeholders in terms of increased accountability.

DOI: 10.4018/978-1-4666-2922-6.ch017

INTRODUCTION

Any technological development in an organization initiates changes at both individual and organizational levels. The demanding responsibilities during an Information Technology (IT) intervention tend to create mental unrest, fear and friction in the minds of the workforce, and more so in a developing country like India (Tarafdar & Roy, 2003). Enterprise Resource Planning (ERP) packages are claimed to be generic, off-the- shelf solutions, but the issues arising out of implementation in different country-specific cultures often hinder smooth implementation. There is growing evidence of social and cultural differences not being addressed during implementation, leading to delays and failures in ERP adaptation. Change practices would need to be customized, along with precautions and planning on how to deal with the ensuing cultural challenges. This conceptual paper emphasizes the need to address essential change elements and factors keeping in mind the country specific context and its socio-cultural differences. It also attempts to provide evidence through case studies of organizations that have identified social and cultural challenges at the outset and proactively managed resistance to change through differential treatment within user groups.

THEORETICAL FRAMEWORK

The theoretical framework of this paper includes:

1. Literature review on Enterprise Resource Planning and Evaluation issues.
2. Case analysis of organizations that have implemented ERP across countries.

CONTEXT AND BACKDROP

Enterprise Resource Planning (ERP): ERP is a tool that integrates information flowing into and out of a company's interdepartmental incompatible database(s) to provide smooth customer service through integration of functions such as marketing, sales, production, financials, materials, and human resources, leading to reduced inventory write-offs and increased profitability in the long run (Davenport, 2000).

Despite the huge need for ERP, the expected benefits are often not forthcoming, with many ERP projects failing to deliver the intended results. The reasons for failure are many and could range from fitment issues in terms of the processes not being mapped to the packages, lack of experience with structured change processes, lack of business integration with the ERP packages, or frequent technical challenges. For instance, by nature, an enterprise system imposes its own logic, best practices and structure on the client company's strategy, culture, organization of departments and structure and hence pushes for and demands a huge amount of change management which client companies may not expect or plan for. These systems force firms to do business in ways that conflict with their vested interests or agenda, and hence could lead to a chaotic scenario. For most organizations, a change management initiative such as an ERP implementation involves the highest cost, time and effort possible in the organization's history (Davenport, 2000). Top managements at times decide to persist with the implementation, thrusting the entire responsibility on the Information technology department without realizing that the issues are to be dealt with elsewhere. Davenport suggests some essential steps for better ERP implementation success which include clarifying strategy in advance, changing organizational structure, putting the right people in place and installing an enterprise system gradually.

ERP IMPLEMENATION IN INDIA

In the context of ERP, there are unique issues faced in India, which could be rooted in its history of Information communication technology (ICT) development in purely Indian origin organizations. The IT growth in this context has hitherto been unbridled and fast paced, taking the workforce by surprise in a very short span of time. From basic rudimentary IT systems in the country, these organizations have moved to sophisticated packages such as Enterprise resource planning (ERP) to address the business pressures, information integration issues and expectations arising out of competition with the multinationals operating in the same space. In this context, it becomes important to test and evaluate the required elements in a change program to ensure better outcome. Indian organizations typically introduce Enterprise Resource Planning to address business issues such as information integration, customer satisfaction, and shared services. Unfortunately, despite the heavy investments, the implementation process of ERP in Indian organizations has been fraught with problems related to technological issues and organizational change aspects. There is thus a need to study and analyze the issues related to both these major aspects of ERP implementation. Most of the studies undertaken till date address overall critical factors that an ERP implementer could face and a few technical issues have been analyzed. However, a detailed study on change management aspects need to be undertaken in the context of ERP of Indian origin organizations, especially in the milieu of the current cultural and social context in Indian organizations during the implementation stage (Tarafdar & Roy, 2003). The change factors and their outcome at every stage of change, the extent to which the implementation of a program is undertaken, in a set of fully owned Indian organization vis-a-vis western origin organizations need to be studied.

Literature on the Indian ERP experience has largely been confined to the popular press in the nature of business articles, journalistic literature published in business magazines, industry reports on individual companies (Tarafdar & Roy, 2003). Host country nationals (HCNs) in developing countries employed with multinational companies are accustomed to the latest IT infrastructure, newer systems, and automation, and hence are less prone to culture shocks arising from uniform change management practices adopted by MNCs across all their subsidiaries. However, organizations based outside India would be expected to have cultural differences, which could affect the change practices. In order to address the social and cultural issues, a structured outcome-based evaluation of change program might need to be looked at. Given this, a comparative study between Indian and western origin organizations is expected to yield interesting insights.

The ERP literature also recommends a comprehensive list of 'Critical success factors" (CSF) deemed essential at a strategic and tactical level for successful implementation of ERP (Somers & Nelson, 2004) (see Table 1).

These factors are segregated into two major areas, viz. (1) Technical aspects, and (2) Change Management and Organizational factors. Our study primarily focuses on Change Management

Table 1. Critical Success factors (CSF) (adapted from Somers & Nelson, 2004)

Strategic and Tactical critical success factors (CSF)
Top management
The project champion
The steering committee
Implementation consultants
The project team
Vendor–customer partnership
Vendors' customization tools
Vendor support
User training and education
Management of expectations
Careful package selection
Project management
Customization
Data analysis and conversion
Business process reengineering
Architecture choices
Dedicating resources
Change management
Clear goals and objectives
Education on new business processes
Interdepartmental communication
Interdepartmental cooperation

aspects that include a few variables from a) the Critical Success Factors, (Somers & Nelson, 2004), b) recommendations from the change management literature and models used by ERP vendors, and 3) our findings of the case studies provided below. Based on the above, the following change elements have been identified as relevant in the context of an ERP implementation:

1. Change factors at the pre-change stage: Shared vision and objective setting, organizational readiness and management of change resistance, change strategy and blue print, communication plan, leadership support and commitment.
2. Implementation / transition factors: Training, Communication.
3. Post-change stage change factors: Tracking systems, Incentivisation.

These change processes can be evaluated through an impact study. The ERP literature provides a model for impact of Information Systems (Gable, Sedera, & Chan, 2003). The authors have

suggested a four-impact-factor matrix, including Systems Quality, Information Quality, Individual Impact and Organizational Impact, as shown in Table 2.

It is proposed to broadly use the factors of Individual Impact, but restricted to the impact on an individual level due to the change management aspects. It is not proposed to include all the Critical Success Factors (e.g., technical factors) in the scope of this study. In addition to the individual impact recommended above, the change processes will be evaluated through various other outcome measures. These outcome measures are proposed as per a structured program evaluation methodology. To select an appropriate methodology, an evaluation research study on program evaluation was undertaken as described below.

EVALUATION RESEARCH

The concept of evaluation began with evaluation of training and development. In its broadest

Table 2. Impact factor index (adapted from Gable et al., 2003)

Individual Impact	Organizational Impact	System Quality	Information Quality
II4 Individual productivity	OI2 Staff requirements	SQ1 Data accuracy	IQ1 Importance
II2 Awareness / Recall	OI3 Cost reduction	SQ2 Data currency	IQ2 Availability
II3 Decision effectiveness	OI4 Overall productivity	SQ3 Database contents	IQ3 Usability
	OI5 Improved outcome	SQ4 Ease of use	IQ4 Understandability
	OI6 Increased capacity	SQ5 Ease of learning	IQ5 Relevance
	OI7 e-government	SQ6 Access	IQ6 Format
	OI8 Business Process Change	SQ7 User requirements	IQ7 Content Accuracy
		SQ8 System features	IQ8 Conciseness
		SQ9 System accuracy	IQ9 Timeliness
		SQ10 Flexibility	IQ10 Uniqueness
		SQ11 Reliability	
		SQ12 Efficiency	
		SQ13 Sophistication	
		SQ14 Integration	
		SQ15 Customization	

definition, evaluation includes all efforts to place value on events, things, processes, or people. Evaluation is normally conducted to determine if program objectives have been achieved or if the achievement of program objectives has resulted in enhanced individual performance.

EVALUATION MODELS

Approximately 22 evaluation frameworks and models were studied. The models could be clustered under a similar framework, primarily for use of quantitative or qualitative aspects of evaluation. The quantitative models report financial outcomes, a mix of measures, or more subjective measures of program success. The qualitative models essentially deal with learning, applications, reactions or feedback on programs. The models include CBA (Cost Benefits Analysis), CIRO (Context, input, reaction, outcome), Program objective, content, implementation and outcome, evaluation of knowledge and skills, business impact, and success case evaluation (Stufflebeam, Madaus, & Kellaghan, 2001; Phillips, 2003). Based on available literature on evaluation research, an adaptation of the Kirkpatrick's model was found to be appropriate for evaluation of the change management program studied in this paper.

KIRKPARTRICK'S FOUR LEVEL FRAMEWORK

This evaluation framework describes four levels of evaluation (Blanchard, Thacker, & Way, 2000).

Level 1: Reaction: At the reaction level, Kirkpatrick's framework focuses on participant perceptions about the program.

Level 2: Learning: Learning evaluation is concerned with measuring the extent to which learning is acquired.

Level 3: Job Behavior: Also known as application, it is defined as whether or not the participant uses what was learned in the program on the job.

Level 4: Results: This is concerned with the program influence on organizational effectiveness. Here, questions regarding improvements in organization effectiveness are answered.

The first level of evaluation mentioned above (reaction and feedback) has been based on the change management process, while the other three levels of the evaluation methodology of Kirkpatrick (1975) have been utilized to measure the impact at an individual level of the change program in the ERP context. It is also assumed that an impact on an individual will eventually effect and influence the organization, which can provide scope for further research in this area.

PROPOSITIONS

Based on literature evidence, the impact and influence of various factors on change practices in the context of ERP is studied through the propositions given below:

- **P1:** The social and cultural issues arising during an ERP implementation require differential management in countries adopting similar ERP packages.
- **P2:** Elements of change processes in the context of ERP implementation are associated with social and cultural issues.

In certain cases, there could be evidences of the current economic scenario (e.g., recession) superseding socio-cultural influential factors in these countries. The study has attempted to identify evidences validating the above proposition(s), based on data from multiple case studies from different countries.

METHODOLOGY

The case studies were undertaken on-site at an ERP consulting organization based in India. The consulting organization has undertaken various large scale ERP implementations for their large multi-national and national clients, implementing both Oracle and SAP's R/3 packages.

DATA COLLECTION

The data was collected by way of case recordings based on open ended questions. Further, personal interviews of respondents were undertaken to elicit qualitative information that could support the questionnaire responses. In addition to direct interviewing, company documents, e-mails, and group discussions were used to shape and triangulate our findings. Finally, data collected were combined and synthesized into a thematic structure and collated into a tabular format to provide a holistic data corpus for further case analysis.

CRIRTERIA USED FOR RIGOUR OF THE STUDY

- **Construct validity:** utilization of multiple interviews with multiple stakeholders at different points of the project, usage of other modes of interaction such as e-mail, telephonic conversations and additional evaluation through documentary evidence of existing practices, change strategies, communication plans and project plans
- **Reliability:** Creating and maintaining a case study database, case study notes, case study documents, summary tables, case narratives and responses to a set of questionnaires.

COMPARTIVE STUDY OF IMPLEMENTATION IN SIX COUNTRIES

Case outline: Overview of the implementation

BRIEF BACKDROP/ CHARACTERISTICS OF SAMPLE FIRMS

The brief background of the sample firms is as provided in Table 3.

Modeling region wise analysis of underlying social and challenges and change practices to address the issues (The case analysis matrix): The analysis covers the social and cultural challenges faced by the sample firms, challenges faced due to change resistance at individual and organizational levels, managing and overcoming resistance to change, the challenges faced from top management, change mechanisms followed and the details of the process owners, details of change tracks planned, formal change readiness assessment undertaken, grievance handling or socialization mechanisms adopted, tracking mechanism of complaints, grievances and open ended issues, status of change plans implemented during each phase, and any unique change process that had to be introduced due to the challenges faced or customization of existing change practices undertaken. The analysis matrix is provided in Table 4.

CHANGE OUTCOME ENVISAGED, KEY PERFORMANCE INDICATORS (KPIS)

The key performance indicators were agreed upon before the commencement of the ERP implementation. While this was the practice in most of the countries, it was not followed in India. The matrices were used to evaluate effectiveness of change if any. In Indian organizations, KPIs are

Table 3. Brief backdrop and characteristics of sample firms

Organization characteristics	Case Studies		
	Belgium	USA	India
Presence in countries	presence in Latin America, Europe (East), Russia, North America, Canada, China, South Korea	USA & Canada	India
Base location of the organization	Belgium (operations in 30 countries)	Maryland, USA	Hospet, Karnataka
Organization facts: Brief backdrop, history, LOB, Manpower strength, type of industry, presence in countries, technology usage prior to ERP implementation.	Beer manufacturer,.	The organization is a large computer software consultancy organization handling highway toll processing for government transportation department, has its presence in USA	Manufacturing, castings of car engine, 1500 crore, KFIL, technology – Informix database taking care of distribution, tally – no integrated ERP, manual excel based
Industry	Manufacturing	Information communication Technology/ Software	Manufacturing
Manpower Strength	70,000 employees	50,000 employees	20,000 employees
Brief history of ICT implementation in the client site:	Across different countries they had from basic spread sheets to people soft and SAP, not standardized.	The client was using legacy database, and basic programming language to develop their application, prior to implementing ERP	No ERP, pockets of HRMS payroll, distribution, finance
Motivations / Reasons for an ERP implementation:	Business issues / problems faced: Need for global reporting,	Current application built on legacy technology, resulting in higher cost of ownership	a) Integration of data, communication.
	Huge number of expats moving across countries,	need to migrate to newer technology that can support upto the next decade	b) Centralization of decision making
	no common processes,	Leverage web technologies for accessing the application from remote areas	c) optimization of resources
	Need to cut HR costs, number of HR FTE (full time equivalent) count was to be reduced, HR to employee ratio was high, move to shared services	Use data warehousing features for analytical reporting	
	The driver was to standardize the processes and data, implement SAP, move to a shared services center per zone (Europe, America and 6 other zones)		
Expected benefits from the initiative	Standardization, global reporting, cost effectiveness (AMC to one vendor), FTE cost reduction	Use an ERP that is configurable so that it can be deployed for multiple clients with minimum configuration	Quicker decision making, good MIS reporting, financial reporting, integration of busi functions
Year of ERP implementation:	2007 (implemented in 7 countries)	2008	2007
Name of ERP Vendor/Consultant	SAP	Siebel CRM application	Oracle

continued on following page

Table 3. Continued

ERP Modules/Applications/ platforms implemented:	ECC 6.0, PA, organization management, time reporting, benefits, e-recruiting, learning solutions, employee and manager self service	Sales, Order Management, Analytics Platform - Oracle / Unix	Manufacturing, order management, purchasing, financials, HRMS, payroll
Broad phases of ERP implementation: basis on which phases were planned	Process standardization, requirements, blue printing, build and test phase and deployment – standard ERP implementation phases	Requirement study, Functional design, Technical design, development, testing, data migration, roll out, post go-live support	108 days of requirement study, BPR exercise, conference room pilot, user training, user acceptance testing, go live and post production support

generally not fixed or monitored. A sample of the KPI index is provided in Table 5.

The steps for creation and modification of a change management plan is provided in Figure 2.

FORMULATION OF A STANDARDIZED CHANGE MANAGEMENT PROGRAM

A rigorous change management program was designed and implemented by the organizations. Specific to the countries, the change management plan was modified and different versions were implemented keeping in mind the nuances of practices required as per the social and cultural issues identified at the outset. A sample of the change management plan is provided in Table 6. Deployment of the detailed plan would be based on the complexity of the ERP implementation exercise.

The transition flow chart explains the stages of change from the outset of the ERP implementation exercise to the final stage of adoption of the new systems in Figure 3.

Discussion: Summarizing the Findings

Based on our analysis of the cultural and region specific challenges of change processes, we recommend strategies which, we feel, could be useful for ERP implementers while addressing culture-specific challenges during change implementation.

Eastern Europe, i.e., Russia and Ukraine being communist countries have challenges similar to developing countries where the Information Technology infrastructure is fairly rudimentary and IT systems need to mature, while the ERP systems are very sophisticated and may cause resistance to change. Care should be taken to gauge this resistance at the beginning of the implementation, with organizations addressing them and a detailed Organization and Individual change readiness assessment pursued. The effectiveness of transfer of technology is also affected by the fact that in the case of Eastern Europe, the technology transfer happened in a developing recipient country (Ukraine) while the parent company (Belgium) from a developed country had different challenges, the learning from which were of little use in the Ukrainian context. This difference has been highlighted in an earlier study which states that the effectiveness of transfer of technology is most affected by variations in societal cultures of the two nations when one of them is an advanced industrialized nation (e. g., West Germany or Japan) and the recipient is a developing country (e.g., India or Brazil) (Kedia & Bhagat, 1988).

Similarly, it is seen that top-down communication works very well in these countries. This could be explained by the fact that Eastern European countries also rate high on power distance, which measures the extent to which less powerful members of a society accept the unequal distribution of power and rewards as normal features of their society. Hofstede (1980) has shown that different cultures possess different distributions of power in their organizational and social hierarchies. Hence,

Table 4. The case analysis matrix

Theme	Western Europe	Eastern Europe	Canada	USA	India
Social and cultural challenges	a) The unit in France had a slow pace of work, b) did not face many challenges c) did not have much resistance, d) had to spell out requirement often due to change in teams, e) employees were generally low profile, f) there was a need to be careful on wording negative remarks g) passive following, h) they need a lot of planning in advance i) Hierarchical setup, too many levels of approval, j) Detail and analysis oriented.	a) Eastern European i.e., Ukraine, Russia, old communist countries consider issues with suspicion. b) They are vocal about their problems. c) Familiar with the systems, d) resistance is primarily from the operational layer, e) comfortable with existing systems, f) since it is a global system, resistance was more, g) faced a lot of language issues, training manuals had to be translated, data was to be loaded and added in the local language so language supporting systems were to be used.	a) Canadian employees were more open to feedback b) open to change, c) losing support of old systems, d) looking forward to new systems, e) more enthusiastic f) language is not an issue	a)Employees not familiar with working with an offshore vendor b) Fear of job loss prevented employees from providing clear and precise requirements	a)Employees feared losing their jobs because of automation, b) labor intensive processes, c) Employees were not communicative d) lack of transparency in sharing information e) Senior and older employees found adoption of change difficult due to operational issues like not used to a basic usage of a mouse. f) Language issues, employees spoke kannada or hindi (local language issues)
	Project team was drawn from HR team and had heavy incentive based targets so they were pushing hard to implement the ERP Western Europe revenues were stagnating, employees had to increase their responsibilities otherwise they could be dispensable	Eastern Europe was on a growth path	In Canada most of the activities are considered to be the line manager's responsibilities		a) Requirements – mental block of sharing information, jobs at stake, around 70% of info was shared with great difficulty. b) Conference room pilot – showcases the ERP, resistance felt as they were used to legacy way of working, not wanting to adapt to newer systems, dos based. Tendency to compare with current way of functioning. comparison with the older formats they were used to. c) Training – oracle processes were different from the home grown system, extra screens and efforts being rejected, language issues – communication, the interaction levels were low, shy of asking questions. d) User acceptance testing – sign off stage, responsibility and ownership moves to the customer and faced challenges during that period, any changes beyond go live would be additional cost so try to prolong the sign off period. e) Go live – success of go live depends on earlier stages, communication gaps
Challenges faced due to change resistance: individual level, organizational level (separately)	Top driven organization, some resistance was felt during standardization workshop, pushing what they were currently doing, conflict in terms of role of people manager and the line manager Individual level – targets to implement ERP were high but employees who did not carry these targets in their scorecard refused to cooperate.	Organization level – In Eastern Europe, employees were comfortable with the current systems hence they did not feel the need to replace with an ERP just because global or corporate level board wanted a change in the system	Employees didn't not have any major issue implementing these systems due to high computer literacy so they were not very cynical	Resistance to change as the employees were not familiar with the technology. There was a push to go in for ERP and mandate to offshore work which posed a lot of opposition	Used to the old way of working, b) Ownership factor c) no buy in within the organization d) advantages and disadvantages of an ERP not explained in advance hence higher resistance e) The Shollapur office had more senior officers, more intra organizational issues and clashes were being faced

continued on following page

Table 4. Continued

Managing and overcoming resistance to change	a) In France as the resistance was not much, planned change management activities went smoothly b) employees took their own initiatives c) wanted higher responsibilities as revenues were low d) employee communication was pertinent e) ESS, MSS forms were built to enhance communication	a) took a lot of persistence b) issues were raised to the top management, c) HR was made to talk to the resistors d) a Russian speaking person was made a change leader who used to mediate between the consultant, the employees and management in addition to organizing change related activities. e) Many workshops were conducted to sell the ERP by way of educating on user friendliness, meeting needs, showing a bigger picture, f) Interactions were high g) a leader who was active respected and who could drive things internally was appointed	a) Dependency on communication dockets, detailing of plans was higher b) project team churning and turnover was high hence due to lesser continuity consultants had to pitch often	At individual level, the people were provided training on the new technology and product. an assurance of job continuation was given	a) Every time the management was pulled in and made them to participate, internal escalations. During the business process sign off it was to loop it by using the equations of the top management with the employees. Extensive top down efforts b) Respect for an ERP implementation seeing value, adhering to schedules etc is an issue in the Indian scenario, value for the vendor or consultant's time. c) In an Indian context if meeting requests are sent by an email it is generally ignored, personal relationships or personal invitation is a must, rapport building is required, so to organize communication or training programs was found to be cumbersome, d) Process orientation was also low
challenges faced due to the top management	Resource starved, cost conscious company, investments on resources was difficult, teams to be allocated Steer course meetings were fairly smooth in Western Europe, project team would have briefed them in detail	Eastern Europe top management wanted to dig deep into what was the progress	a) In Canada replacement took time and difficult, teams fairly stable in other countries. resource commitment was a challenge. b) Getting the top management to participate and attend meetings was a challenge	a) Top management was not familiar with the limitations of the product b) Commitments towards completion date, cooperation not provided	Top management had not sold the ideas as effectively as it could have been sold and internal issues were not addressed
Change mechanism and process owner details – (change agents – internal, consultant)	Western Europe engaged an external consultant	In Eastern Europe the change agents, Leaders were appointed from within the employees		Change management was controlled by Steering committee comprising of the Project manager and other stakeholders. b) the project manager was the process owner, the business analyst was the change leader c) The project manager and the consultants were responsible for communicating the changes to business users across various locations	The IT director was the change agent – sponsor, The team of managers under IT director with local language skills was also responsible as change catalysts. IT director was part of the board – he was on the vendor side, had to tread a difficult path of managing the set of users and vendors. Politically correct. Their strength was managing issues and decisions taken fast, few things not enforced. Manufacturing head was strong in his business process, internal customers were adamant about their processes
Details of change tracks planned:	a) Communication, track included preparation of dockets of communication, stakeholder holder analysis, development of change management plans, determination of the key stakeholders. Unions had to be informed, process of communication was determined for the same, frequency of informing the people manager, level of messaging, keep track of progress during each HR conference, request for slots in these conferences. b) Training - Identifying the training, target audience and delivery of training	a) In Russia the workshops were increased b) target audience was changed as per the employees who needed a buy in. c) HR needed to be convinced first and then other teams were covered. d) People who could negatively influence were talked to and convinced		Training and communication workshops for each application was undertaken covering all end users	Communication, training, escalation path from the lower level and project manager to the director

continued on following page

Table 4. Continued

Formal change readiness assessment undertaken	In western Europe, after the stage of stakeholders analysis a checklist of readiness, an "as is" and "gap" analysis was prepared, managers and trade unions participated in this study. Issues addressed during readiness assessment included the level of training required, openness to use the system,	The project charter detailed the project team size, roles and responsibilities and the cost was discussed with the top management		a) Change readiness was periodically assessed and necessary steps were taken to improve communication to the user community b) Intranet, websites and unit magazines carried details of the change progress	During the proposal stage and before sign off of the contract readiness assessment was done. As per the Gant chart it was a smooth transaction except the hardware ordered was not received with the right configuration, not ready in terms of the technical aspect of readiness
Grievance handling or socialization mechanisms	No need felt for grievance handling mechanisms			none planned	Escalation path as above
Any tracking mechanism of complaints, grievances, open ended issues	a) two months of training, b) issues were handled during and after trainings c) email id was set up for Q&A d) Issue trackers and help desk, frontline contact center setup and tickets logged in. e) Preliminary activities had a script FAQs to the contact centers – level 1 support. f) Beyond level 1 support logged in as a ticket and directed to the consultant			a)all issues were tracked in the quality center, b) rapid response team was organized for immediate response to priority issues c) post deployment, changes were recorded, analyzed and implemented d) weekly tracking plan for change management	Issue log – weekly presentation to the IT director
Status of change plans implemented during each phase:		Transition was for a long duration, in Eastern Europe the stabilization took a longer time			Difficulties in transition stage was the highest, planning stage was comparatively smooth, post implementation was the toughest, all hidden issues, attitude problems was brought up during this stage.100% undertaken
Any unique change process that had to be introduced due to the challenges faced or did a customization of existing change practices suffice	a) Train the trainer workshop: a common workshop planned for all the countries with two representatives from each organization. b) best practices from each location was shared between countries, c) cost effective change management in three zones d) The hired external consultants did not make much of an impact as their had a mandate to deliver, emerging of an internal leader would have been preferred			Every change had to be reviewed by business user, followed by the business analyst and technical analyst, before it was prioritized by the Project manger	during automation and BPR exercise quite a few additional training sessions were undertaken, training were far more than planned as the understanding of ERP user concepts even after go live on basic concepts was found to be low, user manuals were not being used, preference was for personal interaction
The extent to which expected benefits were realized	70% of the benefits was realized	Russia and Ukraine didn't receive benefits as they were running it internally, they refused to move towards a business shared services as planned by the organization		change management process was well streamlined and helped address all business requirements	a)Integrated reporting, productivity improvement – 10 hours to 3 hours, distribution system had 3 manpower for 24 hours, after implementation, the same job was undertaken by 1 person and 2 employees on job rotation hence there was an element of cost saving due to 3 shifts being managed by 2 persons) Due to attitude issues there were layoffs which also helped to reduce costs) Payroll implemented but not being used, HR head did not agree to use it. d)80% overall implementation is being utilized

continued on following page

Table 4. Continued

Understanding employees' reactions to the change process (satisfaction, morale, commitment, engagement.)	a)survey of usage, usefulness of system, information sharing, documentation required made available but restricted to the HR team and not to wider employee population b)post learning of transaction training, report running, self desk for managers and employees, tests were conducted to understand learning	Application level in Russia – despite repeated training, learning was good but data entered as they were not filling in required information for pay roll processing hence the system was modified to stop transaction if mandatory fields were not filled, exception message was sent to the participants, systematic checking. Transaction took longer		post roll out, there were multiple levels of meetings held to understand the level of satisfaction amongst the users	Informal sessions were conducted as employees were not found to be process oriented. No rating or surveys for support by vendors, no structured surveys, no quantitative techniques to understand the issues were utilized.
Main learning from the process (people related, business related)	Cultural issues to be taken care of differently, need adequate time to understand these issues			a)Learning were around handling user expectation, assuring security and bridging cultural differences between employees, management and the vendor representatives b)Provide ample opportunities for interaction at all employee levels	a)Communication, attitude, adaption issues to be anticipated, planned and implemented better but business can be lost due to cost or opportunity cost b)Encourage the top management to sell the advantages more effectively, remove fear of losing jobs, organization to be prepared, board should not take independent decisions, benefits should communicated to the last user
Open ended issues that remain to be addressed despite undertaking all measures	Number of tickets is a barometer, if it comes down it is considered good and in this case the system is stabilizing but few open ended issues are there.			additional scope expansions need to be addressed	none

messages communicated in a top down fashion tend to be well-accepted and followed.

Further, Russia ranked number one on value ratings on conservatism, hierarchy and harmony, as compared to western European countries. Hierarchy values, when defined in terms of its emphasis on the legitimacy of the hierarchical allocation of fixed roles and of resources, were found to be more important in Eastern Europe (Schwartz & Bardi, 1997). This being the case, the communication in a change process could well be top-down.

Challenges were also experienced in Project Management, Business Process Re-engineering, contractual scope expansion and technical aspects. Hence, detailed schedules for each of these processes should be drawn in advance and monitored frequently. In relation to this aspect, a similar study has found that such cultures tend to be "other-directed" (Kedia & Bhagat, 1988) making monitoring a key necessity. To manage the resistance and for smooth change transitions, frequent training and communication workshops need to be planned and deployed.

In the cases studied, it was found that in the Western European organizations, specifically employees based in the unit in France had a slightly laid-back attitude. In such cases, incentivisation would work well along with specific Key Performance Indicators internally formulated for each team. The employees who were primarily found to be non-cooperative could be given charge of small parts of the project. Western Europe required a lot of detailing in every aspect of the project, leading to delays in project delivery. In these cultures, external interference is not encouraged. Hence, the change agents need to be groomed internally.

Developed countries like USA and Canada face far less resistance to change due to their familiarity with information technology and sophisticated systems. In fact, they look forward to newer systems. In Hofstede's (1980) study,

Table 5. Sample change management plan

#	Activity		Location	Resource Responsible	Start date	End date
	Consolidated Plan (country specific version)					
1.	Change Program Management					
	1.1.	Decide on types of CM workshops and dates			06/23/2008	06/27/2008
	1.2.	Stakeholders Analysis			06/09/2008	06/21/2008
	1.3.	Develop Change Management plan (detailed version)			06/09/2008	06/21/2008
	1.4.	Evaluation approach for Change Management effectiveness			06/23/2008	06/27/2008
	1.5.	Decide structure for CM liaisons & Role of Internal Communication			06/23/2008	07/04/2008
	1.6.	Formalize and Communicate Change Management Team Roles and Expectations			06/16/2008	06/27/2008
	1.7.	Change Workshop #1 (2 days): Intro to Managing Change; Workshops: Impact Analysis and Action Plan, Communication Planning, Training Strategy			06/23/2008	07/04/2008
	1.8.	Change Workshop #2: Train the Trainer Session to SMEs				
	1.9.	Change Workshop #3 (0,5 day): Training Results Follow-up and Readiness Preparation				
	1.10.	Change Workshop #4 (0,5 day): Go-Live Preparation				
2.	Communication and Mobilization					
	2.1.	Put forward immediate communication needs			06/16/2008	06/27/2008
	2.2.	Change Workshop #1: Managing Change for Communication Planning Activity			06/23/2008	07/04/2008
	2.3.	Define communication approach			06/23/2008	07/04/2008
	2.4.	Build communication plan for country			06/30/2008	07/11/2008
	2.5.	Build communication materials			06/30/2008	10/31/2008
	2.6.	Conduct initial presentations for Key Stakeholders such as MANCOMM, DIRECTORS, PEOPLE MANAGERS			07/07/2008	07/18/2008
	2.7.	Release initial communications for SILVER - Newsletter, e-mail, on PUB			07/14/2008	08/01/2008
	2.8.	Release additional communications - Newsletter, e-mail, on /Iknow			07/21/2008	10/31/2008
	2.9	Define communication feedback criteria and tools			07/21/2008	08/01/2008
	2.10.	Collect information and report on communication feedback and actions			08/11/2008	09/27/2008

continued on following page

Table 5. Continued

	2.11.	Final communication on SILVER deployment (prepare Go-Live communication plan)			09/01/2008	09/27/2008
3.	**Impact Analysis**					
	3.1.	Collect current Process Flows and Documentation			06/02/2008	06/27/2008
	3.2.	Prepare initial SILVER process catalog and level of changes			06/23/2008	06/27/2008
	3.3.	Prepare To-be-process flows			06/23/2008	06/27/2008
	3.4.	Change Workshop #1: Impact Analysis and Change Action Plan			06/23/2008	07/04/2008
	3.5.	Identify changes to SOP			04/07/2008	08/01/2008
	3.6.	Consolidate Impact Analysis and Action Plan			04/07/2008	08/01/2008
	3.8.	Execute Action Plan			08/04/2008	10/31/2008
4.	**Training**					
	4.1.	**Change Workshop #1: Define Training Strategy**			06/23/2008	07/04/2008
	4.2.	**Plan and Invite**				
	4.2.1.	Identify training requirements (build training course inventory)			06/30/2008	07/04/2008
	4.2.2.	Identify Super Users and target audience for training			06/23/2008	07/11/2008
	4.2.3.	Identify training instructors and training infrastructure (NOTE: Revalidate roles for training)			06/23/2008	07/11/2008
	4.2.4.	Identify resources (rooms, projectors, pcs)			06/23/2008	07/11/2008
	4.2.5.	Build training plan for SMEs, Super Users and Trainers			04/07/2008	07/18/2008
	4.2.6.	Invite SMEs, Super Users and Trainers			07/14/2008	07/25/2008
	4.2.9.	Build training plan for People and Payroll, Managers and Employees			04/07/2008	07/25/2008
	4.2.10.	Invite People and Payroll			07/28/2008	07/31/2008
	4.2.11.	Invite Managers and Employees			07/28/2008	07/31/2008
UKI	**4.3.**	**Execute Training**				
UKI	4.3.2.	Train SMEs			08/04/2008	08/29/2008
UKI	4.3.3.	Train Super Users and Trainers + Train the Trainers			08/18/2008	08/29/2008
UKI	4.3.4.	Super Users and Trainers prepare to perform End Users training			08/25/2008	09/05/2008
UKI	4.3.5.	Train People and Payroll			09/01/2008	09/12/2008
UKI	4.3.6.	Train Managers and Employees			09/01/2008	09/26/2008
UKI	4.3.7.	Additional Trainings for Readiness Improvement			09/22/2008	10/30/2008
UKI	**4.4.**	**Training Material**				
UKI	4.4.1.	Collect existing training material from other zones			06/16/2008	06/27/2008

continued on following page

Table 5. Continued

UKI	4.4.2.	Define structure of training manuals			06/30/2008	07/04/2008
UKI	4.4.3.	Prepare training material - First Version			07/21/2008	08/08/2008
UKI	4.4.4.	Adapt First Version and build final training material			07/21/2008	08/08/2008
UKI	4.4.5.	Review final training material			08/18/2008	08/22/2008
UKI	4.4.6.	Print final training material / Deploy to intranet			08/18/2008	08/29/2008
FR	**4.5.**	**Execute Training FR**				
FR	4.5.1.	Train People Team			08/25/2008	08/29/2008
FR	4.5.2.	Train Payroll			09/01/2008	09/10/2008
FR	4.5.3.	Train Managers			09/01/2008	09/10/2008
FR	4.5.6.	Train Employees			09/10/2008	09/26/2008
FR	4.5.7.	Additional Trainings for Readiness Improvement			09/15/2008	09/26/2008
FR	**4.6.**	**Training Material FR**				
FR	4.6.1.	Collect existing training material from other zones			06/16/2008	06/27/2008
FR	4.6.2.	Define structure of training manuals			06/30/2008	07/04/2008
FR	4.6.3.	Prepare training material - First Version			07/21/2008	08/08/2008
FR	4.6.4.	Adapt First Version and build final training material			07/21/2008	08/08/2008
FR	4.6.5.	Identify needs and resources for any language translation			08/04/2008	08/22/2008
FR	4.6.6.	Validate translated version and build final translated material			08/18/2008	08/22/2008
FR	4.6.7.	Review final training material			08/18/2008	08/22/2008
FR	4.6.8.	Print final training material / Deploy to intranet			08/18/2008	08/29/2008
	4.7.	**Logistics**				
	4.7.1.	Organize training infrastructure / Organize training logistics - rooms, projectors, network			06/23/2008	07/18/2008
	4.8.	**Training Environment Preparation**				
	4.8.1.	Define training environment			07/21/2008	08/01/2008
	4.8.2.	Identify needs for exercises data			07/21/2008	08/01/2008
	4.8.3.	Define data refresh needs and strategy			07/21/2008	08/01/2008
	4.8.4.	Populate training environment with data for exercises			08/18/2008	08/29/2008
	4.8.5.	Request training users			07/07/2008	08/01/2008
	4.8.6.	Create training users			07/14/2008	08/15/2008
	4.9.	**Manage Training**				
	4.9.1.	Prepare training evaluation/ attendances forms			07/07/2008	07/18/2008
	4.9.2.	Prepare template for report of training results			07/07/2008	07/18/2008
	4.9.5.	Present report on training results and follow-up actions			08/18/2008	09/26/2008

continued on following page

Table 5. Continued

5.		**Change Readiness**					
	5.1.	Define readiness criteria and setup evaluation forms				07/21/2008	08/08/2008
	5.2.	Review initial feedback from the stakeholders on deployment readiness, include on communication feedback				07/28/2008	08/08/2008
	5.3.	Evaluate and report on readiness of SMEs				07/28/2008	09/26/2008
	5.5.	Evaluate and report on readiness End Users				09/01/2008	09/26/2008
	5.6.	Conduct discussions on readiness of external providers				09/01/2008	09/26/2008
6.		**Help-Desk and End-User Support**					
	6.1.	Define End-User Support Structure and Procedures				08/04/2008	08/29/2008
7.		**Go Live Preparation Activities**					
	7.4.	Monitor go-live results				10/01/2008	10/31/2008

Table 6. Key performance indicators index

#	Theme	Name	Description	Formula	Data elements	Analysis dimension	Scope (which part of the organization)
1	Effectiveness	Attrition rate	It indicates XYZ organization employees leave due to dismissal (involuntary separation) and resignation (voluntary separation) and also according to specific reasons	Total Attrition Rate = Total headcount leaving in the period / Headcount at beginning of the period AR Dismissal = Headcount dismissed in the period / Headcount at beginning of period AR Resignation = Headcount resigned in the period / Headcount at beginning of period		Geographical area/ Entity Functional area Top Performers(OPR) Talent Pool Years at service	All employees
2	Effectiveness	Internal Staffing rate	Indicates the degree to which vacancies have been staffed internally (as opposed to external staffing). It measures the success of the company in developing the own people and to provide them an internal career path.	Number of vacancies staffed with internal employees during the period / Total number of vacancies staffed during the period		Geographical area/ Entity Functional area Top Performers(OPR) Talent Pool Years at service	All employees

continued on following page

Table 6. Continued

3	Effectiveness	People Performance indicator	It indicates XYZ organization employees at certain performance levels as a % of total headcount	Major gap' indicator:Headcount with Overall Performance equal to 'Major gap' for the period / Total headcount who received a performance appraisal for the period'Progressing' indicator:Headcount with Overall Performance equal to 'Progressing' for the period / Total headcount who received a performance appraisal for the period'Proficient' indicator:Headcount with Overall Performance equal to 'Proficient' for the period / Total headcount who received a performance appraisal for the period'Role Model' indicator:Headcount with Overall Performance equal to 'Role Model' for the period / Total headcount who received a performance appraisal for the period		Geographical area/EntityFunctional areaTop Performers(OPR) Talent PoolYears at service	Managers (Band I-VII)
4	Measure	Performance Appraisal Success Factor	It indicates the # of employees receiving a performance appraisal and is an indicator of the leadership engagement to provide yearly performance feedback to our employees.	Headcount who received a performance appraisal / Total headcount		Geographical area/ Entity Functional area	Managers (Band I-VII)
7	World class	Employee engagement survey	Measures employee engagement (e.g., overall satisfaction; job content; management effectiveness; working environment; remuneration; development & performance management)	Score calculated country by country based on the Employee Engagement Survey		Geographical area/ Entity Functional area	All employees
8	World class	Training cost index	It measure the total training cost against the total compensation (within a zone, country, entity and functional area.)	ZBB Training Sub package Costs / Total Compensation Excl. Travel costs		Geographical area/ Entity Functional area	Managers (Band I-VII)

continued on following page

Table 6. Continued

13	Effec-tiveness	FTE	Number of Full-Time Equivalents as measured from the contrac-tual work hours against legal contract full-time work week (take ZBB/PTT FTE definition)	(Theoretical work hours during the period + overtime during the period - absences during the period) / Planned working time of full time employee during the period		Geographical area/ Entity Functional area	All employees
25	World class	Manage-ment in-vestment factor	Measures money spent on manag-ers per FTE.	Management compensation cost / Total FTE		Geographical area/ EntityFunctional area	All employees
26	Measure	100% Cor-rectness of data within SLA	This measure ensures data entry process. (i.e., con-sistency between form & system)	Audit		Zone	All employees

Table 7. Variable studies

Variable	**Studies**
Organization Readiness	Kwahk (2006)
Awareness of change objective	Ragsdell (2000),Bhatti (2005)
Leadership	Bhatti (2005),Law and Ngai (2007)
Socialization	Fisher (1986; as cited in Poole & Van de Ven, 2004)
Training	Robey, Ross, and Boudreau (2002), Bhatti (2005)
Communication	Aladwani (2001), Bhatti (2005)
Incentives	Robey, Ross, and Boudreau (2002)
Learning	Philips (2003)
Application	Philips (2003)
Change results	Gable et al. (2003)
Size of the Organization/Type of Sector	Nadler and Tushman (1989; as cited in Poole & Van de Ven, 2004), Brown (1991; as cited in Poole & Van de Ven, 2004), Kelly and Amburgey (1991; as cited in Poole and Van de Ven, 2004), Philips (2003)
Reaction and Feedback	Gable et al. (2003)

the highest scores on Uncertainty Avoidance (the extent to which member of an organization or society strive to avoid uncertainty by reliance on social norms, rituals, and bureaucratic practices to alleviate the unpredictability of future events) (Liddell, 2005) were obtained for countries such as Greece, Japan, and most of the Catholic countries in Latin America, whereas low scores were obtained for Hong Kong, Singapore, and the Scandinavian countries (Kedia & Bhagat, 1988). Further, being at the individualistic end of the individualism- collectivism dimension, ties between individuals tend to be loose and people look after their own self- interests in both work

Figure 1. Conceptual framework

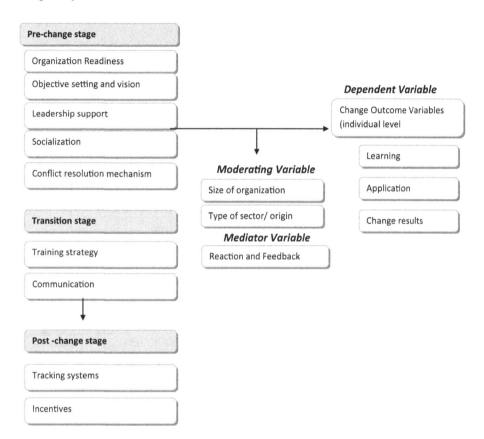

and non- work domains. This being the case, an initiation into the new ERP related process is all that is required, as the onus of learning (being "inner- directed") lies with the individual concerned. Language is also not an issue in these countries, facilitating the ERP implementation.

India also belongs to the south Asian cluster, which in a study of the 9 dimensions of societal practices rated high on performance orientation, group collectivism, power distance and human orientation. Hence, this cluster is distinguished as highly group- oriented, humane, male-dominated and hierarchical in nature (Liddell, 2005). On societal values, this cluster rates high on group collectivism and humane orientation; hence, issues such as loss of jobs for peers could be taken very seriously in India. In India, family owned and managed organizations are generally not

familiar or comfortable with key performance indicators (KPI) indices but these should form a critical part of the change process. For example, despite the existence of a rather sophisticated system of technical core and educated personnel, the ownership structure of Indian industry strongly influences the absorption of technology. The impact of regulatory policies and stringent laws on the acquisition and absorption of foreign technology by both public and private-sector companies makes it difficult to introduce technologies that might affect traditional management and control-related processes (Kedia & Bhagat, 1988). The change outcome needs to be defined and monitored. It is important to introduce tracking systems as culturally, India is high on uncertainty avoidance, and such monitoring initiatives could help increase the predictability and success rate

Figure 2. Creation/modification of change management plan

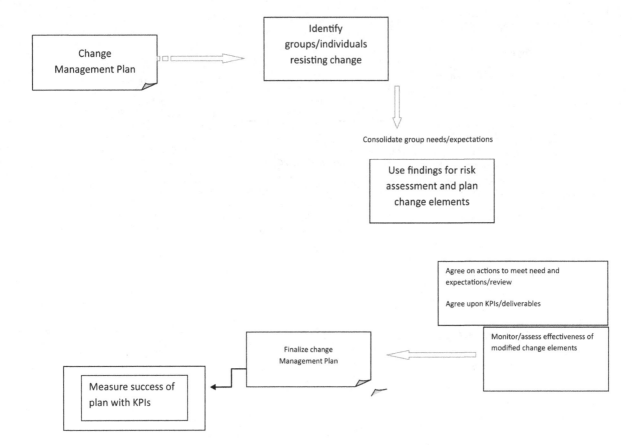

of such critical initiatives. Language could be a problem; hence change agents would need to be conversant with the local language. Finally, in India, resistance to change could be due to various reasons. Organizations might lack interest in pre-change arrangements and plunge into action without change readiness assessments, resource planning, deployment plan, communication and training plans. Hence it is recommended to focus on the pre-change stage as much as the transition stage. Incentivisation would also facilitate employee involvement. Kedia and Bhagat (1988) found that countries high in power distance also had significant differences in rewards, privileges, and opportunities among various levels of management. Hence, rewards and incentives are likely to be highly motivating.

It may thus be summarized that India and some countries in Latin America (e.g., Peru, Colombia, Venezuela) are more associative than European countries in general (Glenn & Glenn, 1981). A closer examination of the recent literature on technology transfers reveal that they are less effective in absorbing imported technologies in their organizational contexts (Marton, 1986; Sachs, 1986)

CONCLUSION

In light of the above literature on ERP and Evaluation with the case analysis highlighting the social and cultural challenges faced in different countries, the following research framework is proposed

Figure 3. Transition flowchart

Existing Scenario

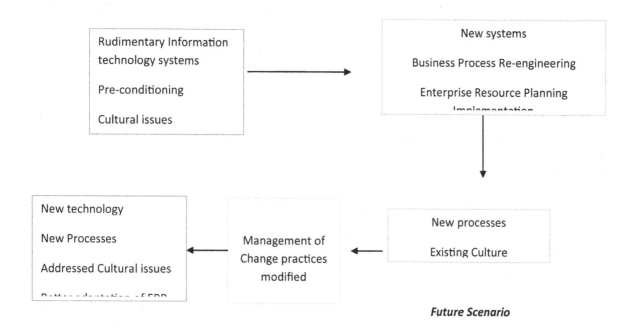

Future Scenario

suggesting a focus on certain key elements of change processes and strategies for addressing these challenges. The framework would include select variables of change processes and its outcome at an individual impact level to be studied in detail, in order to take a holistic view on how effectively an organization can deal with these identified cultural challenges.

In view of the socio-cultural challenges faced and the "change management and organizational factors" based on the list of the Critical Success Factors and their impact, relevant change variables and their individual impact are to be selected and their relationship with the change outcome needs to be tested. This study could be undertaken in both Indian and western origin organization and a discriminant analysis could provide a distinction on their respective outcome and relationships. The independent variables selected are the change variables, the moderating variables would include size of organization, type of sector, the mediating

variable would be the first level of evaluation - 'reaction and feedback'- as the outcome might be enhanced if there is a positive reaction to the change processes in an organization. The variables and studies are as listed in the Table 7.

The conceptual framework proposed in Figure 1.

LIMITATIONS OF THE STUDY

Though the sample firms represent the country specific culture and range throughout the entire spectrum of developed and developing countries, the sample firms and cases recorded may not be sufficient from point of view of generalizability. The cases have been analyzed essentially with a view to highlight the need for a study of the social and cultural scenario in each of the countries and organizations and to address the same through detailed change processes.

ACKNOWLEDGMENT

The authors would like to thank the Editor-in-Chief of the journal, Professor John Wang, and the anonymous reviewers for their inputs and patience.

REFERENCES

Aladwani, A. M. (2001). Change management strategies for successful ERP implementation. *Business Process Management Journal, 7*, 266–275. doi:10.1108/14637150110392764

Bhatti, T. R. (2005, September, 26). *Critical success factors for the implementation of enterprise resource planning (ERP) empirical validation.* Paper presented at the Second International Conference on Innovation in Information Technology, Dubai, India.

Blanchard, P. N., Thacker, J. W., & Way, S. A. (2000). Training evaluation: Perspectives and evidence from Canada. *International Journal of Training and Development, 4*, 295–304. doi:10.1111/1468-2419.00115

Davenport, T. H. (1998). Putting the enterprise into the enterprise system. *Harvard Business Review*, 3–13.

Gable, G. G., Sedara, D., & Chan, T. (2003). Enterprise systems success: A measurement model. In *Proceedings of the Twenty Fourth International Conference on Information Systems*, Seattle, WA (pp. 576-591).

Glenn, E. S., & Glenn, C. G. (1981). *Man and mankind: Conflict and communication between cultures.* Norwood, NJ: Ablex.

Hofstede, G. (1980). *Culture's consequences: National differences in thinking and organizing.* Thousand Oaks, CA: Sage.

Kedia, B. L., & Bhagat, R. S. (1988). Cultural constraints on transfer of technology across nations: Implications for research in international and comparative management. *Academy of Management Review, 13*(4), 559–571.

Kirkpatrick, D. L. (1975). Evaluating training programs. *American Society of Training & Development Report*, 119-120.

Kwahk, K. (2006, January 4-7). ERP acceptance: Organizational change perspective. In *Proceedings of the 39th Hawaii International Conference on System Sciences*, Kauai, HI.

Law, C. C. H., & Ngai, E. W. T. (2007). An investigation of the relationships between organizational factors, business process improvement, and ERP success. *Benchmarking: An International Journal, 14*, 387–406. doi:10.1108/14635770710753158

Lidell, W. W. (2005). Macroeconomic processes and regional economies management, project GLOBE: A large scale cross-cultural study of leadership. *Problems and Perspectives in Management, 3*, 5–9.

Marton, K. (1986). *Multinationals, technology, and industrialization.* Lexington, MA: Heath.

Philips, P. P. (2003). *Training evaluation in the Public sector.* Unpublished doctoral dissertation, Graduate School of the University of Southern Mississippi, Hattiesburg, MS.

Poole, M. S., & Van de Ven, A. H. (2004). *Handbook of organizational change and innovation.* New York, NY: Oxford University Press.

Ragsdell, G. (2000). Engineering a paradigm shift? A holistic approach to organizational change management. *Journal of Organizational Change Management, 13*, 104–120. doi:10.1108/09534810010321436

Robey, D., Ross, J. W., & Boudreau, M. C. (2002). Learning to implement enterprise systems: An exploratory study of the dialectics of change. *Journal of Management Information Systems, 19*, 17–46.

Sachs, J. (1986). A tale of Asian winners and Latin losers. *Economist,* 63–64.

Schwartz, S. H., & Bardi, A. (1997). Influences of adaptation to communist rule on value priorities in eastern Europe. *Political Psychology, 18*(2), 385–410. doi:10.1111/0162-895X.00062

Somers, T. M., & Nelson, K. G. (2004). A taxonomy of players and activities across the ERP project life cycle. *Information & Management, 41,* 257–278. doi:10.1016/S0378-7206(03)00023-5

Stufflebeam, D. L., Madaus, G. F., & Kellaghan, T. (2001). *Evaluation models.* Boston, MA: Kluwer Academic.

Tarafdar, M., & Roy, R. K. (2003). Analyzing the adoption of ERP systems in Indian organizations: A process framework. *Journal of Global Information Technology Management, 6*, 31–51.

This work was previously published in the International Journal of Information Systems and Social Change, Volume 2, Issue 4, edited by John Wang, pp. 44-67, copyright 2011 by IGI Publishing (an imprint of IGI Global).

Compilation of References

Abraham, C., Junglas, I., Watson, R. T., & Boudreau, M. (2009). Studying the role of human nature in technology acceptance. In *Proceeding of the Thirtieth International Conference of Information Systems*, Phoenix, AZ.

Ackerman, M., & Halverson, C. (2004). Organizational memory as objects, processes, and trajectories: An examination of organizational memory in use. *Computer Supported Cooperative Work*, *13*(2), 155–189. doi:10.1023/B:COSU.0000045805.77534.2a

Afriat, S. (1972). Efficiency estimation of production functions. *International Economic Review*, *13*, 568–598. doi:10.2307/2525845

Agarwal, S., Handschuh, S., & Staab, S. (2003). Surfing the service web. In *Proceedings of the 2nd International Semantic Web Conference* (pp. 211-226).

Ägerfalk, P. J. (2003). *Information systems actability: Understanding information technology as a tool for business action and communication*. Unpublished doctoral dissertation, Linköping University, Linköping, Sweden.

Ägerfalk, P. J., Sjöström, J., Eliason, E., Cronholm, S., & Goldkuhl, G. (2002). Setting the scene for actability evaluation – Understanding information systems in context. In *Proceedings of 9ᵗʰ European Conference on Information Technology Evaluation*, Paris, France.

Ägerfalk, P. J. (2004). Investigating actability dimensions: A language/action perspective on criteria for information systems evaluation. *Interacting with Computers*, *16*(5), 957–988. doi:10.1016/j.intcom.2004.05.002

Ägerfalk, P. J., & Eriksson, O. (2004). Action-oriented conceptual modelling. *European Journal of Information Systems*, *13*(1), 80–92. doi:10.1057/palgrave.ejis.3000486

Ajzen, I. (1991). The theory of planned behavior. *Organizational Behavior and Human Decision Processes*, *50*, 179–211. doi:10.1016/0749-5978(91)90020-T

Ajzen, I., & Fishbein, M. (1980). *Understanding attitudes and predicting social behavior*. Upper Saddle River, NJ: Prentice Hall.

Akiyama, T., & Furukawa, Y. (2009). Intellectual property rights and appropriability of innovation. *Economics Letters*, *103*(3), 138–141. doi:10.1016/j.econlet.2009.03.006

ALAsswad, M. M. (2011) Semantic Information Systems Engineering: A Query-based Approach for Semi-automatic Annotation of Web Services, Brunel University.

Aladwani, A. M. (2001). Change management strategies for successful ERP implementation. *Business Process Management Journal*, *7*, 266–275. doi:10.1108/14637150110392764

Albrechtslund, A. (2007). House 2.0: Towards an ethics for surveillance in intelligent living and working environments. In *Proceedings of the Computer Ethics Philosophical Enquiry*.

Alchian, A. A. (1965). Some economics of property rights. *II Politico*, *30*, 816–829.

Aldrich, H. (1972). *An organization-environment perspective on cooperation and conflict between organizations in the manpower training system*. Kent, OH: The Kent State University Press.

Almeida, L. D. A., Neris, V. P. A., Miranda, L. C., Hayashi, E. C. S., & Baranauskas, M. C. C. (2009). Designing inclusive social networks: A participatory approach. In *Proceedings of the 3ʳᵈ International Conference on Online Communities and Social Computing* (pp. 653-662).

ALT-C 2009 conference, Manchester, UK. (2009, September 8-10). *The VLE is Dead*. Retrieved December 9, 2009, from http://www.alt.ac.uk/altc2009/index.html

Andersen, P. B. (2001). What semiotics can and cannot do for HCI. *Knowledge-Based Systems, 14*(8), 419–424. doi:10.1016/S0950-7051(01)00134-4

Anderson, T. (2003). Getting the mix right again: An updated and theoretical rationale for interaction. *The International Review of Research in Open and Distance Learning, 4*(2). Retrieved December 9, 2009, from http://www.irrodl.org/index.php/irrodl/article/view/149/230

Arrow, K. J. (1962). Economic welfare and the allocation of resources for invention. In Nelson, R. R. (Ed.), *The rate and direction of inventive activity: Economic and social factors* (pp. 609–625). Princeton, NJ: Princeton University Press.

Australian Bureau of Statistics. (2003). *Australian social trends*. Retrieved from http://www.abs.gov.au/AUS-STATS/abs@.nsf/DetailsPage/4102.02003

Australian Institute of Health and Welfare. (2003). *Australia's welfare*. Retrieved from http://www.aihw.gov.au/

Bajraktarevic, N., Hall, W., & Fullick, P. (2003). Incorporating learning styles in hypermedia environment: Empirical evaluation. In *Proceedings of the Workshop on Adaptive Hypermedia and Adaptive Web-Based Systems* (pp. 41-52).

Balog, K., & Rijke, M. D. (2008). Combining candidate and document models for expert search. In *Proceedings of the Seventeenth Text REtrieval Conference (TREC 2008)*.

Balog, K., Azzopardi, L., & de Rijke, M. (2006). Formal models for expert finding in enterprise corpora. In *Proceedings of the 29th annual international ACM SIGIR conference on Research and development in information retrieval* (p. 50). New York: ACM.

Balog, K., Soboroff, I., Thomas, P., Csiro, A., & Craswell, N. (2008). *Overview of the TREC 2008 Enterprise Track*.

Banker, R. D., Chang, H., & Cooper, W. W. (1996). Equivalence and implementation of alternative methods for determining returns to scale in data envelopment analysis. *European Journal of Operational Research, 89*, 473–481. doi:10.1016/0377-2217(95)00044-5

Banker, R. D., Charnes, A., & Cooper, W. W. (1984). Some models for estimating technical and scale inefficiencies in data envelopment analysis. *Management Science, 30*, 1078–1092. doi:10.1287/mnsc.30.9.1078

Baranauskas, M. C. C., Bonacin, R., & Liu, K. (2002). Participation and signification: Towards cooperative system design. In *Proceedings of the Anais do V Simpósio sobre Fatores Humanos em Sistemas Computacionais* (pp. 3-14).

Baranauskas, M. C. C., Schimiguel, J., Simoni, C. A. C., & Medeiros, C. M. B. (2005). Guiding the process of requirements elicitation with a semiotic approach – A case study. In *Proceedings of the 11th International Conference on Human Computer Interaction* (pp. 100-110).

Bartis, E., & Mitev, N. (2008). A multiple narrative approach to information systems failure: A successful system that failed. *European Journal of Information Systems, 17*(2), 112–124. doi:10.1057/ejis.2008.3doi:10.1057/ejis.2008.3

Basu, S., & Fernald, J. G. (1997). Returns to scale in US production: Estimates and implications. *The Journal of Political Economy, 105*, 249–283. doi:10.1086/262073

Baumol, W. J., Panzar, J. C., & Willig, R. D. (1998). *Contestable markets and the theory of industry structure*. New York, NY: Harcourt Brace Jovanovich.

Beamon, B., & Balcik, B. (2008). Performance measurement in humanitarian relief chains. *International Journal of Public Sector Management, 21*(1), 4–25. doi:10.1108/09513550810846087

Beaudin, J., Intille, S., & Tapia, E. M. (2004). Lessons learned using ubiquitous sensors for data collection in real homes. In *Proceedings of the CHI Extended Abstracts on Human Factors in Computing*.

Benamati, J., & Rajkumar, T. M. (2002). The application development outsourcing decision: An application of the technology acceptance model. *Journal of Computer Information Systems, 42*(4), 35–43.

Benamati, J., & Rajkumar, T. M. (2008). An outsourcing acceptance model: An application of TAM to application development outsourcing decisions. *Information Resources Management Journal, 21*, 80–102. doi:10.4018/irmj.2008040105

Benbasat, I., Goldstein, D., & Mead, M. (1987). The case research strategy in studies of information systems source. *MIS Quarterly*, [REMOVED HYPERLINK FIELD]*11*(3), 369-386.

Bennett, J. (1995). *Meeting needs: NGO coordination in practice*. London, UK: Earthscan.

Bennett, J. (1994). *NGO coordination at field level: A handbook*. Oxford, UK: ICVA.

Bereiter, C., & Scardamalia, M. (1993). *Surpassing ourselves: An inquiry into the nature and implications of expertise*. Chicago: Open Court.

Berg, B. (1989). *Qualitative research methods for social sciences*. Boston, MA: Allyn & Bacon.

Berger, A. N., & Humphrey, D. (1992). Measurement and efficiency issues in commercial banking. In Griliches, Z. (Ed.), *Output measurement in the service sector*. Chicago, IL: Chicago University Press.

Berger, A. N., & Mester, J. L. (1997). Beyond the black box: What explains differences in the efficiencies of financial institutions? *Journal of Banking & Finance, 21*, 87–98. doi:10.1016/S0378-4266(97)00010-1

Bergman, M. K. (2001). The deep web: Surfacing hidden value. *Journal of Electronic Publishing, 7*(1). doi:10.3998/3336451.0007.104

Bhattacharyya, A., Lovell, C. A. K., & Sahay, P. (1997). The impact of liberalization on the productive efficiency of Indian commercial banks. *European Journal of Operational Research, 98*, 332–345. doi:10.1016/S0377-2217(96)00351-7

Bhatti, T. R. (2005, September, 26). *Critical success factors for the implementation of enterprise resource planning (ERP) empirical validation*. Paper presented at the Second International Conference on Innovation in Information Technology, Dubai, India.

Biswas, H., & Hasan, M. (2007). Using Publications and Domain Knowledge to Build Research Profiles. In *Information and Communication Technology (ICICT'07)* (pp. 82-86).

Blanchard, P. N., Thacker, J. W., & Way, S. A. (2000). Training evaluation: Perspectives and evidence from Canada. *International Journal of Training and Development, 4*, 295–304. doi:10.1111/1468-2419.00115

Bloom, B. S. (1964). *Taxonomy of educational objectives*. New York, NY, USA: David McKay Company Inc.

Blumer, T. (2009). *Story on new GM, with AP help, buries news about financial non-disclosure, unique risks*. Retrieved from http://newsbusters.org/blogs/tomblumer/2009/08/10/ap-story-new-gm-ap-help-buriesnews-about-financial-non-disclosure-uniqu

Boldrin, M., & Levine, D. K. (2008). Perfectly competitive innovation. *Journal of Monetary Economics, 55*(3), 435–453. doi:10.1016/j.jmoneco.2008.01.008

Boldrin, M., & Levine, D. K. (2009). A model of discovery. *The American Economic Review, 99*(2), 337–342. doi:10.1257/aer.99.2.337

Boldrin, M., & Levine, D. K. (2009). Does intellectual monopoly help innovation? *Review of Law & Economics. Berkeley Electronic Press, 5*(3), 1211–1219.

Boldrin, M., & Levine, D. K. (2009). Market structure and property rights in open source industries. *Washington University Journal of Law and Policy, 30*, 325–363.

Boldrin, M., & Levine, D. K. (2009). Market size and intellectual property protection. *International Economic Review, 50*(3), 855–881. doi:10.1111/j.1468-2354.2009.00551.x

Bonacin, R., Rodrigues, M. A., & Baranauskas, M. C. C. (2009). An agile process model for inclusive software development. In *Proceedings of the 11th International Conference on Enterprise Information Systems* (pp. 807-818).

Bonacin, R., Simoni, C. A. C., Melo, A. M., & Baranauskas, M. C. C. (2006). Organisational semiotics: Guiding a service-oriented architecture for e-government. In *Proceedings of the 9th International Conference on Organisational Semiotics* (pp. 47-58).

Bonacin, R., Melo, A. M., Simoni, C. A. C., & Baranauskas, M. C. C. (2009). Accessibility and interoperability in e-government systems: outlining an inclusive development process. *Universal Access in the Information Society, 9*(1), 17–33. doi:10.1007/s10209-009-0157-0

Borau, K., Ullrich, C., Geng, J., & Shen, R. (2009). Microblogging for Language Learning: Using Twitter to Train Communicative and Cultural Competence. In *Proceedings of the ICWL 2009 conference* (pp. 78-87). Retrieved December 9, 2009, from http://www.carstenullrich.net/pubs/Borau09Microblogging.pdf

Borland, J., & Yang, X. (1995). Specialization, product development, evolution of the institution of the firm, and economic growth. *Journal of Evolutionary Economics*, *5*, 19–42. doi:10.1007/BF01199668

Botan, C., & Vorvoreanu, M. (2005). What do employees think about electronic surveillance at work? In Weckert, J. (Ed.), *Electronic monitoring in the workplace: Controversies and solutions* (pp. 123–144). Hershey, PA: IGI Global.

Bouhnik, S., Golany, B., Passy, S., Hackman, S. T., & Vlatsa, D. A. (2001). Lower bound restrictions on intensities in data envelopment analysis. *Journal of Productivity Analysis*, *16*, 241–261. doi:10.1023/A:1012510605812

Briec, W., Kerstens, K., & Vanden Eeckaut, P. (2004). Non-convex technologies and cost functions: Definitions, duality and nonparametric tests of convexity. *Journal of Economics*, *81*, 155–192. doi:10.1007/s00712-003-0620-y

Briggs, R. O., Adkins, M., Mittleman, D., Kruse, J., Miller, S., & Nunamaker, J. F. (1999). A technology transition model derived from field investigation of GSS use aboard the U.S.S. Coronado. *Journal of Management Information Systems*, *15*(3), 151–195.

Brin, S., & Page, L. (1997). *The anatomy of a large-scale hypertextual web search engine.* Retrieved from http://infolab.stanford.edu/~backrub/google.html

Broberg, H. (2008). Understanding IT-systems in practice: Investigating the potential of activity and actability theory. In *Proceedings of the 5th International Conference on Action in Language, Organisations and Information Systems*, Venice, Italy.

Brooks, F. P. (1995). *The mythical man-month* (Anniversary ed.). Reading, MA: Addison-Wesley.

Brown, E., Brailsford, T., Fisher, T., & Moore, A. (2009). Evaluating learning style personalization in adaptive systems: Quantitative methods and approaches. *IEEE Transactions on Learning Technologies*, *2*(1), 10–22. doi:10.1109/TLT.2009.11

Brusilovsky, P., & Millan, E. (2007). User models for adaptive hypermedia and adaptive educational systems. In P. Brusilovsky, A. Kobsa, & W. Neidl (Eds.), *The Adaptive Web: Methods and Strategies of Web Personalization* (LNCS 4321, pp. 3-53). New York: Springer.

Brusilowsky, P. (2001). Adaptive hypermedia. *Journal of User Modeling and User-Adapted Interaction*, *11*, 87–110. doi:10.1023/A:1011143116306

Brusilowsky, P., & Vassileva, J. (2003). Course sequencing techniques for large-scale web-based education. *Journal of Continuing Engineering Education and Life-long Learning*, *13*, 75–94.

Bryman, A. (1989). *Research methods and organization studies*. London, UK: Unwin Hyman. doi:10.4324/9780203359648doi:10.4324/9780203359648

Buckland, M., & Gey, F. (1994). The relationship between recall and precision. *Journal of the American Society for Information Science American Society for Information Science*, *45*(1), 12–19. doi:10.1002/(SICI)1097-4571(199401)45:1<12::AID-ASI2>3.0.CO;2-L

Bui, T., Cho, S., Sankaran, S., & Sovereign, M. (2000). A framework for designing a global information network for multinational humanitarian assistance/disaster relief. *Information Systems Frontiers*, *1*(4), 427–442. doi:10.1023/A:1010074210709

Burbridge, L. C., & Nightingale, D. S. (1989). *Local coordination of employment and training services to welfare recipients*. Washington, DC: The Urban Institute.

Burden, K., & Atkinson, S. (2008). Evaluating pedagogical affordances of media sharing Web 2.0 technologies: A case study. In *Proceedings of the ascilite*, Melbourne, Australia. Retrieved December 14, 2009, from http://www.ascilite.org.au/conferences/melbourne08/procs/burden-2.pdf

Burgress, D. F. (1974). A cost minimization approach to import demand equations. *The Review of Economics and Statistics*, *56*, 224–234.

Buss, D. (2009). Museums: Rise and fall of a car town. *Wall Street Journal (Eastern Edition)*, p. D7.

Callon, M., & Muniesa, F. (2003). Les marchés économiques comme dispositifs collectifs de calcul. *Reseaux*, *122*, 189–233.

Canzi, A., Folcio, A., Milani, M., Radice, S., Santangelo, E., & Zanoni, E. (2003). *The management of flows of distance communication between tutors and students in the context of an English course in blended learning held at the Università degli Studi of Milan*. Retrieved December 9, 2009, from http://www.ctu.unimi.it/pdf/Edmedialucerna_2003.pdf

Carver, C. A., Howard, R. A., & Lane, W. D. (1999). Enhancing student learning through hypermedia courseware and incorporation of student learning styles. *IEEE Transactions on Education*, *42*, 33–38. doi:10.1109/13.746332

Castano, S., Ferrara, A., & Montanelli, S. (2006). Matching ontologies in open networked systems: Techniques and applications. *Journal on Data Semantic*, *5*, 25–63.

Centrality. (n.d.). In *Wikipedia, The Free Encyclopedia*.

CETIC.BR. (2009). *Proportion of Brazilian households with Internet access*. Retrieved from http://www.cetic.br/usuarios/tic/2008-total-brasil/rel-geral-04.htm

CETIC.BR. (2009). *Proportion of Brazilian households with computers*. Retrieved from http://www.cetic.br/usuarios/tic/2008-total-brasil/rel-geral-01.htm

CETIC.BR. (2009). *Proportion of Brazilian individuals who used cell phone in the last three months*. Retrieved from http://www.cetic.br/usuarios/tic/2008-total-brasil/rel-semfio-01.htm

Cha, H. J., Kim, Y. S., Lee, J. H., & Yoon, T. B. (2006). An adaptive learning system with learning style diagnosis based on interface behaviors. In *Proceedings of Intl. Conf. E-learning and Games (Edutainment)*, Hangzhou, China.

Chai, C. S., & Tan, S. C. (2005). Fostering learning communities among teachers and students: potentials and issues. In Looi, C. K., (Eds.), *Towards sustainable and scalable educational innovations informed by the learning sciences*. Amsterdam, The Netherlands: IOS Press.

Chamberlin, E. H. (1933). *The theory of monopolistic competition: A reorientation of the theory of value*. Cambridge, MA: Harvard University Press.

Chang, R. L., Stern, L., Sondergaard, H., & Hadgraft, R. (2009). Places for learning engineering: A preliminary report on informal learning spaces. In *Proceedings of the Research in Engineering Education Symposium*, Palm Cove, QLD. Retrieved December 14, 2009, from http://rees2009.pbworks.com/f/rees2009_submission_86.pdf

Checkland, P., & Scholes, J. (1991). *Soft systems methodology in action*. New York, NY: John Wiley & Sons.

Che, H., & Seetharaman, P. (2009). Speed of Replacement: Modeling brand loyalty using last-move data. *JMR, Journal of Marketing Research*, *46*(4), 494–505. doi:10.1509/jmkr.46.4.494

Cheng, Y., & Ku, H. (2009). An investigation of the effects of reciprocal peer tutoring. *Computers in Human Behavior*, 25.

Cherchye, L., Kuosmanen, T., & Post, T. (2000a). What is the economic meaning of FDH? A reply to Thrall. *Journal of Productivity Analysis*, *13*, 259–263. doi:10.1023/A:1007827126369

Cherchye, L., Kuosmanen, T., & Post, T. (2000b). *Why convexify? An assessment of convexity axioms in DEA*. Helsinki, Finland: Helsinki School of Economics and Business Administration.

Chifu, V. R., Salomie, I., & Chifu, E. S. (2007). Taxonomy learning for semantic annotation of web services. In *Proceedings of the 11th WSEAS International Conference on Computers*, Crete, Greece (pp. 300-305).

Christensen, L. R., & Green, W. H. (1976). Economies of scale in U.S. electric power generation. *The Journal of Political Economy*, *84*, 655–676. doi:10.1086/260470

Christensen, L. R., Jorgenson, D. W., & Lau, L. J. (1971). Conjugate duality and the transcendental logarithmic production function. *Econometrica*, *39*, 255–256.

Christensen, L. R., Jorgenson, D. W., & Lau, L. J. (1973). Transcendental logarithmic production frontiers. *The Review of Economics and Statistics*, *55*, 28–45. doi:10.2307/1927992

Clements-Croome, D., Noy, P., & Liu, K. (2006). *Occupant behaviour analysis.* Retrieved from http://www.serg.soton.ac.uk/idcop/outcomes/IDCOP_Scientific_Report_OccupantBehaviour.pdf

Coffield, F., Moseley, D., Hall, E., & Ecclestone, K. (2004). *Learning styles and pedagogy in post-16 learning. A systematic and critical review.* London: Learning and Skills Research Centre.

Cohen, W. M., & Levin, R. C. (1989). Empirical studies of innovation and market structure. In Schmalensee, R., & Willig, R. D. (Eds.), *Handbook of industrial organization* (*Vol. 2*, pp. 1059–1107). Amsterdam, The Netherlands: Elsevier.

Cohen, W. M., & Levinthal, D. A. (1990). Absorptive capacity: A new perspective on learning and innovation. *Administrative Science Quarterly*, *35*(1), 128–152. doi:10.2307/2393553

Cole, J. H. (2001). Patents and copyrights: Do the benefits exceed the costs? *The Journal of Libertarian Studies*, *15*(4), 79–105.

Comfort, L. K. (1990). Turning conflict into co-operation: organizational designs for community response in disasters. *International Journal of Mental Health*, *19*(1), 89–108.

Comfort, L. K. (1993). Integrating information technology into international crisis management and policy. *Journal of Contingencies and Crisis Management*, *1*(1), 15–26. doi:10.1111/j.1468-5973.1993.tb00003.x

Comfort, L. K., & Kapucu, N. (2006). Inter-organizational coordination in extreme events: The World Trade Center attacks, September 11, 2001. *Natural Hazards*, *39*(2), 309–327. doi:10.1007/s11069-006-0030-x

Comfort, L. K., Sungu, Y., Johnson, D., & Dunn, M. (2001). Complex systems in crisis: Anticipation and resilience in dynamic environments. *Journal of Contingencies and Crisis Management*, *9*(3), 144–158. doi:10.1111/1468-5973.00164

Conner, M. (2009). *Marcia Conner Web site.* Retrieved December 12, 2009, from http://marciaconner.com/

Conrad, R. M. (2009). Assessing collaborative learning. In Rogers, P., (Eds.), *Encyclopedia of distance learning* (pp. 89–93). Hershey, PA: IGI Global.

Cooper, W. W., Thompson, R. G., & Thrall, R. M. (1996). Introduction: Extensions and new developments in DEA. *Annals of Operations Research*, *66*, 3–46. doi:10.1007/BF02125451

Costa, P. D. P., De Martino, J. M., & Nagle, E. J. (2009). Speech synchronized image-based facial animation. In *Proceedings of the International Workshop on Telecommunications*, São Paulo, Brazil (pp. 235-241).

Costa, C. (2010). Lifelong learning in Web 2.0 environments. *International Journal of Technology Enhanced Learning*, *2*(3), 275–284. doi:10.1504/IJTEL.2010.033582

Costello, R. L. (2006). *The camera ontology.* Retrieved from http://protege.cim3.net/file/pub/ontologies/camera/

Cotler, J., & Rizzo, J. (2010). Designing value sensitive social networks for the future. *Journal of Computing Sciences in Colleges*, *25*(6), 40–46.

Cranmer, S. (2006). Children and young people's uses of the internet for homework' learning. *Media and Technology*, *31*(3), 301–315. doi:10.1080/17439880600893358 doi:10.1080/17439880600893358

Creswell, W. J. (1998). *Quality inquiry and research design.* Thousand Oaks, CA: Sage.

Cronen, V. (2001). Practical theory, practical art, and the pragmatic-systemic account of inquiry. *Communication Theory*, *11*(1), 14–35. doi:10.1111/j.1468-2885.2001.tb00231.x

Cronholm, S., & Goldkuhl, G. (2002). Actable information systems - Quality ideals put into practice. In *Proceedings of the 11th International Conference on Information Systems Development*, Riga, Latvia.

Cronholm, S., & Goldkuhl, G. (2003). Strategies for information systems evaluation – Six generic types. *Electronic Journal of Information Systems Evaluation*, *6*(2).

Crooks, A. (2001). Professional development and the JET program: Insights and solutions based on the Sendai City Program. *JALT journal*, *23*, 31-46.

Crowston, K. (1994). *A taxonomy of organizational dependencies and coordination mechanisms.* Cambridge, MA: MIT Center for Coordination Science.

Crowston, K. (1997). A coordination theory approach to organizational process design. *Organization Science*, *8*(2), 157–175. doi:10.1287/orsc.8.2.157

Cunningham, B., Nikolai, L., & Bazley, J. (2004). *Accounting information for business decisions* (2nd ed.). Australia: Thomson.

Curbera, F., Duftler, M., Khalaf, R., Nagy, W., Mukhi, N., & Weerawarana, S. (2002). Unraveling the web services web: An introduction to SOAP, WSDL, and UDDI. *IEEE Internet Computing*, *6*(2), 86–93. doi:10.1109/4236.991449

Daly, H. E., & Cobb, J. B. (1989). *For the common good*. Boston, MA: Beacon Press.

Damanpour, F. (1991). Organizational innovation: a meta-analysis of effects of determinants and moderators. *Academy of Management Journal*, *34*, 555–590. doi:10.2307/256406

DAML. (2003). *Ontology of book*. Retrieved from http://www.daml.org

Dandapani, K. (2004). Success and failure in web-based financial services. *Communications of the ACM*, *47*(5), 31–33. doi:10.1145/986213.986233doi:10.1145/986213.986233

Davenport, T. H. (1998). Putting the enterprise into the enterprise system. *Harvard Business Review*, 3–13.

Davis, F. D. (1986). *Technology acceptance model for empirically testing new end-user information systems theory and results*. Unpublished doctoral dissertation, MIT, Cambridge, MA.

Davis, F. D. (1989). Perceived usefulness, perceived ease of use, and user acceptance of information technology. *Management Information Systems Quarterly*, *13*(3), 319–340. doi:10.2307/249008doi:10.2307/249008

Davis, F. D., Bagozzi, R. P., & Warshaw, P. R. (1989). User acceptance of computer technology: A comparison of two theoretical models. *Management Science*, *35*, 36–48. doi:10.1287/mnsc.35.8.982

Davis, F. D., Bagozzi, R. P., & Warshaw, P. R. (1992). Extrinsic and intrinsic motivation to use computers in the workplace. *Journal of Applied Social Psychology*, *22*, 1111–1132. doi:10.1111/j.1559-1816.1992.tb00945.x

Davis, G. B. (2002). Anytime/anyplace computing and the future of knowledge work. *Communications of the ACM*, *45*(12), 67–73. doi:10.1145/585597.585617

Dawes, S., Cresswel, A., & Cahan, B. (2004). Learning from crisis: Lessons in human and information infrastructure from the world trade center response. *Social Science Computer Review*, *22*(1), 52–66. doi:10.1177/0894439303259887

DCMI. (2009). *Dublin Core Metadata Initiative*. Retrieved from http://dublincore.org

De Alessi, L. (1980). The economics of property rights: A review of the evidence. In Richard, O. Z. (Ed.), *Research in law and economics: A research annual* (*Vol. 2*, pp. 1–47). Greenwich, CT: JAI Press.

De Bruijn, H. (2006). One fight, one team: The 9/11 commission report on intelligence, fragmentation and information. *Public Administration*, *84*(2), 267–287. doi:10.1111/j.1467-9299.2006.00002.x

De Marsico, M., Sterbini, A., & Temperini, M. (2010). Tunnelling Model between Adaptive e-Learning and Reputation-based Social Activities. Proc *21st International Conference on Database and Expert Systems Applications*, DEXA 2010, *3rd International Workshop on Social and Personal Computing for Web-Supported Learning Communities*, SPeL 2010, Aug.30th-Sept.3 2010, Bilbao, Spain, A Min Tjoa, R. Wagner (Eds.), IEEE Computer Society.

De Marsico, M., Sterbini, A., & Temperini, M. (2011). The Definition of a Tunneling Strategy between Adaptive Learning and Reputation-based Group Activities. In Proc. *11th IEEE International Conference on Advanced Learning Technologies* (ICALT 2011), July 6-8, 2011, Athens, Georgia, USA. IEEE Computer Society, Los Alamitos.

de Moor, A., & Weigand, H. (2002). Towards a semiotic communication quality model. In Liu, K., Clarke, R. J., Andersen, P. B., & Stamper, R. K. (Eds.), *Orgnizational semiotics: Evolving a science of information systems* (pp. 275–285). Amsterdam, The Netherlands: Kluwer Academic.

De Souza, C. S. (2005). *The semiotic engineering of human-computer interaction*. Cambridge, MA: MIT Press.

De Vreede, G., Jones, N., & Mgaya, R. J. (1999). Exploring the application and acceptance of group support systems in Africa. *Journal of Management Information Systems*, *15*(3), 197–234.

DeLone, W. H., & McLean, E. R. (1992). Information systems success: The quest for the dependent variable. *Information Systems Research*, *3*(1), 60–95. doi:10.1287/isre.3.1.60doi:10.1287/isre.3.1.60

DeLone, W. H., & McLean, E. R. (2003). The DeLone and McLean model of information systems success: A ten-year update. *Journal of Management Information Systems*, *19*(4), 9–30.

DeLone, W., & McLean, W. (1992). Information systems success: The quest for the dependant variable. *Information Systems Research*, *3*(1), 60–95. doi:10.1287/isre.3.1.60

Deprins, D., Simar, L., & Tulkens, H. (1984). Measuring labor efficiency in post offices. In Marchand, M., Pestieau, P., & Tulkens, H. (Eds.), *The performance of public enterprises: Concepts and measurements*. Amsterdam, The Netherlands: North-Holland.

Despotis, D. K. (2005). A reassessment of the human development index via data envelopment analysis. *The Journal of the Operational Research Society*, *56*, 969–980. doi:10.1057/palgrave.jors.2601927

Devereux, M. B., Head, A. C., & Lapham, B. J. (1996). Monopolistic competition, increasing returns, and the effects of government spending. *Journal of Money, Credit and Banking*, *28*, 233–254. doi:10.2307/2078025

Dewar, R. J. (2009). *A savage factory: An eyewitness account of the auto industry's self-destruction* (pp. 74–98). Bloomington, IN: AuthorHouse.

Diewert, W. E., & Fox, K. J. (2008). On the estimation of returns to scale, technical progress and monopolistic markups. *Journal of Economics*, *145*, 174–193.

Dixton, B., Garcia, P., & Anderson, M. (1987). Usefulness of pretests for estimating underlying technologies using dual profit functions. *International Economic Review*, *28*, 623–633. doi:10.2307/2526570

Doctorow, C. (2003) *Down and out in the magic kingdom*. Tor Books. ISBN 0765304368. Under licence Creative Commons, Attribution-NonCom-mercial-ShareAlike, 2004. Retrieved March, 5th 2010, from http://www/craphound.com

Donini, A., & Niland, N. (1999). *Rwanda, lessons learned: A report on the coordination of humanitarian activities*. Retrieved from http://www.grandslacs.net/doc/2404.pdf

Donini, A. (1996). The Bureaucracy and the free spirits: Stagnation and innovation in the relationship between the UN and NGOs. In Weiss, T. G., & Gordenker, L. (Eds.), *NGOs, the UN and Global Governance*. London, UK: Lynne Rienner Publishers.

Downes, S. (2005). E-Learning 2.0. *Stephen's Web*. Retrieved December 9, 2009, from http://www.downes.ca/post/31741

Dumai, S. T., & Nielsen, J. (1992). Automating the assignment of submitted manuscripts to reviewers. In *Proceedings of the Annual ACM Conference on Research and Development in Information Retrieval*.

Duo, Z., Juan-Zi, L., & Bin, X. (2005). Web service annotation using ontology mapping. In *Proceedings of the IEEE International Workshop on Service-Oriented System Engineering* (pp. 235-242).

Duval, E. (2007). *Snowflakes Effect: Open learning without boundaries*. Retrieved December 9, 2009, from http://ariadne.cs.kuleuven.be/mediawiki2/index.php/SnowflakeEffect

e-Cidadania. (2008). *E-cidadania project: System and methods for the constitution of a culture mediated by information and communication technology*. Retrieved from http://www.nied.unicamp.br/ecidadania

Edmund, M. (2009). The big three: Will a bailout be enough? *Quality Progress*, *42*(1), 14–15.

Ehrig, M., & Euzenat, J. (2005). Relaxed precision and recall for ontology matching. In *Proceedings of the K-CAP Workshop on Integrating Ontologies* (pp. 25-32).

Ehrig, M., & Sure, Y. (2005). *Test ontologies*. Retrieved from http://www.aifb.uni-karlsruhe.de/WBS/meh/foam/ontologies.htm

Elbeltagi, I., McBride, N., & Hardaker, G. (2005). Evaluating the factors affecting DSS usage by senior managers in local authorities in Egypt. *Journal of Global Information Management, 13*(2), 42–65. doi:10.4018/jgim.2005040103doi:10.4018/jgim.2005040103

Elkiss, A., Shen, S., Fader, A., States, D., & Radev, D. (2008). Blind men and elephants: what do citation summaries tell us about a research article. *Journal of the American Society for Information Science and Technology, 59.*

Ellis, A., & Calvo, A. (2007). Minimum Indicators to Assure Quality of LMS-supported Blended Learning. *Educational Technology and Society, 10*(2), 60-70. Retrieved December 14, 2009, from http://www.ifets.info/journals/10_2/6.pdf

Emrouznejad, A., Parker, B. R., & Tavares, G. (2008). Evaluation of research in efficiency and productivity: A survey and analysis of the first 30 years of scholarly literature in DEA. *Socio-Economic Planning Sciences, 42,* 151–157. doi:10.1016/j.seps.2007.07.002

Enjolras, B. (2008). A governance-structure approach to voluntary organizations. *Nonprofit and Voluntary Sector Quarterly, 20*(10).

Epstein, L., & Martin, A. (2005). *Coding variables.* London, UK: Academic Press.

Euzenat, J. (2004). An API for ontology alignment. In S. McIlraith, D. Plexousakis, & F. van Harmelen (Eds.), *Proceedings of the Third International Semantic Web Conference* (LNCS 3298, pp. 698-712).

Euzenat, J., & Shvaiko, P. (2007). *Ontology matching.* New York, NY: Springer.

Evaghorou, E. L. (2010). The state strategy in today's global economy: A reset position in the theory of international political economy. *International Journal of Society Systems Science, 2*(3), 297–311. doi:10.1504/IJSSS.2010.033496

Fanselow, J. F. (1994). JET as an exercise in program analysis. In Wada, M., & Cominos, A. (Eds.), *Studies in team teaching* (pp. 201–216). Tokyo, Japan: Kenkyusha.

Faraj, S., & Xiao, Y. (2006). Coordination in fast-response organizations. *Management Science, 52*(8), 155–189. doi:10.1287/mnsc.1060.0526

Färe, R., & Grosskopf, S. (1985). A nonparametric cost approach to scale efficiency. *The Scandinavian Journal of Economics, 87,* 594–604. doi:10.2307/3439974

Färe, R., Grosskopf, S., & Lovell, C. A. K. (1985). *The measurement of efficiency of production.* Boston, MA: Kluwer Academic.

Färe, R., Grosskopf, S., & Lovell, C. A. K. (1988). Scale elasticity and scale efficiency. *Journal of Institutional and Theoretical Economics, 144,* 721–729.

Farrell, J. N. (2000). Long live c-learning: The advantages of the classroom. *Training & Development, 54*(9), 43–46.

Farrell, M. J. (1959). Convexity assumptions in theory of competitive markets. *The Journal of Political Economy, 67,* 377–391. doi:10.1086/258197

Feier, C., Roman, D., Polleres, A., Domingue, J., & Fensel, D. (2005). Towards intelligent web services: The web service modeling ontology. In *Proceedings of the International Conference on Intelligent Computing,* Hefei, China. (pp. 1-10).

Felder, R. M., & Silverman, L. K. (1988). Learning and teaching styles in engineering education. *Engineering Education, 78*(7), 674-681. Retrieved from http://www4.ncsu.edu/unity/lockers/users/f/felder/public/Papers/LS-1988.pdf

Fenton-Smith, B. (2000). Foreign teachers in Japanese secondary schools: Why aren't they happier? *Kanda university of international studies, 12,* 409-426.

Fernandez, G., Sterbini, A., & Temperini, M. (2007). On the specification of learning objectives for course configuration. In *Proc. Int. Conf. on Web-Based Education (WBE07).*

Ferrier, G., & Lovell, C. A. K. (1990). Measuring cost efficiency in banking: Econometric and linear programming evidence. *Journal of Econometrics, 46,* 229–245. doi:10.1016/0304-4076(90)90057-Z

Ferry, B., Kiggins, J., Hoban, G., & Lockyer, L. (2000). Using computer-mediated communication to form a knowledge-building community with beginner teachers. *Journal of Educational Technology & Society, 3*(3).

Fishbein, M., & Ajzen, I. (1975). *Belief, attitude, intention and behavior: An introduction to theory and research.* Reading, MA: Addison-Wesley.

Fisher, C. W., & Kingma, D. R. (2001). Criticality of data quality as exemplified in two disasters. *Information & Management, 39*(2), 109–116. doi:10.1016/S0378-7206(01)00083-0

Førsund, F. R. (1996). On the calculation of scale elasticity in DEA models. *Journal of Productivity Analysis, 7,* 283–302. doi:10.1007/BF00157045

Førsund, F. R., & Hjalmarsson, L. (2004). Calculating scale elasticity in DEA models. *The Journal of the Operational Research Society, 55,* 1023–1038. doi:10.1057/palgrave.jors.2601741

Førsund, F. R., Hjalmarsson, L., Krivonozhko, V., & Utkin, O. B. (2007). Calculation of scale elasticities in DEA models: Dsirect and indirect approaches. *Journal of Productivity Analysis, 28,* 45–56. doi:10.1007/s11123-007-0047-5

Foster-Fishman, P. G., Salem, D. A., & Allen, N. A. (2001). Facilitating inter-organization collaboration: the contribution of inter-organizational alliances. *American Journal of Community Psychology, 29*(6), 875–905. doi:10.1023/A:1012915631956

Friedman, B. K. (2006). Value sensitive design and information systems. *Human-Computer Interaction and Management Systems, 6,* 348–372.

Frisch, R. (1965). *Theory of production.* Dordrecht, The Netherlands: D. Reidel Publishing.

Fukuyama, H. (2000). Returns to scale and scale elasticity in data envelopment analysis. *European Journal of Operational Research, 125,* 93–112. doi:10.1016/S0377-2217(99)00200-3

Fukuyama, H. (2001). Returns to scale and scale elasticity in Farrell, Russell and additive models. *Journal of Productivity Analysis, 16,* 225–239. doi:10.1023/A:1012558521742

Fukuyama, H. (2003). Scale characterizations in a DEA directional technology distance function framework. *European Journal of Operational Research, 144,* 108–127.

Furnas, G., Landauer, T., Gomez, L., & Dumais, S. (1987). The vocabulary problem in human-system communication. *Communications of the ACM, 30*(11), 971. doi:10.1145/32206.32212

Furukawa, Y. (2007). The protection of intellectual property rights and endogenous growth: Is stronger always better? *Journal of Economic Dynamics & Control, 31*(11), 3644–3670. doi:10.1016/j.jedc.2007.01.011

Gable, G. G., Sedara, D., & Chan, T. (2003). Enterprise systems success: A measurement model. In *Proceedings of the Twenty Fourth International Conference on Information Systems,* Seattle, WA (pp. 576-591).

Galbraith, J. R. (1977). *Organization design.* Reading, MA: Addison-Wesley.

Gallini, N., & Scotchmer, S. (2001). Intellectual property: When is it the best incentive system? In Jaffe, A., Lerner, J., & Stern, S. (Eds.), *Innovation policy and the economy (Vol. 2).* Cambridge, MA: MIT Press.

Gardner, H. (1995). Reflections on multiple intelligences: Myths and messages. *Phi Delta Kappan, 77*(3), 200–209.

Gauld, R. (2007). Public sector information system project failures: Lessons from a New Zealand hospital organization. *Government Information Quarterly, 24*(1), 102–114. doi:10.1016/j.giq.2006.02.010doi:10.1016/j.giq.2006.02.010

Gefen, D. (2003). TAM or just plain habit: A look at experienced online shoppers. *Journal of End User Computing, 15*(3), 1–13. doi:10.4018/joeuc.2003070101doi:10.4018/joeuc.2003070101

Gibson, J. J. (1979). *The ecological approach to visual perception.* Boston, MA: Houghton Mifflin.

Gillis-Furutaka, A. (1994). Pedagogical preparation for JET programme teachers. In Wada, M., & Cominos, A. (Eds.), *Studies in team teaching* (pp. 29–41). Tokyo, Japan: Kenkyusha.

Girju, R., Moldovanb, D., Tatub, M., & Antoheb, D. (2005). On the semantics of noun compounds. *Computer Speech & Language, 19*(4), 479–496.

Giunchiglia, F., Yataskevich, M., Avesani, P., & Shvaiko, P. (2009). A large dataset for the evaluation of ontology matching. *The Knowledge Engineering Review*, *24*(2), 137–157. doi:10.1017/S026988890900023X

Glenn, E. S., & Glenn, C. G. (1981). *Man and mankind: Conflict and communication between cultures*. Norwood, NJ: Ablex.

Goldkuhl, G. (2008a). Practical inquiry as action research and beyond. In *Proceedings of the 16th European Conference on Information Systems*, Galway, Ireland.

Goldkuhl, G. (2008b). Actability theory meets affordance theory: Clarifying HCI in IT usage situations. In *Proceedings of the 16th European Conference on Information Systems*, Galway, Ireland.

Goldkuhl, G., Cronholm, S., & Sjöström, J. (2004). User interfaces as organisational action media. In *Proceedings of the 7th International Workshop on Organisational Semiotics*, Setúbal, Portugal.

Goldkuhl, G. (2004). Design theories in information systems – a need for multi-grounding. *Journal of Information Technology Theory and Application*, *6*(2), 59–72.

Goldkuhl, G. (2009). Information systems actability - tracing the theoretical roots. *Semiotica*, (175): 379–401. doi:10.1515/semi.2009.054

Goldkuhl, G., & Ågerfalk, P. J. (2002). Actability: A way to understand information systems pragmatics. In Liu, K., Clarke, R. J., Andersen, P. B., & Stamper, R. K. (Eds.), *Coordination and communication using signs: Studies in organisational semiotics – 2*. Boston, MA: Kluwer Academic.

Goldsmiths University London Blog. Learning Technologist jottings at Goldsmiths Blog. (2009). *The VLE is Dead. Debate at ALT-C 2009*. Retrieved December 9, 2009, from http://celtrecord.wordpress.com/2009/09/08/the-vle-is-dead-debate-at-alt-c-2009/

Goodwin, N., Nelson, J. A., Ackerman, F., & Weisskopf, T. (2005). *Microeconomics in context*. Boston, MA: Houghton Mifflin.

Graf, S., Lan, C. H., Liu, T. C., & Kinshuk. (2009). Investigations about the effects and effectiveness of adaptivity for students with different learning styles. In *Proceedings of ICALT 2009* (pp. 415-419). Washington, DC: IEEE Computer Society Press.

Grandori, A. (1997). An organizational assessment of inter-firm coordination modes. *Organization Studies*, *18*(6), 897–925. doi:10.1177/017084069701800601

Grassé, P. P. (1959). La reconstruction du nid et les coordinations interindividuelles chez bellicositermes natalensis et cubitermes sp. la théorie de la stigmergie: Essai d'interprétation du comportement des termites constructeurs. *Insectes Sociaux*, *6*(1). doi:10.1007/BF02223791

Gratton, L., & Erickson, T. (2007). Eight ways to build collaborative teams. *Harvard Business Review*, 100–109.

Gray, B. (2004). Informal Learning in an Online Community of Practice. *Journal of Distance Education*, *19*(1), 20–35.

Greenfield, G., & Rohde, F. (2009). Technology acceptance: Not all organisations or workers may be the same. *International Journal of Accounting Information Systems*, *10*, 263–272. doi:10.1016/j.accinf.2009.10.001

Griliches, Z., & Ringstad, V. (1971). *Economies of scale and the form of the production function*. Amsterdam, The Netherlands: North-Holland.

Guilkey, D. K., & Lovell, C. A. K. (1980). On the flexibility of the translog approximation. *International Economic Review*, *21*, 137–147. doi:10.2307/2526244

Guo, C., & Acar, M. (2005). Understanding collaboration among nonprofit organizations: Combining resource dependency, institutional, and network perspectives. *Nonprofit and Voluntary Sector Quarterly*, *34*(3), 340–361. doi:10.1177/0899764005275411

Hadjicostas, P., & Soteriou, A. C. (2006). One-sided elasticities and technical efficiency in multi-output production: A theoretical framework. *European Journal of Operational Research*, *168*, 425–449. doi:10.1016/j.ejor.2004.05.008

Halonen, R., Thomander, H., & Laukkanen, E. (2010). DeLone & McLean IS Success Model in Evaluating Knowledge Transfer in a Virtual Learning Environment. [IJISSC]. *International Journal of Information Systems and Social Change, 1*(2).

Hanoch, G. (1970). Homotheticity in joint production. *Journal of Economic Theory, 2,* 423–426. doi:10.1016/0022-0531(70)90024-4

Hanoch, G., & Rothschild, M. (1972). Testing assumptions in production theory: A nonparametric approach. *The Journal of Political Economy, 80,* 256–275. doi:10.1086/259881

Harpviken, K. B., Millard, A. S., Kjellman, K. E., & Strand, A. (2001). *Sida's contributions to humanitarian mine action: Final report* (Tech. Rep. No. 01/06). Stockholm, Sweden: Swedish International Development Cooperation System.

Hartshorne, C., & Weiss, P. (1960). *Collected papers of C. S. Peirce (1931-1935)*. Cambridge, MA: Harvard University Press.

Hartung, A. (2009). *The fall of GM: What went wrong and how to avoid its mistakes* (pp. 1-15). Upper Saddle River, NJ: Financial Times Press.

Hartvigsen, D. (1999). The Conference Paper-Reviewer Assignment Problem. *Decision Sciences, 30*(3), 865–876. doi:10.1111/j.1540-5915.1999.tb00910.x

Hasencamp, G. (1976). A study of multiple-output production functions. *Journal of Econometrics, 4,* 253–262. doi:10.1016/0304-4076(76)90036-1

Hayashi, E. C. S., & Baranauskas, M. C. C. (2009). Communication and expression in social networks: Getting the "making common" from people. In *Proceedings of the IEEE Latin American Web Congress, Joint LA-WEB/CLIHC Conference* (pp. 131-137).

Hayes, G. R., Poole, E. S., Iachello, G., Patel, S. N., Grimes, A., & Abowd, G. D. (2007). Physical, social, and experiential knowledge in pervasive computing environments. *Pervasive Computing, 6*(4), 56–63. doi:10.1109/MPRV.2007.82

He, B., Macdonald, C., Ounis, I., Peng, J., & Santos, R. (2008). University of Glasgow at TREC 2008: Experiments in blog, enterprise, and relevance feedback tracks with terrier. In *Proceedings of the 17th text retrieval conference, Gaithersburg,* MD.

Heeks, R., & Bhatnagar, S. (1999). Understanding success and failure in information age reform. In R. Heeks (Ed.), *Reinventing government in the information age* (pp. 49–74). London, UK: Routledge. doi:10.4324/9780203204962doi:10.4324/9780203204962

Helpman, E. (1993). Innovation, imitation, and intellectual property rights. *Econometrica, 61*(6), 1247–1280. doi:10.2307/2951642

Hendler, J., Shadbolt, N., Hall, W., Berners-Lee, T., & Weitzner, D. (2008). Web science: An interdisciplinary approach to understanding the web. *Communications of the ACM, 51*(7), 60–69. doi:10.1145/1364782.1364798

Hepp, M. (2006). Semantic web and semantic web services: Father and son or indivisible twins? *IEEE Internet Computing, 10*(2), 85–88. doi:10.1109/MIC.2006.42

Heß, A., & Kushmerick, N. (2003). Learning to attach semantic metadata to web services. In *Proceedings of the Semantic Web Conference* (pp. 258-273).

Heß, A., Johnston, E., & Kushmerick, N. (2004). ASSAM: A tool for semi-automatically annotating semantic web services. In *Proceedings of the Semantic Web Conference* (pp. 320-334).

Hettich, S., & Pazzani, M. (2006). Mining for proposal reviewers: lessons learned at the national science foundation. In *Proceedings of the 12th ACM SIGKDD international conference on Knowledge discovery and data mining* (pp. 862-871). New York: ACM.

Hjalmarsson, L., Kumbhakar, S. C., & Heshmati, A. (1996). DEA, DFA and SFA: A comparison. *Journal of Productivity Analysis, 7,* 303–328. doi:10.1007/BF00157046

Hofstede, G. (2001). *Culture's consequences* (2nd ed.). Thousand Oaks, CA: Sage.

Holstein, W. (2009). *Why GM matters: Inside the race to transform an American icon.* New York, NY: Walker Books.

Honey, P., & Mumford, A. (2000). *The learning styles helper's guide*. Maidenhead, UK: Peter Honey Publications Ltd.

Howcroft, B., Hamilton, R., & Hewer, P. (2002). Consumer attitude and the usage and adoption of home-based banking in the United Kingdom. *International Journal of Bank Marketing, 20*(2-3), 111–121. doi:10.1108/026523 20210424205doi:10.1108/02652320210424205

Huang, L., Lu, M., & Wong, B. K. (2003). The impact of power distance on email acceptance: Evidence from the PRC. *Journal of Computer Information Systems, 44*(1), 93–101.

Hylton, K. N. (2003). *Antitrust law: Economic theory and common law evolution*. Cambridge, UK: Cambridge University Press. doi:10.1017/CBO9780511610158

Igbaria, M., Iivari, J., & Maragahh, H. (1995). Why do individuals use computer technology? A Finnish case study. *Information & Management, 29*, 227–239. doi:10.1016/0378-7206(95)00031-0

Information Retrieval (*n.d.*). Wikipedia, The Free Encyclopedia.

Ingrassia, P. (2010, April 23). Two cheers for General Motors; Yes, it's paid back $6 billion and is more efficient. It still owes taxpayers about $52 billion. *Wall Street Journal (Online Edition)*, p. A17.

Intille, S. S., Larson, K., Beaudin, J., Tapia, E. M., Kaushik, P., Nawyn, J., et al. (2005). The PlaceLab: A live-in laboratory for pervasive computing research [Video]. In *Proceedings of the Pervasive Video Program*.

Intille, S. S., Tapia, E. M., Rondoni, J., Beaudin, J., Kukla, C., Agarwal, S., et al. (2003). Tools for studying behavior and technology in natural settings. In *Proceedings of the Conference on Ubiquitous Computing*.

Ireland, N., & Stoneman, P. (1986). Technological diffusion, expectations and welfare. *Oxford Economic Papers, 38*(2), 283–304.

Ivanova, M. (2008). Knowledge Building and Competence Development in eLearning 2.0 Systems. In *Proceedings of the I-KNOW '08 conference*, Graz, Austria (pp. 84-91).

Ivanova, M. (2009). Use of Start Pages for Building a Mashup Personal Learning Environment to Suport Self-Organized Learners. *Serdica Journal of Computing*, 227-238.

Jacek, K., Tomas, V., Carine, B., & Joel, F. (2007). SAWS-DL: Semantic annotations for WSDL and XML schema. *IEEE Internet Computing, 11*(6), 60–67. doi:10.1109/MIC.2007.134

Jiang, J., Lu, W., & Zhao, H. (2008). CSIR at TREC 2008 Expert Search Task: Modeling Expert Evidence in Expert Search. In *Proceedings of the 2008 Text REtrieval Conference (TREC 2008)*, Gaithersburg, MD.

Jin, F., Niu, Z., Zhang, Q., Lang, H., & Qin, K. (2008). A user reputation model for DLDE learning 2.0 community. In Buchanan, G., Masoodian, M. and Cunningham, S. J. (Eds) *Proc. 11th Int. Conf. on Asian Digital Libraries: Universal and Ubiquitous Access To information*, Dec. 02 - 05, 2008, Bali, Indonesia. Lecture Notes In Computer Science, 5362, 61-70. Berlin, Heidelberg, Springer-Verlag.

Johnson, L., Levine, A., & Smith, R. (2009). *The Horizon Report*. Austin, TX: The New Media Consortium. Retrieved December 9, 2009, from http://www.nmc.org/pdf/2009-Horizon-Report.pdf

Jokisalo, E., & Riu, A. (2009). *Informal learning in the era of Web 2.0*. Retrieved December 14, 2009, from http://www.elearningeuropa.info/files/media/media19656.pdf

Jones, C. I. (2004). *Growth and ideas*. Cambridge, MA: National Bureau of Economic Research.

Jonsson, K. (2006). The embedded panopticon: Visibility issues of remote diagnostics surveillance. *Scandinavian Journal of Information Systems, 18*(2), 7–28.

Jung, Y., & Lee, A. (2000). Design of a social interaction environment for electronic marketplaces. In *Proceedings of the 3rd Conference on Designing Interactive Systems: Processes, Practices, Methods, and Techniques* (pp. 129-136).

Kamel, S., & Hassan, A. (2003). Assessing the introduction of electronic banking in Egypt using the technology acceptance model. *Annals of Cases on Information Technology, 5*, 1–25.

Kamppinen, M., Vihervaara, P., & Aarras, N. (2008). Corporate responsibility and systems thinking - tools for balanced risk management. *International Journal of Sustainable Society*, *1*(2), 158–171. doi:10.1504/IJS-SOC.2008.022572

Karahanna, E., Ahuja, M., Srite, M., & Galvin, J. (2002). Individual differences and relative advantage: the case of GSS. *Decision Support Systems*, *32*(4), 327–341. doi:10.1016/S0167-9236(01)00124-5

Karimzadehgan, M., White, R., & Richardson, M. (2009). Enhancing expert finding using organizational hierarchies. In *Proceedings of the 31th European Conference on IR Research on Advances in Information Retrieval* (pp. 177-188). New York: Springer.

Kedia, B. L., & Bhagat, R. S. (1988). Cultural constraints on transfer of technology across nations: Implications for research in international and comparative management. *Academy of Management Review*, *13*(4), 559–571.

Keller, M. (1989). *Rude awakening: The rise, fall, and struggle for recovery of General Motors*. New York, NY: William Morrow. ·

Kelly, D., Harper, D. J., & Landau, B. (2008). Questionnaire mode effects in interactive information retrieval experiments. *Information Processing & Management*, *44*(1), 122–141. doi:10.1016/j.ipm.2007.02.007doi:10.1016/j.ipm.2007.02.007

Kessler, M. M. (1963). Bibliographic coupling between scientific papers. *American Documentation*, *14*(1), 10–25. doi:10.1002/asi.5090140103

Khalifa, M., & Ning Shen, K. (2008). Explaining the adoption of transactional B2C mobile commerce. *Journal of Enterprise Information Management.*, *21*, 110–124. doi:10.1108/17410390810851372

Khan, B. H. (2005). *Managing e-learning strategies*. Hershey, PA: Information Science Publishing.

Kiley, D., & Welch, D. (2009). The hard road ahead for government motors. *Business Week, 28*.

Kim, S. N., & Baldwin, T. (2005). Automatic interpretation of noun compounds using WordNet similarity. In *Proceedings of the 2nd International Joint Conference on Natural Language Processing* (pp. 945-956).

Kirkpatrick, D. L. (1975). Evaluating training programs. *American Society of Training & Development Report*, 119-120.

Kirschner, P. A. (2001). Using integrated electronic environments for collaborative teaching/learning. *Research Dialogue in Learning and Instruction*, *2*(1), 1–10. doi:10.1016/S0959-4752(00)00021-9

Klein, D., & Randić, M. (1993). Resistance distance. *Journal of Mathematical Chemistry*, *12*(1), 81–95. doi:10.1007/BF01164627

Kolkman, M. (1993). *Problem articulation methodology*. Unpublished doctoral dissertation, University of Twente, Enschede, The Netherlands.

Kolodinsky, J. M., Hogarth, J. M., & Hilgert, M. A. (2004). The adoption of electronic banking technologies by US consumers. *International Journal of Bank Marketing*, *22*(4-5), 238–259. doi:10.1108/02652320410542536doi:10.1108/02652320410542536

Konomi, S., & Roussos, G. (2007). Ubiquitous computing in the real world: Lessons learnt from large scale RFID deployments. *Personal and Ubiquitous Computing*, *11*(7), 507–521. doi:10.1007/s00779-006-0116-1

Kothandapani, V. (1971). Validation of feeling, belief, and intention to act as three components of attitude and their contribution to prediction of contraceptive behaviour. *Journal of Personality and Social Psychology*, *19*, 321–333. doi:10.1037/h0031448

Kreijns, K., Kirschner, P. A., & Jochems, W. (2003). Identifying the pitfalls for social interaction in computer-supported collaborative learning environments: a review of the research. *Computers in Human Behavior*, *19*, 335–353. doi:10.1016/S0747-5632(02)00057-2

Kuosmanen, T. (2003). Duality theory of non-convex technologies. *Journal of Productivity Analysis*, *20*, 273–304. doi:10.1023/A:1027399700108

Kushima, C. (2008). *Research on the system design for supporting ALT's job preparation: a global learning community ALTs and JTEs are building*. Unpublished doctoral dissertation, Hokkaido University, Sapporo, Japan.

Kushima, C., & Nishihori, Y. (2006). Reconsidering the role of the ALT: Effective preparation for ALTs based on the questionnaire survey. *Annual review of English language education in Japan, 17*, 221-230.

Kushima, C., Obari, H., & Nishihori, Y. (2008). Fostering global teacher training: the design and practice of a web-based discussion forum as a knowledge building community. In. *Proceedings of WorldCALL, 2008*, 236–239. Retrieved from http://www.j-let.org/~wcf/proceedings/proceedings.pdf.

Kwahk, K. (2006, January 4-7). ERP acceptance: Organizational change perspective. In *Proceedings of the 39th Hawaii International Conference on System Sciences*, Kauai, HI.

Laender, A. H. F., Ribeiro-Neto, B. A., da Silva, A. S., & Teixeira, J. S. (2002). A brief survey of web data extraction tools. *SIGMOD Record, 31*(2), 84–93. doi:10.1145/565117.565137

Lagsten, J., & Karlsson, F. (2006). Multiparadigm analysis – Clarity of information systems evaluation. In *Proceedings of 13th European Conference on Information Technology Evaluation*, Genoa, Italy.

Land, K. C., Lovell, C. A. K., & Thore, S. (1993). Chance-constrained data envelopment analysis. *Managerial and Decision Economics, 14*, 541–554. doi:10.1002/mde.4090140607

Langheinrich, M., Coroamă, V., Bohn, J., & Mattern, F. (2005). Living in a smart environment – Implications for the coming ubiquitous information society. *Telecommunications Review, 15*(1).

Lara, R., Roman, D., Polleres, A., & Fensel, D. (2004). A conceptual comparison of WSMO and OWL-s. In *Proceedings of the European Conference on Web Services* (pp. 254-269).

Lauer, M. (1995). Designing statistical language learners: Experiments on noun compounds (Doctoral dissertation, Macquarie University). Cornell University Library, 1, 226.

Lave, J., & Wenger, E. (1999). Legitimate peripheral participation in the communities of practice. In McCormic, R., & Paechter, C. (Eds.), *Leraning and knowledge* (pp. 21–35). Thousand Oaks, CA: Sage.

Law, C. C. H., & Ngai, E. W. T. (2007). An investigation of the relationships between organizational factors, business process improvement, and ERP success. *Benchmarking: An International Journal, 14*, 387–406. doi:10.1108/14635770710753158

Learning Design, I. M. S. (2009). *IMS Learning Design best practice guide* and *information binding* and *information model*. IMSGLOBAL publication, retrieved November 13th 2009 from website http://www.imsglobal.org/learningdesign/.

Lee, Y., Kozar, A. K., & Larsen, K. R. T. (2003). The technology acceptance model: Past present, and future. *Communications for the AIS, 12*, 752–780.

Lee, C. H. M., Cheng, Y. W., Rai, S., & Depickere, A. (2005). What Affect Student Cognitive Style in the Development of Hypermedia Learning System? *Computers & Education, 45*, 1–19. doi:10.1016/j.compedu.2004.04.006

Legris, P., Ingham, J., & Collerette, P. (2003). Why people use information technology? A critical review of the technology acceptance model. *Information & Management, 40*, 191–204. doi:10.1016/S0378-7206(01)00143-4doi:10.1016/S0378-7206(01)00143-4

Lerman, K., Plangrasopchok, A., & Knoblock, C. A. (2006). Automatically labeling the inputs and outputs of web services. In *Proceedings of the National Conference on Artificial Intelligence* (pp. 1363-1368).

Leuf, B. (2006). Enhancing the web. In Leuf, B. (Ed.), *The semantic web: Crafting infrastructure for agency* (pp. 3–30). New York, NY: John Wiley & Sons. doi:10.1002/0470028173.ch1

Levin, R. C. (1988). Appropriability, R&D spending, and technological performance. *The American Economic Review, 78*(2), 424–428.

Levy, B. (1987). A theory of public enterprise behavior. *Journal of Economic Behavior & Organization, 8*, 75–96. doi:10.1016/0167-2681(87)90022-9

Lewis, I., & Talalayevsky, A. (2004). Improving inter-organizational supply chain through optimizing of information flows. *Journal of Enterprise Information Management, 17*(3), 229–237. doi:10.1108/17410390410531470

Li, Q. (2004). Knowledge building community: keys for using online forums. *Research & practice to improve learning, 48*(4), 24-28.

Liao, Z., & Landry, R. (2000). An empirical study on organizational acceptance of new information systems in a commercial bank environment. In *Proceedings of the 33rd Hawaii International Conference on System Sciences* (pp. 2021-2030).

Lidell, W. W. (2005). Macroeconomic processes and regional economies management, project GLOBE: A large scale cross-cultural study of leadership. *Problems and Perspectives in Management, 3*, 5–9.

Lim, J. (2003). A conceptual framework on the adoption of negotiation support systems. *Information and Software Technology, 45*, 469–477.

Limongelli, C., Sciarrone, F., Temperini, M., & Vaste, G. (2011). The Lecomps5 framework for personalized web-based learning: a teacher's satisfaction perspective. *Computers in Human Behavior* 27:4, Elsevier, pp.1285-1466, ISSN 0747-5632

Limongelli, C., Sciarrone, F., Temperini, M., & Vaste, G. (2009). Adaptive learning with the LS-Plan system: A field evaluation. *IEEE Transactions on Learning Technologies, 2*(3), 203–215. doi:10.1109/TLT.2009.25

Lim, T. (2010). Rapid emergence of rechargeable car battery market. *SERI Quarterly, 3*(2), 23–29.

Lin, B., & Hsieh, C. (2001). Web-based teaching and learner control: a research review. *Computers & Education, 37*.

Linden, G., Smith, B., & York, J. (2003). *Amazon.com recommendations*. Retrieved from http://www.computer.org/portal/web/csdl/doi/10.1109/MIC.2003.1167344

Liu, X. (2001). *Employing measure methods for business process reengineering in China*. Unpublished doctoral dissertation, University of Twente, Enschede, The Netherlands.

Liu, K. (2000). *Semiotics in information systems engineering*. Cambridge, UK: Cambridge University Press. doi:10.1017/CBO9780511543364

Livingstone, D. (1999). Exploring the icebergs of adult learning: Findings of the first Canadian survey of informal learning practices. *CJSAE, 13*(2), 49–72.

Lochbaum, K., & Streeter, L. (1989). Comparing and combining the effectiveness of latent semantic indexing and the ordinary vector space model for information retrieval. *Information Processing & Management, 25*(6), 665–676. doi:10.1016/0306-4573(89)90100-3

London Knowledge Lab. (2008). *Education 2.0? Designing the web for teaching and learning*. Retrieved December 9, 2009, from http://www.tlrp.org/pub/documents/TELcomm.pdf

Lozano, S., & Gutierrez, E. (2008). Data envelopment analysis of the human development index. *International Journal of Society Systems Science, 1*, 132–150. doi:10.1504/IJSSS.2008.021916

Luján-Mora, S. (2006). *A Survey of Use of Weblogs in Education. Formatex*. Retrieved December 14, 2009, from http://www.formatex.org/micte2006/pdf/255-259.pdf

Lyneis, J. M. (1999). System dynamics for business strategy: A phased approach. *System Dynamics Review, 15*(1), 37–70. doi:10.1002/(SICI)1099-1727(199921)15:1<37::AID-SDR158>3.0.CO;2-Z

Maitland, C., Ngamassi, L., & Tapia, A. (2009, May). *Information management and technology issues addressed by humanitarian relief coordination bodies*. Paper presented at the 6th International ISCRAM Conference, Göteborg, Sweden.

Malone, T. (1987). Modeling coordination in organizations and markets. *Management Science, 33*(10), 1317–1332. doi:10.1287/mnsc.33.10.1317

Marchiori, M. (1997). *The quest for correct information on the web: Hyper search engines*. Retrieved from http://www.w3.org/People/Massimo/papers/quest_hypersearch.pdf

Marco Dorigo, T. S. (2004). *From real to artificial ants: Ant colony optimization*. Cambridge, MA: MIT Press.

Marlow, C., Naaman, M., Boyd, D., & Davis, M. (2006). *Tagging paper, taxonomy, flickr, academic article, to read*. Paper presented at the Conference on Hypertext and Hypermedia, Odense, Denmark.

Marsh, L., & Onof, C. (2007). *Stigmergic epistemology, stigmergy cognition.* Retrieved from http://mpra.ub.uni-muenchen.de/10004/

Martin, D., Burstein, M. H., McDermott, D., McIlraith, S. A., Paolucci, M., & Sycara, K. (2007). Bringing semantics to web services with OWL-S. *World Wide Web (Bussum), 10*(3), 243–277. doi:10.1007/s11280-007-0033-x

Marton, K. (1986). *Multinationals, technology, and industrialization.* Lexington, MA: Heath.

Maybury, M. (2006). *Expert finding systems (Tech. Rep.).* MITRE Corporation.

McCall, J. J. (1967). Competitive production for constant risk utility functions. *The Review of Economic Studies, 34*, 417–420. doi:10.2307/2296560

McCoy, S., Everard, A., & Jones, B. M. (2005). An examination of the technology acceptance model in Uruguay and the US: A focus on culture. *Journal of Global Information Technology Management, 8*(1), 27–45.

McCoy, S., Galletta, D., & King, W. (2007). Applying TAM across cultures: The need for caution. *European Journal of Information Systems, 16*, 81–90. doi:10.1057/palgrave.ejis.3000659

McIlraith, S. A., Son, T. C., & Zeng, H. (2001). Semantic web services. *IEEE Intelligent Systems, 16*(2), 46–53. doi:10.1109/5254.920599

McLoughlin, C., & Lee, M. J. W. (2008). Future learning landscapes: Transforming pedagogy through social software. *Innovate Journal of Online Education, 4*(5). Retrieved December 9, 2009, from http://www.jeffrudisill.com/Content/Student%20Centered%20Learning/Future%20Learning%20Landscapes%20Exhibit%202.pdf

Mead, G. H. (1938). *Philosophy of the act.* Chicago, IL: University of Chicago Press.

Mean Reciprocal Rank (*n.d.*). Wikipedia, The Free Encyclopedia.

Miller, G. A. (1995). WordNet: A lexical database for English. *Communications of the ACM, 38*(11), 39–41. doi:10.1145/219717.219748

Miranda, L. C., Hornung, H. H., Solarte, D. S. M., Romani, R., Weinfurter, M. R., Neris, V. P. A., & Baranauskas, M. C. C. (2007). Laptops educacionais de baixo custo: Prospectos e desafios. In *Proceedings of the Anais do 18ᵗʰ Simpósio Brasileiro de Informática na Educação* (pp. 358-367).

Miranda, L. C., Piccolo, L. S. G., & Baranauskas, M. C. C. (2008). Artefatos físicos de interação com a TVDI: Desafios e diretrizes para o cenário brasileiro. In *Proceedings of the Anais do 8ᵗʰ Simpósio Brasileiro de Fatores Humanos em Sistemas Computacionais* (pp. 60-69).

Miyake, N. (1997). *The Internet children.* Tokyo, Japan: Iwanami Shoten.

Moodle. (2009). Retrieved from http://moodle.org

Morrison, C. J. (1992). Unraveling the productivity growth slowdown in the United States, Canada and Japan: The effects of subequilibrium, scale economies and markups. *The Review of Economics and Statistics, 74*, 381–393. doi:10.2307/2109482

Moss, M., & Townsend, A. (2006, May). Disaster forensics: Leveraging crisis information systems for social science. In F. B. Van de Walle & M. Turoff (Eds.), *Proceedings of the 3rd International ISCRAM Conference,* Newark, NJ.

Mulford, C. L. (1984). *Inter-organizational relations: Implication for community development.* New York, NY: Human Science Press.

Mulford, C. L., & Rogers, D. L. (1982). *Definitions and models.* Ames, IA: Iowa State University Press.

Muller, M. (1997). Participatory practices in the software lifecycle. In Helander, M., Landauer, T. K., & Prabhu, P. V. (Eds.), *Handbook of human-computer interaction* (pp. 255–297). Amsterdam, The Netherlands: Elsevier Science.

Mumford, E. (1964). *Living with a computer.* London, UK: Institute of Personnel Management.

Mumford, E., & Henshall, D. (1979). *A participative approach to computer systems design: A case study of the introduction of a new computer system.* London, UK: Associated Business Programmes.

Nadin, M. (1988). Interface design: A semiotic paradigm. *Semiotica*, *69*(3-4), 269–302. doi:10.1515/semi.1988.69.3-4.269

Nanba, H., & Okumura, M. (1999). Towards multi-paper summarization using reference information. In *Proceedings of IJCAI* (pp. 926-931).

Neris, V. P. A., & Baranauskas, M. C. C. (2009a). Designing e-government systems for all – a case study in the Brazilian scenario. In *Proceedings of the IADIS International Conference on WWW/Internet* (pp. 1-8).

Neris, V. P. A., & Baranauskas, M. C. C. (2009b). Interfaces for all - a tailoring-based approach. In *Proceedings of the 11th International Conference on Enterprise Information Systems*, Milan, Italy (pp. 928-939).

Neris, V. P. A., Almeida, L. D. A., de Miranda, L. C., Hayashi, E. C. S., & Baranauskas, M. C. C. (2009a). Towards a socially-constructed meaning for inclusive social network systems. In *Proceedings of the 11th International Conference on Informatics and Semiotics in Organisations* (pp. 247-254).

Neris, V. P. A., Hornung, H. H., Miranda, L. C., Almeida, L. D. A., & Baranauskas, M. C. C. (2009b). Building social applications with an agile semio-participatory approach. In *Proceedings of the IADIS International Conference on WWW/Internet* (pp. 1-8).

Nguyen, D. H., & Mynatt, E. D. (2002). *Privacy mirrors: Understanding and shaping socio-technical ubiquitous computing systems*. Atlanta, GA: Georgia Institute of Technology.

Nielsen, J. (1993). *Usability engineering*. San Diego, CA: Academic Press.

Nishihori, Y. (2007). The design of web-based collaborative environments for global teacher training: A knowledge building community for future assistant language teachers in Japan. In *Proceedings of the 12th Conference of Pan-Pacific Association of Applied Linguistics* (pp. 168-171).

Niskanen, W. (1971). Bureaucrats and politicians. *The Journal of Law & Economics*, *18*, 617–643. doi:10.1086/466829

Nordhaus, W., & Tobin, J. (1972). *Is growth obsolete?* New York, NY: Columbia University Press.

Norman, D. A. (1988). *The psychology of everyday things*. New York, NY: Basic Books.

Norman, D. A. (2008). Signifiers, not affordances. *Interaction*, *15*(6), 18–19. doi:10.1145/1409040.1409044

Notari, M. (2006, August 21-23). How to use a wiki in education:"Wiki based effective constructive learning". In *Proceedings of the WikiSym'06*, Odense, Denmark. Retrieved December 14, 2009, from http://www.wikisym.org/ws2006/proceedings/p131.pdf

Nunnally, J. (1967). *Psychometric theory*. New York, NY: McGraw-Hill.

O'Reilly, T. (2005). *What is Web 2.0. Design Patterns and Business Models for the Next Generation of Software*. Retrieved December 9, 2009, from http://oreilly.com/pub/a/web2/archive/what-is-web-20.html?page=1

O'Donoghue, T., & Zweimuller, L. (2004). Patents in a model of endogenous growth. *Journal of Economic Growth*, *9*(1), 81–123. doi:10.1023/B:JOEG.0000023017.42109.c2

Oosterhof, A., Conrad, R. M., & Ely, D. P. (2008). *Assessing learners online*. Upper Saddle River, NJ: Pearson Education.

Ou, Y., Massrour, B., & Noormohamed, N. (2009). Putting the pedal to the metal: Forces driving the decision-making process toward American-made vehicles by consumers in Taiwan, China, and Thailand. *Competition Forum*, *7*(2), 343–353.

Palloff, R., & Pratt, K. (2005). *Collaborating online: learning together in community*. San Francisco, CA: Jossey-Bass.

Pan, G. S. C. (2005). Information systems project abandonment: A stakeholder analysis. *International Journal of Information Management*, *25*(2), 173–184. doi:10.1016/j.ijinfomgt.2004.12.003doi:10.1016/j.ijinfomgt.2004.12.003

Panitz T. (1997). Collaborative versus cooperative learning: comparing the two definitions helps understand the nature of interactive learning. *Cooperative Learning and College Teaching*, 8(2).

Panzar, J. C., & Willig, R. D. (1977). Economies of scale in multioutput production. *The Quarterly Journal of Economics*, *41*, 481–493. doi:10.2307/1885979

Paolucci, M., Kawamura, T. R., Payne, T. R., & Sycara, K. (2002). Importing the semantic web in UDDI. In *Proceedings of the International Workshop on Web Services, E-business and the Semantic Web*, (pp. 815-821).

Papanikolaou, K. A., Grigoriadou, M., Kornilakis, H., & Magoulas, G. D. (2003). Personalizing the interaction in a Web-based educational hypermedia system: the case of INSPIRE. *User Modeling and User-Adapted Interaction*, *13*, 213–267. doi:10.1023/A:1024746731130

Parunak, H. V. D. (2005). *Expert assessment of human-human stigmergy*. Ann Arbor, MI: Altarum Institute.

Patil, A., Oundhakar, S., Sheth, A., & Verma, K. (2004). Meteor-s web service annotation framework. In *Proceedings of the 13th International Conference on World Wide Web* (pp. 553-562).

Philips, P. P. (2003). *Training evaluation in the Public sector*. Unpublished doctoral dissertation, Graduate School of the University of Southern Mississippi, Hattiesburg, MS.

Phillips, L. A., Calantone, R., & Lee, M. (1994). International technology adoption: behavior, structure, demand certainty and culture. *Journal of Business and Industrial Marketing*, *9*(2), 16–28. doi:10.1108/08858629410059762doi:10.1108/08858629410059762

Pikkarainen, T., Pikkarainen, K., Karjaluoto, H., & Pahnila, S. (2004). Consumer acceptance of online banking: An extension of the technology acceptance model. *Internet Research*, *14*(3), 224–235. doi:10.1108/10662240410542652doi:10.1108/10662240410542652

Pineno, C., & Tyree, M. (2009). The changing public reports by management and the auditors of publicly held corporations: An updated comparative study of General Motors Corporation and Ford Motor Company. *Competition Forum*, *7*(2), 465–472.

Podinovski, V. V., Førsund, F. R., & Krivonozhko, V. E. (2009). A simple derivation of scale elasticity in data envelopment analysis. *European Journal of Operational Research*, *197*, 149–153. doi:10.1016/j.ejor.2008.06.015

Poole, D., Mackworth, A., & Goebel, R. (1998). *Computational Intelligence: A Logical Approach*. Oxford, UK: Oxford University Press.

Poole, M. S., & Van de Ven, A. H. (2004). *Handbook of organizational change and innovation*. New York, NY: Oxford University Press.

Popescu, E., & Badica, C. (2009). Providing personalized courses in a Web-supported learning environment. In *Proceedings of WI-IAT 2009, Workshop SPeL* (pp. 239-242). Washington, DC: IEEE Computer Society Press.

Popescu, E., Badica, C., & Moraret, L. (2009). WELSA: An intelligent and adaptive Web-based educational system. In *Proceedings of IDC 2009* (SCI 237, pp. 175-185). New York: Springer.

Popescu, E., Trigano, P., Badica, C., Butoi, B., & Duica, M. (2008a). A course authoring tool for WELSA adaptive educational system. In *Proceedings of ICCC 2008* (pp. 531-534).

Popescu, E. (2010a). A Unified Learning Style Model for Technology-Enhanced Learning: What, Why and How? *International Journal of Distance Education Technologies*, *8*(3).(pp. 65-81)

Popescu, E. (2010b). Adaptation Provisioning with respect to Learning Styles in a Web-Based Educational System: An Experimental Study. *Journal of Computer Assisted Learning*, *26*(4).(pp.243-257) doi:10.1111/j.1365-2729.2010.00364.x

Popescu, E., Badica, C., & Trigano, P. (2008b). Learning objects' architecture and indexing in WELSA adaptive educational system. *Scalable Computing: Practice and Experience*, *9*(1), 11–20.

Potencier, F. (2008). The symfony 1.3 & 1.4 reference guide. Sensio Labs Books, ISBN: 9782918390145. License Creative Commons Attribution-Share Alike 3.0 Unported. Retrieved March 10th 2010 from http://www.symfony-project.org.

Prekop, P. (2007). *Supporting Knowledge and Expertise Finding within Australia's Defence Science and Technology Organisation*. HICSS.

Quarantelli, E. L. (1982). Social and organizational problems in a major emergency. *Emergency Planning Digest*, *9*, 7–10.

Quarantelli, E. L. (1997). Problematical aspects of the information/communication revolution for disaster planning and research: Ten non-technical issues and questions. *Disaster Prevention and Management, 6*(2), 94–106. doi:10.1108/09653569710164053

Quirmbach, H. C. (1986). The diffusion of new technology and the market for an innovation. *The Rand Journal of Economics, 17*(1), 33–47. doi:10.2307/2555626

Raghavan, S., & Garcia-Molina, H. (2000). *Crawling the hidden web.* Retrieved from http://ilpubs.stanford.edu:8090/456/1/2000-36.pdf

Ragsdell, G. (2000). Engineering a paradigm shift? A holistic approach to organizational change management. *Journal of Organizational Change Management, 13,* 104–120. doi:10.1108/09534810010321436

Rajasekaran, P., Miller, J., Verma, K., & Sheth, A. (2005). Enhancing web services description and discovery to facilitate composition. In *Proceedings of the First International Workshop on Semantic Web Services and Web Process Composition* (pp. 55-68).

Randall, D., Hughes, J., O'Brien, J., Rouncefield, M., & Tolmie, P. (2001). 'Memories are made of this': Explicating organisational knowledge and memory. *European Journal of Information Systems, 10,* 113–121. doi:10.1057/palgrave.ejis.3000396

Rankine, S. Malfroy, & Ashford-Rowe. (2009). Benchmarking across universities: A framework for LMS analysis. In *Proceedings of the ascilite Auckland 2009.* Retrieved December 9, 2009, from http://www.ascilite.org.au/conferences/auckland09/procs/rankine.pdf

Redecker, C. (2009). *Review of Learning 2.0 Practices: Study on the Impact of Web 2.0 Innovations on Education and Training in Europe* (Tech. Rep.). JRC Scientific. Retrieved December 9, 2009, from http://ftp.jrc.es/EURdoc/JRC49108.pdf

Reinhard, S. R., Lovell, C. A. K., & Thijssen, G. J. (2000). Environmental efficiency with multiple environmentally detrimental variables: Estimated with SFA and DEA. *European Journal of Operational Research, 121,* 287–303. doi:10.1016/S0377-2217(99)00218-0

Robey, D., Ross, J. W., & Boudreau, M. C. (2002). Learning to implement enterprise systems: An exploratory study of the dialectics of change. *Journal of Management Information Systems, 19,* 17–46.

Robinson, J. (1933). *The economics of imperfect competition.* London, UK: Macmillan.

Roethlisberger, F. J., & Dickson, W. J. (1939). *Management and the worker.* Cambridge, MA: Harvard University Press.

Roman, D., Keller, U., Lausen, H., de Bruijn, J., Lara, R., & Stollberg, M. (2005). Web service modeling ontology. *Applied Ontology, 1*(1), 77–106.

Rose, G., & Straub, D. (1998). Predicting general IT use: Applying TAM to the Arabic world. *Journal of Global Information Management, 6*(3), 39–46.

Rosenberg, N. (1963). Technological change in the machine tool industry, 1840-1910. *The Journal of Economic History, 23,* 414–443.

Rosenberg, N. (1981). Why in America? In Mayr, O., & Post, R. C. (Eds.), *Yankee enterprise, the rise of the American system of manufactures.* Washington, DC: Smithsonian Institution Press.

Röstlinger, A., & Cronholm, S. (2009). Design criteria for public e-services. In *Proceedings of the 17th European Conference on Information Systems*, Verona, Italy.

Rule, J., & Brantley, P. (1992). Computerized surveillance in the workplace: Forms and distributions. *Sociological Forum, 7*(3), 405–423. doi:10.1007/BF01117554

Saab, D., Maldonado, E., Orendovici, R., Ngamassi, L., Gorp, A., Zhao, K., et al. (2008). Building global bridges: Coordination bodies for improved information sharing among humanitarian relief agencies. In F. Fiedrich & B. Van de Walle (Eds.), *Proceedings of the 5th International ISCRAM Conference*, Washington, DC (pp. 471-483).

Sachs, J. (1986). A tale of Asian winners and Latin losers. *Economist,*63–64.

Saen, R. E., & Azadi, M. (2009). The use of super-efficiency analysis for strategy ranking. *International Journal of Society Systems Science, 1,* 281–292. doi:10.1504/IJSSS.2009.022819

Sahoo, B. K., Mohapatra, P. K. J., & Trivedi, M. L. (1999). A comparative application of data envelopment analysis and frontier translog production function for estimating returns to scale and efficiencies. *International Journal of Systems Science*, *30*, 379–394. doi:10.1080/002077299292335

Sahoo, B. K., Sengupta, J. K., & Mandal, A. (2007). Productive performance evaluation of the banking sector in India using data envelopment analysis. *International Journal of Operations Research*, *4*, 1–17.

Sahoo, B. K., & Tone, K. (2009a). Decomposing capacity utilization in data envelopment analysis: An application to banks in India. *European Journal of Operational Research*, *195*, 575–594. doi:10.1016/j.ejor.2008.02.017

Sahoo, B. K., & Tone, K. (2009b). Radial and non-radial decompositions of profit change: With an application to Indian banking. *European Journal of Operational Research*, *196*, 1130–1146. doi:10.1016/j.ejor.2008.04.036

Saidel, J., & Cour, S. (2003). Information technology and the voluntary sector workplace. *Nonprofit and Voluntary Sector Quarterly*, *32*, 5–24. doi:10.1177/0899764002250004

Sakakibara, M. (2002). Formation of R&D consortia: Industry and company effects. *Strategic Management Journal*, *23*(11), 1033–1050. doi:10.1002/smj.272

Salm, J. (1999). Coping with globalization: A profile of the northern NGO sector. *Nonprofit and Voluntary Sector Quarterly*, *28*(4s), 87. doi:10.1177/089976499773746447

Sandmo, A. (1971). On the theory of the competitive firm under price uncertainty. *The American Economic Review*, *61*, 65–73.

Sangineto, E., Capuano, N., Gaeta, M., & Micarelli, A. (2008). Adaptive course generation through learning styles representation. *Universal Access in the Information Society*, *7*(1), 1–23. doi:10.1007/s10209-007-0101-0

Santana, V. F., Solarte, D. S. M., Neris, V. P. A., Miranda, L. C., & Baranauskas, M. C. C. (2009). Redes sociais online: desafios e possibilidades para o contexto brasileiro. In *Proceedings of the Anais do XXIX Congresso da Sociedade Brasileira de Computação* (pp. 339-353).

Saracho, A. I. (1996). The diffusion of a durable embodied capital innovation. *Economics Letters*, *54*(1), 45–50. doi:10.1016/S0165-1765(97)00003-7

Scardamalia, M., & Bereiter, C. (1994). Computer support for knowledge-building communities. *Journal of the Learning Sciences*, *3*(3), 265–283. doi:10.1207/s15327809jls0303_3

Scarf, H. E. (1981a). Production sets with indivisibilities part I: Generalities. *Econometrica*, *49*, 1–32. doi:10.2307/1911124

Scarf, H. E. (1981b). Production sets with indivisibilities part II: The case of two activities. *Econometrica*, *49*, 395–423. doi:10.2307/1913318

Scarf, H. E. (1986). Neighborhood systems for production sets with indivisibilities. *Econometrica*, *54*, 507–532. doi:10.2307/1911306

Scarf, H. E. (1994). The allocation of resources in the presence of indivisibilities. *The Journal of Economic Perspectives*, *8*, 111–128.

Schank, D. W. T. (2005). Approximating clustering coefficient and transitivity. *Journal of Graph Algorithms and Applications*, *9*(2).

Schirrer, A., Doerner, K., & Hartl, R. (2007). Reviewer assignment for scientific articles using memetic algorithms. *Metaheuristics, progress in complex optimization systems*, *39*, 113-134.

Scholefield, W. F. (1996). What do JTEs really want? *JALT journal*, *18*, 7-25.

Schuler, D., & Namioka, A. (1993). *Participatory design: Principles and practices*. Mahwah, NJ: Lawrence Erlbaum.

Schwartz, S. H., & Bardi, A. (1997). Influences of adaptation to communist rule on value priorities in eastern Europe. *Political Psychology*, *18*(2), 385–410. doi:10.1111/0162-895X.00062

Searle, J. R. (1969). *Speech acts: An essay in the philosophy of language*. Cambridge, UK: Cambridge University Press.

Seidel, J. (1998). *Qualitative data analysis*. Retrieved from http://www.scribd.com/doc/7129360/Seidel-1998-Qualitative-Data-Analysis

Sengupta, J. K., & Sahoo, B. K. (2006). *Efficiency models in data envelopment analysis: Techniques of evaluation of productivity of firms in a growing economy*. London, UK: Palgrave Macmillan.

Sharp, B. (2009). Detroit's real problem: It's customer acquisition, not loyalty. *Marketing Research, 21*(1), 26–27.

Shen, H., Wang, L., Bi, W., Liu, Y., & Cheng, X. (2008). Research on Enterprise track of TREC 2008. In *Proceedings of the 2008 Text REtrieval Conference (TREC 2008)*.

Sheng, H., Nah, F., & Siau, K. (2006). An experimental study on u-commerce adoption: Impact of personalization and privacy concerns. In *Proceedings of the Fifth Annual Workshop on HCI Research in MIS*.

Shih, H. (2004). An empirical study on predicting user acceptance of e-shopping on the Web. *Information & Management, 41*(3), 351–368. doi:10.1016/S0378-7206(03)00079-Xdoi:10.1016/S0378-7206(03)00079-X

Shi, H., & Yang, X. (1995). A new theory of industrialization. *Journal of Comparative Economics, 20*, 171–189. doi:10.1006/jcec.1995.1008

Shneiderman, B. (1998). *Designing the user interface: Strategies for effective human-computer interaction* (3rd ed.). Reading, MA: Addison-Wesley.

Silva, F. B., & Baranauskas, M. C. C. (2009). Celcidadania: Uma aplicação para celulares integrada a redes sociais inclusiva. In *Proceedings of the 17th Congresso Interno de Iniciação Científica da UNICAMP*.

Singhal, A. (2001). Modern information retrieval: A brief overview. *A Quarterly Bulletin of the Computer Society of the IEEE Technical Committee on Data Engineering, 24*(4), 35–43.

Sivashanmugam, K., Sheth, A., Miller, J., Verma, K., Aggarwal, R., & Rajasekaran, P. (2003). *Metadata and semantics for web services and processes* (pp. 1–19). Datenbanken Und Informationssysteme. [Databases and Information Systems]

Sjöström, J. (2008). *Making sense of the IT artefact – a socio-pragmatic inquiry into IS use qualities*. Unpublished licentiate thesis, Linköping University, Linköping, Sweden.

Sjöström, J., & Goldkuhl, G. (2004). The semiotics of user interfaces – A socio-pragmatic perspective. In Liu, K. (Ed.), *Virtual, distributed and flexible organisations: Studies in organisational semiotics*. Dordrecht, The Netherlands: Kluwer Academic.

Slavin, R. (1990). *Cooperative learning: theory, research, and practice*. Prentice Hall.

Small, H. (1973). Co-citation in the scientific literature: A new measure of the relationship between two documents. *JASIS, 24*, 265–269. doi:10.1002/asi.4630240406

Social network Analysis Software (*n.d.*). Wikipedia, The Free Encyclopedia.

Sokoloff, K. L. (1988). Inventive activity in early industrial America: Evidence from patent records. *The Journal of Economic History, 48*, 813–850. doi:10.1017/S002205070000663X

Somers, T. M., & Nelson, K. G. (2004). A taxonomy of players and activities across the ERP project life cycle. *Information & Management, 41*, 257–278. doi:10.1016/S0378-7206(03)00023-5

Spence, M. (1984). Cost reduction, competition, and industry performance. *Econometrica, 52*(1), 101–121. doi:10.2307/1911463

Staab, S. (2003). Web services: Been there, done that? *IEEE Intelligent Systems, 18*(1), 72–85. doi:10.1109/MIS.2003.1179197

Stamper, R. (1988). Analysing the cultural impact of a system. *International Journal of Information Management, 8*, 107–122. doi:10.1016/0268-4012(88)90020-5

Stamper, R. (1996). Signs, norms, and information systems. In Holmqvist, B., Andersen, P. B., Klein, H., & Posner, R. (Eds.), *Signs at work* (pp. 349–397). Berlin, Germany: Walter de Gruyter.

Stamper, R. K. (2000). New directions for systems analysis and design. In Filipe, J. (Ed.), *Enterprise information systems* (pp. 14–39). Amsterdam, The Netherlands: Kluwer Academic.

Stamper, R. K., Althaus, K., & Backhouse, J. (1988). MEASUR: Method for eliciting, analyzing and specifying user requirements. In *Computerized assistance during the information systems life cycle*. Amsterdam, The Netherlands: Elsevier Science.

Stamper, R. K., Liu, K., Hafkamp, M., & Ades, Y. (2000). Understanding the roles of signs and norms in organizations - A semiotic approach to information systems design. *Behaviour & Information Technology*, *19*(1), 15–27. doi:10.1080/014492900118768

Stamper, R., & Kolkman, M. (1991). Problem articulation: A sharp-edged soft systems approach. *Journal of Applied System Analysis*, *18*, 69–76.

Stanton, J. M. (2000). Reactions to employee performance monitoring: Framework, review and research directions. *Human Performance*, *13*(1), 85–113. doi:10.1207/S15327043HUP1301_4

Starrett, D. A. (1977). Measuring returns to scale in the aggregate and the scale effect of public goods. *Econometrica*, *45*, 1439–1455. doi:10.2307/1912310

Stash, N. (2007). *Incorporating cognitive/learning styles in a general-purpose adaptive hypermedia system*. Unpublished doctoral dissertation, Eindhoven University of Technology, The Netherlands.

Stash, N., Cristea, A., & De Bra, P. (2005). Explicit intelligence in adaptive hypermedia: Generic adaptation languages for learning preferences and styles. In *Proceedings of the Workshop CIAH2005, Combining Intelligent and Adaptive Hypermedia Methods/Techniques in Web Based Education Systems, in conjunction with HT'05* (pp. 75-84).

State Services Commission. (2008). *Factors for successful coordination: A framework to help state agencies coordinate effectively*. Wellington, New Zealand: State Services Commission. Retrieved from http://www.ssc.govt.nz/upload/downloadable_files/successful-coordination-framework.pdf

Stephenson, S. (2008). What went wrong with GM took half a century. *Motor Age*, *124*(8), 2.

Sterbini, A., & Temperini, M. (2007). Good students help each other: improving knowledge sharing through reputation systems. In *Proc. 8th IEEE Conference on Information Technology Based Higher Education and Training (ITHET07)*.

Sterbini, A., & Temperini, M. (2008). Learning from peers: motivating students through reputation systems. In *Proc. Int. Symp. on Applications and the Internet, Social and Personal Computing for Web-Supported Learning Communities. Turku, Finland*.

Sterbini, A., & Temperini, M. (2009). Collaborative projects and self evaluation within a social reputation-based exercise-sharing system. In *Proceedings of the IEEE/WIC/ACM International Conferences on WI and IAT, 2nd International Workshop on SPeL* (Vol. 3, pp. 243-246).

Sterbini, A., & Temperini, M. (2009a). Adaptive construction and delivery of web-based learning paths. *39th ASEE/IEEE Frontiers in Education Conference, Oct. 18-21 2009, San Antonio, TX, USA*.

Sterbini, A., & Temperini, M. (2011). SOCIALX: reputation based support to social collaborative learning through exercise sharing and project teamwork. *Journal of Information Systems and Social Change, IGI Global*, *1*(2), 64–79. doi:10.4018/jissc.2011010105

Stiles, M. (2007). Death of the VLE?: A challenge to a new orthodoxy. *The Journal for the Serials Community*, *20*(1), 31–36. doi:10.1629/20031doi:10.1629/20031

Straub, D. (1994). The effect of culture on IT diffusion e-mail and fax in Japan and the U.S. *Information Systems Research*, *5*(1), 23–47. doi:10.1287/isre.5.1.23doi:10.1287/isre.5.1.23

Straub, D., Keil, M., & Brenner, W. (1997). Testing the technology acceptance model across cultures: A three country study. *Information & Management*, *33*, 1–11. doi:10.1016/S0378-7206(97)00026-8doi:10.1016/S0378-7206(97)00026-8

Stufflebeam, D. L., Madaus, G. F., & Kellaghan, T. (2001). *Evaluation models*. Boston, MA: Kluwer Academic.

Sueyoshi, T. (1997). Measuring efficiencies and returns to scale in Nippon telegraph and telephone in production and cost analysis. *Management Science*, *43*, 779–796. doi:10.1287/mnsc.43.6.779

Sueyoshi, T., & Sekitani, K. (2007a). Measurement of returns to scale using non-radial DEA model: A range-adjusted measure approach. *European Journal of Operational Research, 176*, 1918–1946. doi:10.1016/j.ejor.2005.10.043

Sueyoshi, T., & Sekitani, K. (2007b). The measurement of returns to scale under a simultaneous occurrence of multiple solutions in a reference set and a supporting hyperplane. *European Journal of Operational Research, 181*, 549–570. doi:10.1016/j.ejor.2006.05.042

Summers, R. (1965). A capital intensive approach to the small sample properties of various simultaneous equation estimates. *Econometrica, 33*, 1–41. doi:10.2307/1911887

Sycara, K., Paolucci, M., Ankolekar, A., & Srinivasan, N. (2003). Automated discovery, interaction and composition of semantic web services. *Web Semantics: Science. Services and Agents on the World Wide Web, 1*(1), 27–46. doi:10.1016/j.websem.2003.07.002

Tajino, A., & Tajino, Y. (2000). Native and non-native: What can they offer? *ELT Journal, 54*(1), 3–11. doi:10.1093/elt/54.1.3

Takeuchi, M., Hayashi, Y., Ikeda, M., & Mizoguchi, R. (2006). A collaborative learning design environment to integrate practice and learning based on collaborative space ontology and patterns. In *Proceedings of the 8th international conference on intelligent tutoring systems* (pp. 187-196).

Talton, J. (2010, April 21). *GM's challenge isn't repaying debt but making sexy cars; worry about the second half of '10.* Retrieved from http://seattletimes.nwsource.com/html/soundeconomywithjontalton/2011662521_gms_challenge_isnt_replaying_d.html

Tan, C., & Kwok, P. (2005). *Knowledge building in interschool learning communities: Reflections from a case on project learning in Hong Kong. Towards sustainable and scalable educational innovations informed by the learning sciences.* Amsterdam, The Netherlands: IOS Press.

Tapia, A., Maitland, C., Maldonado, E., & Ngamassi, L. (2010, August 12-15). *Crossing borders, organizations, hierarchies and sectors: IT collaboration in international humanitarian and disaster relief.* Paper presented at the 16th Americas Conference on Information Systems, Lima, Peru.

Tarafdar, M., & Roy, R. K. (2003). Analyzing the adoption of ERP systems in Indian organizations: A process framework. *Journal of Global Information Technology Management, 6*, 31–51.

Taylor, A. III. (2010). *Sixty to zero: An inside look at the collapse of General Motors--and the Detroit auto industry* (pp. 31–34). New Haven, CT: Yale University Press.

Thompson, D. (1967). *Organizations in action.* New York, NY: McGraw-Hill.

Thompson, F. J., Frances, J., & Mitchell, J. (1991). *Markets, hierarchy and networks: The coordination of social life.* London, UK: Sage.

Thompson, R. L., Higgins, C. A., & Howell, J. M. (1991). Personal computing: Toward a conceptual model of utilization. *Management Information Systems Quarterly, 15*, 124–143. doi:10.2307/249443

Thrall, R. M. (1999). What is the economic meaning of FDH? *Journal of Productivity Analysis, 11*, 243–250. doi:10.1023/A:1007742104524

TimeCruiser Computing Corporation. (2008). *LMS 2.0: How to Select an Advanced Learning System.* Retrieved December 12, 2009, from http://www.timecruiser.com/timecruiser/admin/UserFiles/White_Paper_LMS_43008.pdf

Tone, K., & Sahoo, B. K. (2003). Scale, indivisibilities and production function in data envelopment analysis. *International Journal of Production Economics, 84*, 165–192. doi:10.1016/S0925-5273(02)00412-7

Tone, K., & Sahoo, B. K. (2004). Degree of scale economies and congestion: A unified DEA approach. *European Journal of Operational Research, 158*, 755–772. doi:10.1016/S0377-2217(03)00370-9

Tone, K., & Sahoo, B. K. (2005). Evaluating cost efficiency and returns to scale in the life insurance corporation of India using data envelopment analysis. *Socio-Economic Planning Sciences, 39*, 261–285. doi:10.1016/j.seps.2004.06.001

Tone, K., & Sahoo, B. K. (2006). Re-examining scale elasticity in DEA. *Annals of Operations Research, 145*, 69–87. doi:10.1007/s10479-006-0027-6

Tong, C., & Tong, L. (2009). GM: Problems, solutions and lessons. *Competition Forum, 7*(1), 136–141.

Treiber, M., Truong, H. L., & Dustdar, S. (2009). SOAF - Design and implementation of a service-enriched social network. In *Proceedings of the 9th International Conference on Web Engineering* (pp. 379-393).

Triantafillou, E., Pomportsis, A., Demetriadis, S., & Georgiadou, E. (2004). The value of adaptivity based on cognitive style: an empirical study. *British Journal of Educational Technology, 35*(1), 95–106. doi:10.1111/j.1467-8535.2004.00371.x

Tulkens, H. (1993). On FDH analysis: Some methodological issues and applications to retail banking, courts and urban transit. *Journal of Productivity Analysis, 4*, 183–210. doi:10.1007/BF01073473

Tulkens, H., & Vanden Eeckaut, P. (1995). Non-parametric efficiency, progress and regress measures for panel data: methodological aspects. *European Journal of Operational Research, 80*, 474–499. doi:10.1016/0377-2217(94)00132-V

Twain, M. (1876). *The adventures of Tom Sawyer* (2003rd ed.). Bath, UK: Paragon.

Ullrich, C. (2005). The Learning-Resource-Type is dead, long live the Learning-Resource-Type! *Learning Objects and Learning Designs, 1*, 7–15.

UNDP. (2002). *Human development report 2002: Deepening democracy in a fragmented world*. New York, NY: Oxford University Press.

Uvin, P. (1999). *The influence of aid in situations of violent conflict: A synthesis and commentary on the lessons learned from case studies on the limit and scope of the use of development assistance incentives and disincentives for influencing conflict situations*. Paris, France: OECD. Retrieved from http://www.ndu.edu/itea/storage/610/Impact%20of%20Aid%20Uvin.pdf

Van Brabant, K. (1999). *Opening the black box: An outline of a framework to understand, promote and evaluate humanitarian coordination*. London, UK: Humanitarian Policy Group.

Van De Ven, A. H., Delbecq, A. L., & Koenig, R. Jr. (1976). Determinate of coordination modes within organizations. *American Sociological Review, 41*(2), 322–338. doi:10.2307/2094477

Van Gorp, A., Ngamassi, L., Maitland, C., Saab, D., Tapia, A., Maldonado, A., et al. (2008, June 24-27). *VSAT deployment for post-disaster relief and development: Opportunities and constraints for inter-organizational coordination among international NGOs*. Paper presented at the 17th Biennial Conference of the International Telecommunications Society Montreal, QC, Canada.

Varian, H. R. (1984). The nonparametric approach to production analysis. *Econometrica, 52*, 279–297. doi:10.2307/1913466

Varian, H. R. (1992). *Microeconomic analysis* (3rd ed.). New York, NY: W.W. Norton and Company.

Venkatesh, M. (2009). The constitutive and the instrumental in social design. In *Proceedings of the 4th International Conference on Design Science Research in Information Systems and Technology*.

Venkatesh, V., & Davis, F. (2000). A theoretical extension of the technology acceptance model: Four longitudinal field studies. *Management Science, 46*(2), 186–204. doi:10.1287/mnsc.46.2.186.11926doi:10.1287/mnsc.46.2.186.11926

Venkatesh, V., & Morris, M. G. (2000). Why don't men ever stop to ask for directions? Gender, social influence, and their role in technology acceptance and usage behavior. *Management Information Systems Quarterly, 24*(1), 115–139. doi:10.2307/3250981doi:10.2307/3250981

Venkatesh, V., & Davis, F. (2000). A theoretical extension of the technology acceptance model: Four longitudinal field studies. *Management Science, 46*, 186–204. doi:10.1287/mnsc.46.2.186.11926

Venkatesh, V., Morris, M., Davis, G., & Davis, F. (2003). User acceptance of information technology: Toward a unified view. *Management Information Systems Quarterly, 27*, 425–478.

Verma, K., & Sheth, A. (2007). Semantically annotating a web service. *IEEE Internet Computing, 11*(2), 83–85. doi:10.1109/MIC.2007.48

Vernick, J., Rutkow, L., & Salmon, D. (2007). Availability of litigation as a public health tool for firearm injury prevention: Comparison of guns, vaccines, and motor vehicles. *American Journal of Public Health*, *97*(11), 1991–1997. doi:10.2105/AJPH.2006.092544

Vijayasarathy, L. R. (2004). Predicting consumer intentions to use on-line shopping: The case for an augmented technology acceptance model. *Information & Management*, *41*(6), 747–762. doi:10.1016/j.im.2003.08.011doi:10.1016/j.im.2003.08.011

Vila na Rede. (2009). *Vila na rede: An inclusive social network.* Retrieved from http://www.vilanarede.org.br

Vlaar, P., Van den Bosch, F., & Volberda, H. (2006). Coping with problems of understanding in inter-organizational relationships: Using formalization as a means to make sense. *Organization Studies*, *27*(11), 1617–1638. doi:10.1177/0170840606068338

Vorvoreanu, M., & Botan, C. H. (2000). Examining electronic surveillance in the workplace: A review of theoretical perspectives and research findings. In *Proceedings of the Conference of the International Communication Association.*

Vygotskij, L. S. (1978). Mind in society: the development of higher psychological processes. (M. Cole, V. John-Steiner, S. Scribner, & E. Souberman, Eds. and Trans.) Cambridge MA: Harvard University Press. (Original material published between 1930 and 1935).

W3C. (2001). *XML linking language (xlink).* Retrieved from http://www.w3.org/TR/xlink/

W3C. (2009). *Web content accessibility guidelines 2.0.* Retrieved from http://www.w3.org/TR/WCAG/

Waddell, W., & Bodek, N. (2005). *Rebirth of American industry - a study of lean management.* Vancouver, WA: PCS Press.

Waldman, M. (1987). Noncooperative entrance, uncertainty, and the free rider problem. *The Review of Economic Studies*, *54*(2), 301–310. doi:10.2307/2297519

Walker, J. (2005). *Feral hypertext: When hypertext literature escapes control.* Paper presented at the Conference on Hypertext and Hypermedia, Salzburg, Austria.

Wang, Y. S., Wang, Y. M., Lin, H. H., & Tang, T. I. (2003). Determinants of user acceptance of Internet banking: An empirical study. *International Journal of Service Industry Management*, *14*(5), 501–519. doi:10.1108/09564230310500192doi:10.1108/09564230310500192

Wang, F., Chen, B., & Miao, Z. (2008). *A survey on reviewer assignment problem (LNCS 5027).* New York: Springer.

Wang, T., Wang, K., & Huang, Y. (2008). Using a style-based ant colony system for adaptive learning. *Expert Systems with Applications*, *34*(4), 2449–2464. doi:10.1016/j.eswa.2007.04.014

Wardrip-Fruin, N. (2004). *What hypertext is.* Paper presented at the Conference on Hypertext and Hypermedia, Santa Cruz, CA, USA.

Watanabe, S., Ito, T., Ozono, T., & Shintani, T. (2005). *A Paper Recommendation Mechanism for the Research Support System Papits.* Washington, DC: IEEE.

Weber, J. (2010). Assessing the "Tone at the Top": The moral reasoning of CEOs in the automobile industry. *Journal of Business Ethics*, *92*(2), 167–182. doi:10.1007/s10551-009-0157-2

Webster, J., & Watson, Watson, R. T. (2002). Analyzing the past to prepare for the future: Writing a literature review. *Management Information Systems Quarterly*, *26*(2), xiii–xxiii.

Wei, W., Lee, J., & King, I. (2007). Measuring credibility of users in an e-learning environment. In Proc.*16th Int. Conf. on World Wide Web*, 1279-1280, New York, NY, USA, ACM.

Weiser, M. (1991). The computer for the 21st century. *Scientific American*, *265*(3), 94–104. doi:10.1038/scientificamerican0991-94

Weller, M. (2007). The distance from isolation: Why communities are the logical conclusion in e-learning. *Computers & Education*, 49.

Wenger, E. (1998). *Communities of practice: Learning, meaning, and identity.* Cambridge Un. Press.

Wentz, L. (2006). *An ICT primer: Information and communication technologies for civil-military coordination in disaster relief and stabilization and reconstruction (Tech. Rep. No. OMB 0704-0188)*. Washington, DC: National Defense University.

Whetten, R. A., & Rogers, D. L. (1982). *Inter-organizational coordination: Theory, research and implementation* (1st ed.). Ames, IA: Iowa State University Press.

White, D., & Harary, F. (2001). The cohesiveness of blocks in social networks: Node connectivity and conditional density. *Sociological Methodology*, 305–359. doi:10.1111/0081-1750.00098

Wilson, E. O. (1975). *Sociobiology: The new synthesis*. Cambridge, MA: Belknap Press of Harvard University Press.

Winter, S., Chudoba, K., & Gutek, B. (1998). Attitudes toward computers: When do they predict computer use? *Information & Management*, *34*, 275–285. doi:10.1016/S0378-7206(98)00065-2

Witkin, H. A. (1962). *Psychological differentiation: studies of development*. New York: Wiley.

Wu, Z., & Palmer, M. (1994). Verbs semantics and lexical selection. In *Proceedings of the 32nd Annual Meeting on Association for Computational Linguistics*, Las Cruces, NM (pp. 133-138).

Xue, Y., Zhu, T., Hua, G., Zhang, M., Liu, Y., & Ma, S. (2008). THUIR at TREC2008: Enterprise track. trec.nist.gov.

Yang, K.-H., Kuo, T.-L., Lee, H.-M., & Ho, J. M. (2009). A reviewer recommendation system based on collaborative intelligence. In *Web Intelligence and Intelligent Agent Technology, IEEE/WIC/ACM International Conference on* (Vol. 1, pp. 564-567).

Yang, S., Zhao, J., Zhang, X., & Zhao, L. (2009). Application of pagerank technique in collaborative learning. In Leung, E. W., Wang, F. L., Miao, L., Zhao, J. and He, J. (Eds.) *Revised Selected Papers from Advances in Blended Learning: Second Workshop on Blended Learning*, Jinhua, China, August 20-22, 2008. Lecture Notes in Computer Science 5328 (102-109). Berlin, Heidelberg, Springer-Verlag.

Yang, H., & Yoo, Y. (2004). It's all about attitude: Revisiting the technology acceptance model. *Decision Support Systems*, *38*, 19–31. doi:10.1016/S0167-9236(03)00062-9

Yang, X. (1994). Endogeneous vs. exogeneous comparative advantage and economics of specialization vs. economies of scale. *Journal of Economics*, *60*, 29–54. doi:10.1007/BF01228024

Yang, X., & Ng, Y. K. (1993). *Specialization and economic organization*. Amsterdam, The Netherlands: Elsevier Science.

Yang, X., & Rice, R. (1994). An equilibrium model endogenizing the emergence of a dual structure between the urban and rural sectors. *Journal of Urban Economics*, *35*, 346–368. doi:10.1006/juec.1994.1020

Yao, J., Xu, J., & Niu, J. (2008). Using Role Determination and Expert Mining in the Enterprise Environment. In *Proceedings of the 2008 Text REtrieval Conference (TREC 2008)*, Gaithersburg, MD.

Yarowsky, D., & Florian, R. (1999). Taking the load of the conference chairs: towards a digital paper-routing assistant. In *Proceedings of the 1999 Joint SIGDAT Conference on Empirical Methods in NLP and Very-Large Corpora*.

Yimam-Seid, D., & Kobsa, A. (2003). Expert finding systems for organizations: Problem and domain analysis and the demoir approach. In *Sharing Expertise: Beyond Knowledge Management*. Cambridge, MA: MIT Press.

Yin, R. K. (2003). *Case study research: Design and methods*. Thousand Oaks, CA: Sage.

Yi, U., Wu, Z., & Tung, L. L. (2005/2006). How individual differences influence technology usage behaviour? Toward an intergrated framework. *Journal of Computer Information Systems*, *46*, 52–63.

Yousafzai, S., Foxall, G., & Pallister, J. (2007). Technology acceptance: A meta-analysis of the TAM: Part 1. *Journal of Modelling in Management*, *2*, 251–280. doi:10.1108/17465660710834453

Yu, D., & Chen, X. (2007). Supporting collaborative learning activities with IMS LD. In Proc. *International Conference on Advanced Communication Technology* (ICACT2007).

Zhang, M., Duan, Z., & Zhao, C. (2008). Semi-automatically annotating data semantics to web services using ontology mapping. In *Proceedings of the 12th International Conference on Computer Supported Cooperative Work in Design* (pp. 470-475).

Zhao, K., Maitland, C., Ngamassi, L., Orendovici, R., Tapia, A., & Yen, J. (2008, July 14-17). *Emergence of collaborative projects and coalitions: A framework for coordination in humanitarian relief.* Paper presented at the 2nd World Congress Conference on Social Simulation, Washington, DC.

Zhu, J. (2008). The University College London at TREC 2008 enterprise track. In *Proceedings of the 2008 Text REtrieval Conference (TREC 2008)*.

Zweig, D. (2005). Beyond privacy and fairness concerns: Examining psychological boundary violations as a consequence of electronic performance monitoring. In Weckert, J. (Ed.), *Electronic monitoring in the workplace: Controversies and solutions*. Hershey, PA: IGI Global.

Zweig, D., & Webster, J. (2002). Where is the line between benign and invasive? An examination of psychological barriers to the acceptance of awareness monitoring systems. *Journal of Organizational Behavior, 23*(5), 605–633. doi:10.1002/job.157

About the Contributors

John Wang is a Professor in the Department of Information & Operations Management at Montclair State University, USA. Having received a scholarship award, he came to the USA and completed his PhD in Operations Research from Temple University. Due to his extraordinary contributions beyond a tenured full Professor, Dr. Wang has been honored with a special range adjustment in 2006. He has published over 100 refereed papers and seven books. He has also developed several computer software programs based on his research findings. He is the Editor-in-Chief of *International Journal of Applied Management Science*, *International Journal of Operations Research and Information Systems*, and *International Journal of Information Systems and Supply Chain Management*. He is the Editor of *Data Warehousing and Mining: Concepts, Methodologies, Tools, and Applications* (six-volume) and the Editor of the *Encyclopedia of Data Warehousing and Mining*, 1st (two-volume) and 2nd (four-volume). His long-term research goal is on the synergy of operations research, data mining, and cybernetics.

* * *

Mohammad Mourhaf AL Asswad is a PhD candidate at the School of Information Systems Computing and Mathematics, Brunel University. He holds an MRES (Master of Research) Degree from Brunel University and a bachelor of Control and Electronic Engineering from Aleppo University, Syria. His research interests focus on the adoption of ontologies for the development of flexible and adaptable software and Web applications. Moreover, he is interested in Linguistic analysis and its employment in software applications. Mohammad has published and presented his work in several international conferences. Mohammad is a program committee member of the WEBIST International Conference.

Leonelo Dell Anhol Almeida is a PhD. candidate on Awareness in Collaborative Systems at the Institute of Computing, University of Campinas, Brazil. He obtained his master's degree in Computer Science from Federal University of Paraná. His research interests involve awareness in collaborative systems, Human-Computer Interaction, Universal Design, and Accessibility. He is engaged in multidisciplinary projects that investigate the barriers faced by people while using computers considering scenarios of vast economical and social diversity as the Brazilian. From the participation in those projects and from his doctoral research he recently published works in international conferences (*e.g.* International Conference on Informatics and Semiotics in Organisations'09, HCI International'09), in national journal (Brazilian Journal of Computer Science in Education), and a book chapter focused on web accessibility.

Costin Bădică received the M.Sc. and Ph.D. in Computer Science from University of Craiova, Romania in 1990 and 1999, respectively. In 2006 he also received the title of Professor of Computer Science from University of Craiova. He is currently with the Department of Software Engineering, Faculty of Automatics, Computers and Electronics of the University of Craiova, Romania. During 2001 and 2002 he was Post-Doctoral Fellow with the Department of Computer Science, King's College London, United Kingdom. His current research interests are at the intersection of Artificial Intelligence, Distributed Systems and Software Engineering. He authored and co-authored more than 100 publications related to these topics as journal articles, book chapters and conference papers. He has prepared special journal issues and co-edited 4 books in Springer's Studies in Computational Intelligence series. He co-initiated and he is co-organizing the Intelligent Distributed Computing – IDC series of conferences that is being held yearly. He is member of the editorial board of 4 international journals. He also served as PC member of many international conferences.

M. Cecília C. Baranauskas is Professor at the Institute of Computing, UNICAMP, Brazil. She received a B.Sc. and M.Sc. in Computer Science and a Ph.D. in Electrical Engineering at UNICAMP, Brazil. She spent a sabbatical year in the UK, as Honorary Research Fellow at the Staffordshire University and as a Visiting Fellow at the University of Reading, working with Prof. Kecheng Liu's Applied Informatics with Semiotics Lab (2001-2002). She also received a Cátedra Ibero-Americana Unicamp-Santander Banespa to study accessibility issues on software engineering at Universidad Politécnica de Madrid (2006-2007). Her research interests have focused on human-computer interaction, particularly investigating different formalisms (including Organizational Semiotics and Participatory Design) in the analysis, design and evaluation of societal systems. She is leading several projects investigating the use of these formalisms in design contexts of e-Citizenship and e-Inclusion. Former IFIP TC13 representative, currently she is member of the BR-CHI (an ACM SIGCHI local chapter) Executive Council and member of the Special Committee for HCI at SBC (Brazilian Computing Society).

Sanghamitra Bhattacharyya is an Associate Professor in Organizational Behaviour and Human Resource Management in the Department of Management Studies, Indian Institute of Technology, Madras. Her areas of interest include, among others, group and team dynamics, management of organizational change, cross cross-cultural management issues, international human resource management practices, and growth strategies of BPO sector in India. She has published widely in various national and international peer reviewed journals and presented papers at conferences both in India and abroad. In addition to her academic activities, Dr. Bhattacharyya has undertaken consultancy projects with various Indian corporate in Chennai and Kolkata. She is a metallurgical engineer from Bengal Engineering College, and obtained her doctoral degree (FPM) from Indian Institute of Management Calcutta in the area of Organizational Behaviour. Prior to joining academics, Dr. Bhattacharyya has had industry experience of nearly ten years at various organizations (both manufacturing and service sectors) in public and private sector undertakings in India.

Vânia Paula de Almeida Neris is a PhD. candidate in Computing Science at the University of Campinas (UNICAMP), Brazil and was a visitor student at the Informatics Research Centre (IRC) at the University of Reading, UK. She graduated in Computer Engineering and received a Master degree in Computer Science at the Federal University of São Carlos (UFSCar), Brazil. Her actual research focus on the design of interactive systems, especially tailorable user interfaces. She has been publishing in several journals (*e.g.* Scientometrics, ACM E-learn magazine, Brazilian Journal of Computer Science in Education) and international conferences (*e.g.* IFIP TC13 Conference on Human-Computer Interaction - INTERACT, Human-Computer Interaction International - HCII, International Conference on Informatics and Semiotics in Organisations - ICISO, International Conference on Enterprise Information Systems - ICEIS).

Sergio de Cesare is currently a Lecturer and Director of Postgraduate Studies in the Department of Information Systems and Computing at Brunel University. Sergio's work focuses on the adoption of ontologies in software engineering as a means to develop and evolve information systems that are increasingly flexible and adaptive to change. His research interests relate to the foundational principles and concepts of semantic modelling, the role of semantic models in making sense of real-world domain knowledge and the development of ontology-based development tools and techniques. Sergio has over 50 peer-reviewed publications in leading journals and conferences and has organised several international events related to modelling in software engineering including a series of ACM OOPSLA workshops on the general theme of 'Semantics in Systems Development'. Sergio is also currently the Managing Editor of the European Journal of Information Systems.

Leonardo Cunha de Miranda is a PhD. candidate in Computer Science at the University of Campinas (UNICAMP), Brazil, He received a Master degree in Computer Science at the Federal University of Rio de Janeiro (UFRJ), Brazil. His actual research focus on Interaction Design in Interactive Digital Television (iDTV), especially building a new physical artifact of interaction with iDTV. He has been publishing in several international conferences, e.g., International Conference on Informatics and Semiotics in Organisations (ICISO), International Conference on Human-Computer Interaction (HCII), International Conference on Interfaces and Human Computer Interaction (IHCI), IADIS International Conference on WWW/Internet (ICWI), and Latin American Conference on Human-Computer Interaction (CLIHC).

José Figueiredo, Professor in the Engineering and Management Department of IST – Lisbon Technology Institute – Technical University of Lisbon, is an Electronic Engineer with an MBA in Information Management and a PhD in Industrial Engineering. Always being involved in the university teaching he started two small companies in the information technologies sector. He currently teaches Project Management and Information Management. He is currently the Vice-President of the research unity of the Engineering and Management Department in IST and the coordinator of the department's master course in industrial management (MEGI). He published several papers in international journals, book chapters, and conference proceedings.

Dieter Gstach studied Economics and Mathematics at the University of Vienna where he received his doctoral degree in 1993 after a postdoctoral education at Vienna Institute of Advanced Studies. Since 1991 he holds the position of an Assistant Professor at Vienna University of Economics and Business Administration where he was tenured 2001. His research, documented in many publications in various international scholarly journals, covers, amongst other areas, methodological and empirical issues in efficiency and productivity measurement, the macroeconomic role of housing markets, input-output analysis and property taxation schemes. This is for most of which he also serves as regular referee. Furthermore, he does consulting for various governmental and semi-private institutions.

Göran Goldkuhl, PhD, is professor in information systems at Linköping University and guest professor at Stockholm University, Sweden. He is the director of the Swedish research group VITS (www.vits.org), consisting of 25 researchers. He has published several books and more than 150 research papers at conferences, in journals and as book chapters. He is currently developing a family of theories and methods, which all are founded on socio-instrumental pragmatism; theories as Workpractice Theory, Business Action Theory, Information Systems Actability Theory; and methods for business process modeling, problem analysis, e-service design, user-interface design, information modeling and IS evaluation. He has a great interest in pragmatic and qualitative (interpretive) research methods and he has contributed to the development of Multi-Grounded Theory, (a modified version of Grounded Theory) and Practical Inquiry (a special kind of action research). He has been active in international research communities such as Language Action Perspective (LAP), Action in Language, Organisations and Information Systems (ALOIS), Enterprise Interoperability, Organisational Semiotics and AIS SIGPrag. He is main responsible for Ph D education in information systems at Linköping University and has been the supervisor for 20 PhD dissertations and 41 Licentiate theses. At the moment he is responsible for and actively working with several e-government research projects.

Geoffrey Greenfield is a doctoral student in Information Systems at the UQ Business School, The University of Queensland. Geoffrey has over 25 year's business experience and applies his knowledge and experience in his volunteer role as the President of a NFP organisation. His research interests include technology acceptance by NFP workers, trust development, attitude development, and computer-mediated communication.

Elaine Cristina Saito Hayashi is a PhD Candidate in Computing Sciences at the Institute of Computing at University of Campinas (UNICAMP), Brazil. She graduated at Mackenzie University and has a specialization in Telecommunications from University of Industrial Engineering (UniFEI), both in the state of São Paulo, Brazil. She received her Master's degree from the University of Campinas (UNICAMP). Her research interests include Information and Communication Technology, interaction design and usability. She is currently member of a research team focusing on promoting digital cultures among the less favored Brazilian citizens. She has presented her research in some international conferences including HCI International, ACM SAC and IFIP World Computer Congress (WCC).

Jeffrey Hsu is an Associate Professor of Information Systems at the Silberman College of Business, Fairleigh Dickinson University. He is the author of numerous papers, chapters, and books, and has previous business experience in the software, telecommunications, and financial industries. His research interests include human-computer interaction, e-commerce, IS education, and mobile/ubiquitous computing. He is Managing Editor of the *International Journal of Data Analysis and Information Systems* (IJDAIS), Associate Editor of the *International Journal of Information and Communication Technology Education* (IJICTE), and is on the editorial board of several other journals. Dr. Hsu received his PhD in Information Systems from Rutgers University, a MS in Computer Science from the New Jersey Institute of Technology, and an MBA from the Rutgers Graduate School of Management.

Malinka Ivanova is a Senior Assistant Professor at Technical University – Sofia, College of Energetics and Electronics, Bulgaria. She obtained her PhD degree in Automation of Engineering Labour and Systems for Automated Design (Exploration and Design of eLearning Platforms) and master's degree in Electronics and Automation both from Technical University - Sofia. In the area of eLearning her research interests include eLearning platforms, the influence of Internet technologies on education, knowledge management, learning design, teaching/learning in 3D environments, personal learning environments. She is author and co-author of more than 50 publications, including journal articles, books, book chapters and conference papers. She received several awards and grants related to her research topics. Her teaching courses are in the fields of Electronics and Computer Science.

T.J. Kamalanabhan is a Ph.D in Organizational Psychology, he has done his M.Phil in Organizational Psychology and P.G.D.M - Post-graduate Diploma in Business Management. He has around 20 Years of experience in academics and consultancy projects. He is has obtained a Fulbright fellowship for research and teaching at the Department of International Business, Washington State University, Pullman, USA in 2002, he has been awarded DAAD Fellowship to visit Germany under (German Academic Exchange Programme) to do research in Entrepreneurship at the Department of Management, University of West Saxony, Zwickau, Germany in 1998, He is a regular visiting faculty at the department of Management and Law, Multimedia University, Melaka, Malaysia, University College of Technology and Management Malaysia, (KUTPM) Shah Alam, Selangor, Malaysia Faculty of Management, Multimedia University, Cyberjaya, Malaysia, University College of Technology and Management Malaysia, (KUTPM) Shah Alam, Selangor, Malaysia and a visiting Research fellow at KNU University, Daegu, South Korea.

Dain Kaplan received a Bachelor's degree from the University of California in Irvine, and a Master's degree from the Tokyo Institute of Technology, where he has also continued on to the Ph.D. program, researching computational linguistics. His research interests include textual entailment, citation analysis, and stylistic analysis.

Melih Kirlidog is an associate professor in the Computer Engineering Department of Marmara University, Istanbul, Turkey. He holds a BSc in Civil Engineering from Middle East Technical University, Turkey. He is also the holder of MBA (in MIS) and PhD degrees from University of Wollongong, Australia. He has worked as an ICT analyst and consultant for over twenty years in Turkey and Australia. His current research interests include intercultural ICT development and implementation, ICT in developing countries, decision support systems, and community informatics. Since November 2002 he works as a full time academic. His articles appeared in Information Technology & People, Online Information Review, IEEE Technology and Society Magazine, and Electronic Library. Aygul Kaynak was born in Corum, Turkey in 1980. She completed her primary and high school educations in several schools in Istanbul. She received her BSc degree in mathematical engineering from Yildiz Teknik University, Istanbul in 2001 and her MSc degree in computer engineering from Marmara University, Istanbul in 2004. She attended several seminars and trainings about internet banking technologies in various subjects. She is currently working in one of the multinational banks in Turkey. She has taken an active role on analyzing internet banking application of the mentioned bank, specifying customers' requirements and working on their development. She is married.

Mark Lycett is a Professor in Information Systems Development at Brunel University, and researches the complex dynamics that couple people, process, information and technology. His research interests relate to the principles and mechanisms of adaptation/evolution in the natural and social realm and how they can be brought to fruition using modern technologies, in particular to guide and improve people's experience with technology. Mark has published his work in a number of leading journals and conferences and is engaged in ongoing research with a number of organizations. Prior to returning to education, he spent a number of years in industry, primarily in project management.

Carleen Maitland is an Associate Professor in the College of Information Sciences and Technology at Penn State University. Her research focuses on inter-organizational coordination across multiple forms of organizations, both non-profit and for-profit, focusing on the technologies that support and hinder such coordination.

Edgar Maldonado received his PhD from the Pennsylvania State University. His research interests are national-level information and telecommunication policies specifically directed at how the development and use of Open Source technologies can positively impact the public, private, and voluntary sectors.

Artur Marques teaches at the ESGTS (Management and Technology School of Santarém) and is a PhD student in Engineering and Management at the IST (Lisbon Technology Institute). He graduated in Computer Engineering, has an MsC in Engineering Policy and Management of Technology, and an MsC in Internet and E-Commerce. His research interests include the social and organizational impacts of information technology, Web user experience, and Internet business models.

Stuart Moran is a researcher at the Informatics Research Centre, University of Reading, UK; where he also obtained his BSc in Computer Engineering and PhD in Informatics. While his initial interests were in hardware design, this soon shifted focus as he began to view the importance of humans and their interaction with computer systems. This led to his current more human-oriented research interests in pervasive computing, technology acceptance and persuasive technologies.

Keiichi Nakata is Reader in Social Informatics at the Informatics Research Centre, Henley Business School, University of Reading, UK. He obtained his PhD in Artificial Intelligence from the University of Edinburgh, UK, and M.Eng. and B.Eng. in Nuclear Engineering from the University of Tokyo. His research interests lie at the interface between technology and people, in the areas of computer-supported collaborative work, cognitive systems engineering, and information systems. His editorial board memberships include New Generation Computing, International Series on Computer Supported Cooperative Work, and Automated Experimentation. His previous appointments include International University in Germany, The University of Tokyo, and GMD.

Yuri Nishihori is a Professor of English at the Faculty of Music, Sapporo Otani University, and an emeritus professor of Hokkaido University, Hokkaido, Japan. She received her M.Phil. in applied linguistics from the University of Reading, U.K. and M.A. in literature from Hokkaido University, Japan. Her research interests include applied linguistics, English teaching methodology and educational technology. She has presented her research at several international conferences, including the International

329

Association of Applied Linguistics (AILA), the International Conference on Computers in Education (ICCE) and the WorldCALL. She has also conducted extensive research in the fields of English language teaching and computer supported collaborative learning. She was an executive member of the JACET (Japan Association of College English Teachers) and the president of its Hokkaido Chapter from 2004 to 2010, as well as undertaking the role of counselor for the Japanese Society for Information and Systems in Education (JSiSE) from 2004 to 2009. Chizuko Kushima is an assistant professor at Tsuda College, Tokyo, Japan. She received her M.A. and Ph.D. in international media and communication studies from Hokkaido University, Japan. Her research interests include corpus linguistics, computer supported collaborative learning, educational technology and development of teaching materials. She presented her research at the WorldCALL 2008. Her articles appeared in the Journal of ARELE (Annual Review of English Language Education in Japan) and the Japan Information-Culturology Society. She is also conducting active research in the field of computer supported teacher training. She received the Excellent Paper Award with co-authors of this paper at the 2008 Conference on Education for Information Processing in Japan.

Toshio Okamoto obtained his PhD from Tokyo Institute of Technology. He is a professor at the University of Electro-Communications, Graduate School of Information Systems, Japan. He is the president of the Japanese Society for Information and Systems on Education, and is the director of e-Learning Center at the University of Electro-Communications. His research interests include e-Learning systems, Computer Supported Collaborative Learning systems, and curriculum development of IT education. He is also a convener of WG2 (Collaborative Technology) of LTSC/ISO SC36 (Learning Technologies Standards Committee).

Elvira Popescu is Assistant Professor at the Software Engineering Department, University of Craiova, Romania and a member of the Multimedia Applications Development Research Center at the same university. She obtained her Ph.D. degree in Information Systems from the University of Craiova, Romania and the University of Technology of Compiègne, France (double degree) in 2008. Her research interests include technology enhanced learning, adaptation and personalization in Web-based systems, intelligent and distributed computing. She authored and co-authored more than 40 publications, including journal articles, book chapters and conference papers. She received several scholarships and participated in several national and international research projects. Dr. Popescu is actively involved in the research community by serving as program committee member and reviewer for numerous conferences, and co-organizing three international workshops in the area of social and personal computing for web-supported learning communities.

Anguelina Popova is a PhD student at the University of Utrecht, the Netherlands. She is researching the educational value of podcasts and their effectiveness for learning with a focus on stimulating students' reflection and deep learning. She also has research interests in using Web 2.0 technologies for learning, mobile learning and personal learning environments. She holds a Ms in Educational Technologies and an Ms in Politics in Europe from the University of Nice-Sophia Antipolis, France. She has presented her research on podcasting on several European and international conferences, including Toward a Philosophy of Telecommunications Convergence, Budapest 2007; Junior Educational Researchers (JURE), Leuven 2008; ED-MEDIA, Vienna 2008; European Conference on Educational Research (ECER), Vienna 2009;

World Conference on Educational Sciences (WCES), Istanbul 2010. She has published an article on podcasting in co-authorship with Dr. Palitha Edirisingha from Beyond Distance Research Alliance, UK, for the STRIDE handbook 2009 and in the Procedia - Social and Behavioral Sciences Journal (ISSN: 1877-0428), Vol. 2, Issue 2, (2010). Anguelina Popova is holder of the prestigious grant of the Huygens Scholarship Programme for excellent students, attributed by the Dutch Ministry of Education, Culture and Science.

Sapna Poti is currently pursuing doctoral program in the Department of Management Studies, Indian Institute of Technology, Madras. She has about 15 years of industry experience in the human resources and business consulting. Her previous stints were in the consulting practice of Deloitte in Human Capital Advisory Services and Hewitt Associates in the Talent Organization Consulting division, primarily in the southern and eastern region of India. Sapna Poti's dissertation topic will be, "Evaluation of a change program in the context of ERP implementation in India". She has published in national and international peer reviewed journals.

Harold Robinson is a PhD student in the College of Information Sciences and Technology at Penn State University. His research focuses on inter-organizational coordination between humanitarian relief, governmental and military organizations.

Fiona H. Rohde, PhD, is an Associate Professor within the UQ Business School, The University of Queensland. Her primary research activities involve the area of outsourcing, and also the area of data quality and information management and its effect of information retrieval. Fiona worked for KPMG in the Computer Audit Division before joining the school approximately 15 years ago.

Neil Rubens is an Assistant Professor at the Graduate School of Information Systems, University of Electro-Communications. He received Ph.D. in Computer Science from Tokyo Institute of Technology, M.Sc. from the University of Massachusetts. His research interests include machine learning, artificial intelligence and heuristics.

Biresh K. Sahoo is Professor of Economics at the Xavier Institute of Management, India. He obtained his doctoral degree from Indian Institute of Technology Kharagpur, India. He has been recipient of three prestigious fellowships: *Postdoctoral Fellowship* awarded by the Japan Society of the Promotion of Science, Japan; *Lise Meitner Fellowship* awarded by the Board of Austrian Science Fund, Austria and *Postdoctoral Fellowship* awarded by the Administration Board of Fundo Regional Da Ciência E Technologia, Portugal. His research on applied production frontier analysis has appeared in several scholarly journals such as *Annals of Operations Research, European Journal of Operational Research, International Journal of Production Economics, International Journal of Systems Science, International Journal of Operations Research, Journal of Operations Research Society of Japan, Journal of Quantitative Economics, Socio-Economic Planning Sciences*, etc. He has co-authored with Jati Sengupta a monograph titled 'Efficiency Models in DEA' published by Palgrave Macmillan, London. He also serves as a regular referee for the most of these aforementioned journals.

Andrea Sterbini, Ph.D., is full-time Researcher in Computer Science at the Computer Science Dept. of Sapienza University of Rome, Italy. He teaches Computer Architectures, C Programming and Robotic courses. His main research interest is the usage of AI techniques and social collaboration in e-learning. He strongly supports Free Software and Open Standards.

Satoko Sugie is a graduate student in the Graduate School of International Media, Communication & Tourism Studies, Hokkaido University, Japan. Her interests include Chinese language education and educational technology. She has worked as a research assistant in the fields of Web design and development of e-learning contents at the Information Initiative Center, Hokkaido University. She is also an experienced teacher of Chinese at Sapporo University and is well-known as a capable interpreter of the Chinese language. With her expertise, she is undertaking particularly unique research in the field of collaborative learning between China and Japan. She has presented her research at the 7th International Conference on New Technologies in Teaching and Learning Chinese.

Andrea H. Tapia is an Associate Professor in the College of Information Sciences and Technology at Penn State University. Her research focuses on the functional and structural elements of organizations engaging in technological change and coordination across boundaries.

Louis-Marie Ngamassi Tchouakeu is a PhD student in the College of Information Sciences and Technology at Penn State University. His research focuses on information and communication technology (ICT) use for inter-organizational coordination and social networks among humanitarian organizations.

Marco Temperini is an Associate Professor at the Department of Computer and System Sciences, Sapienza University of Roma where he also has got a PhD in Computer Science in 1992. Marco teaches 'Programming Techniques' and 'Programming of the Web' in degree courses of the Faculty of Information Technology Engineering at Sapienza. His recent research activity is on theory and technology of web-based distance learning, social and collaborative learning and web-based participatory planning.

Zhongxian Wang is a professor at Montclair State University, New Jersey, USA. Professor Wang teaches Operations Analysis, Production/Operations Management, Decision Support & Expert Systems, Business Statistics, Operations Research, and Management Sciences. He is a member of Institute for Operations Research and the Management Sciences (INFORMS), Information Resources Management Association (IRMA), The Decision Sciences Institute (DSI), The Production and Operations Management Society (POMS).

Ruben Xing received his Ph.D., Master of Science (M.S), and Master of Arts (M.A) in from Columbia University, New York. Having worked for more than 15 years, Dr. Xing has held senior IT management positions at several large financial conglomerates like Merrill Lynch, Citigroup, First-Boston/Credit Suisse in metropolitan New York. His current research interests include Broadband and Wireless Communications, the Internet transformations and security, Disaster Recover/Business Continuity Planning, and Supply Chain Management.

Yuichi Yamamoto is an assistant professor in the Research Division of Media Education at the Information Initiative Center, Hokkaido University, Japan. He received his M.S. and Ph.D. in chemistry from Hokkaido University, Japan. His research interests include educational technology and computational chemistry. He has done extensive research in the fields of distance learning and computer supported collaborative learning. He has presented his research at several international conferences, including the International Conference on Computers in Education (ICCE) and the International Conference on Information Technology Based Higher Education and Training (ITHET). He received the Best Poster Presentation Award at the 15th International Conference on Computers in Education with co-authors, Prof. Nishihori and Dr. Sato. Haruhiko Sato is an assistant professor in the Division of Synergetic Information Science, Graduate School of Information Science and Technology, Hokkaido University, Japan. He received his M.S. and Ph.D. in information science from Hokkaido University, Japan. His research interests include automated theorem proving, formal verification and software engineering. He has presented his research at several international conferences, including IJCAR (International Joint Conference on Automated Reasoning) and RTA (international conference on Rewriting Techniques and Applications). He has been published in the journal of IEICE Transactions on Information and Systems. He was awarded the Best Paper Award of the 2010 IAENG International Conference on Computer Science.

Yanli Zhang is Assistant Professor in the Management and Information Systems Department at Montclair State University. She graduated with a PhD in Management concentrating on strategy and international business from Rutgers University in May 2007. Her research interests are in strategy and decision making, technological innovation, and networks and knowledge. Her paper on "Inter-Firm Networks and Innovation" won the Doug Nigh award for the most innovative paper in the International Management Division at the Academy of Management 2007 Annual Conference. She obtained a BA in Economics from Beijing University, China; and worked as an economic analyst in the Ministry of Foreign Affairs, China, and a management consultant in Accenture, Beijing office.

Kang Zhao is a PhD student in the College of Information Sciences and Technology, the Pennsylvania State University. His current research interests include complex networks, computational social and organizational simulations, social computing, and agent-based systems.

Index